PREFACE AND ACKNOWLEDGEMENTS

This volume contains the papers of the ninth technical conference of the Remote Sensing Society which was the first 'International' occasion organised by the Society. The convenor was both surprised by and grateful for the number of contributions offered. In the majority of cases authors provided camera-ready copy enabling reproduction in time for the conference. The Society is grateful to the contributors and their typing and other support staff for making the publication possible.

No editing has been performed on any of the papers or abstracts and the authors are responsible for the contents of their papers.

The contribution of Mrs Elaine Rosindell to the artwork and in designing the cover is gratefully acknowledged.

Tony Allan School of Oriental & African Studies
Malet Street, London WC1 e 7HP

Papers prepared for publication by D Lynn and J A Allan, London, 1981

Note

The contents of the papers in this volume reflect the opinions of the authors. Officials of the Remote Sensing Society which convened the meeting and of The University of London in which the sessions were held do not necessarily share the views expressed herein.

PROGRAMME AND LIST OF CONTENTS OF PROCEEDINGS

MATCHING REMOTE SENSING TECHNOLOGIES AND THEIR APPLICATIONS

16 - 18 December 1981

at the University of London

(Meeting rooms will be notified at registration)

WEDNESDAY 16 DECEMBER

EXHIBITION: AN EXHIBITION OF EQUIPMENT WILL BE OPEN THROUGHOUT THE CONFERENCE

THURSDAY 17 DECEMBER

SESSION A - A1

6.15-7.30 A N N U A L G E N E R A L M E E T I N G

8.30 C O N F E R E N C E D I N N E R
at which THE BELFOTOP AWARD will be announced

THURSDAY 17 DECEMBER

FRIDAY 18 DECEMBER

FRIDAY 18 DECEMBER

FRIDAY 18 DECEMBER

pm 4.30-5.30 CONCLUDING SESSION

Chairman: J O Thomas (Imperial College)

VIEWS ON THE MAIN THEME AND APPROPRIATE APPLICATIONS SESSIONS

D Stanley (Logica): N Press (Nigel Press Associates)
J R G Townshend (University of Reading)

DISCUSSION

E S Mallett (ESA) and R M Watson (Resource Management and Research, Nairobi and London)

****

MATCHING REMOTE SENSING TECHNOLOGIES AND THEIR APPLICATIONS - POSTERS

LIST OF POSTERS

KEYNOTE PAPERS

DEVISING AND MANAGING AN INTERNATIONAL APPLICATIONS PROGRAMME: THE ESA EXPERIENCE

E S Mallett, European Space Agency

The European Space Agency's first remote sensing satellite, Meteosat, was designed to satisfy meteorologists. There is a coherent and organised grouping of meteorologists in Europe, thus making it relatively easy for ESA to interface with the prospective user.

For other remote sensing satellites, the situation is more difficult, there is a far greater range of prospective needs, and the possible users are not organised so formally.

The methods by which ESA has attempted to choose a payload for its first earth resources satellite, ERS-1, having an oceanographic mission, will be described.

The instruments finally chosen will be outlined, and the way in which the data will be transmitted to users will be described.

DOWN MARKET REMOTE SENSING

R. M. Watson
Resource Management & Research
16B, West Central Street,
London W.C.1

Abstract: Remote sensing is a means of establishing realities concerning, among other things, natural resources and their use by man.
The consultant's market place has an important role in shaping remote sensing methods.
The preculiarities of this market place have profoundly distorted the evolution and application of remote sensing methodologies, so that the role of remote sensing in providing information of usefulness to man is neither efficient nor balanced.

This is particularly marked when one considers the use of remote sensing to supply information for development activities in the third world.

The manner in which the problem to be solved can determine the methodological approach is considered through a general model.

Some examples of down market remote sensing technologies are described for range resource surveys in Somalia, for oil pollution surveys in the North Sea, for forest inventory in Sudan, and for Sodium Flouride prospecting in Kenya.

Because of the criterion of usefulness has been given a dominant position in the conference theme remote sensing technologies should be evaluated in a wider enviroment than the purely technical or scientific one adopted by so many speakers in this meeting. This requires a consideration of marketing, of the practical conditions of a method's application, including financial, legal, logistical, political, and behavioural aspects, and of the means of applying the results of remote sensing to problem solving.

Paper presented at a conference on Matching remote sensing technologies and their application convened in London by the Remote Sensing Society, 16-18 December, 1981.

PREFACE

If I had walked into this hall by mistake and had been asked the guess the conference theme, I regret to say I would have been wrong. I would have got the remote sensing part, but if pressed to detect a theme, I might have said:

"My niche in the Remote Sensing Industry (RSI)".

Luckily I came here with about 20kg of literature advertising an International Conference on:

"Matching Remote Sensing Technologies and their applications".

I therefore intend to describe my niche in matching remote sensing technologies and their applications.

I guarantee not to introduce any acronyms, to discuss only that equipment already owned by the average Playboy reader, and to use a language likely to be understood from Brighton to a line running just south of the wash.

My paper is called "Down Market Remote Sensing". The second paragraph of my abstract must be amended in the light of these proceedings to acknowledge the spectacular growth of academic interest in remote sensing. I will include the implications of this in the paper, although it does not change my basis emphasis.

And as a final prefacing point I would like to thank Dr Tony Allan, and the Remote Sensing Society for their kind invitation, and the speakers who have provided such excellent examples of Up Market Remote Sensing that I am assured my short contribution represents no threat to this highly complex and thriving industry.

1. INTRODUCTION

Definitions:

1. Remote Sensing

I remain puzzled to this day as to how remote sensing is. I am also puzzled by the inference that being on the ground enables the truth to be ascertained. What would Einstein have made of the scientist peering into an electron microscope sited in an orbiting space station? Usage has apparently decreed that this most unsatisfactory term applies to the collection of information about the surface and atmosphere of the earth (and possibly other planets) from flying (aircraft, kites, helicopters), floating (balloons) or orbiting (satellites, space probes) platforms.

2. Matching Applications

Knowledge *per se* and science, the search for this knowledge, have not brought mankind the copious blessings we are sometimes led to believe. This is partly because we ask scientists to solve the wrong problems, and partly because we have evaluated knowledge to become a worthwhile end in itself. The last few decades of this century will reveal in a spectacular way the folly of replacing wisdom with knowledge.

In this paper I will take the following eccentric position:

"The problem to be solved, ie the application, should determine every aspect of the search for a solution", and so I have chosen to interpret the slightly ambigious conference theme in this way:

"The influence of real problems on remote sensing technologies".

I observe a different interpretation in some of the papers presented here. They appear to be saying either:

"Here is something new we can do with remote sensing technology; we can use it to solve (only) these sorts of problems".

(Typically represented by Coulson's paper on high altitude aerial photography, or Barr and Bassarab's paper on an interactive digital image processing system).

or:

"Here is something new we can do with remote sensing technology; are there any problems it will solve?"

(Typically represented by Wilson and Henderson-Seller's paper on three dimensional histograms and quantitative image processing from low resolution sensors, and Welsh's paper on vegatation mapping from sideways looking airborne radar.)

or:

"Here is something new we can do with remote sensing technology".

(Typically represented by Kitchco's paper on an inhouse geographic information system and Singh's paper on an updated catalogue system.)

2. THE PROBLEMS OF THE SOLUTION MARKET

Scientists and technologists such as ourselves are working the "solution market". We bring solutions, or our promise of solutions and exchange them for money.

There are three basic market places: Academia, Technologia and the Real World. Ideally the problems of the real world should have a dominant (or at least a guiding) influence on the solutions offered for sale in the academic and technological market places. That they do not is just one more real world problem beyond the scope of this paper.

This conference demonstrates convincingly the abundance of academic and technological solutions which are being sold. Buyers in this market are sophisticated and well-heeled. They have a keen eye for ingenuity, elegance, and complexity. Not a few of them have undeclared interests. For example the US and Russian Governments support satellite remote sensing studies, but also have to launch large numbers of remotely sensing satellites; industrial concerns manufacture and sell the equipment necessary for the technology; National and Internation agencies have to compete for our taxes; and so on.

It is therefore hardly surprising to find remote sensing technologies moving steadily up market. As this trend continues the descriptions of these methods become increasingly esoteric. I quote:

"The efficiency of classification algorithms applied to remotely sensed imagery is largely dependent on the statistical attributes of the data, via the Modulation Transfer Function (MTF) of the imaging system, and the inherent spatial frequencies of the terrain".

(Harrison, 1981)

"Creation of RISS is an example of a trend in maximising the capabilities of an operational GIS system, for remote sensing applications".

(Kitcho, 1981)

"Following ground truthing by ITE the input maps are being digitized at the Experimental Cartography Unit (ECU) where data, after interactive editing, will be used to generate monochromatic maps on the Unit's flat bed projection plotter".

(Drummond and Fuller, 1981)

A part of this esoterism reflects a real increase in complexity; the rest is sales talk.

One might ask why the real world market has not halted or at least retarded this up market climb. In undeveloped parts of the world one would expect economic pressures to exert a powerful force on remote sensing technologies.

In reality solutions to resource problems in the third world are provided by consultants or international agencies working for governments. I may have been unfortunate in the last 15 years, but must sadly report that the consultant's market I know does not serve the public interest. It is both morally and structurally corrupt.

These charges are not made lightly. At the structural level bilateral aid of most donor countries has to be spent in the main on goods and services from that country, regardless of their qualities.

International agencies are required by their statutory regulations to buy equipment and recruit staff from particular National sources. Anyone who has worked for FAO knows the problems of operating Polish planimeters and Russian 35 mm cameras.

In the "marketing" of consultant proposals undue weight is given to Project Managers' and experts' CVs and few marks are awarded for methodologies, even those offering substantial cost savings.

And the grossest structural corruption is to be observed in the political environment prevailing in the majority of developing, ie. recipient, countries.

Aid really is the movement of money from poor people in rich nations to rich people in poor nations; the widening gap between the developed and developing worlds reflects the validity of the epigram. That gap cannot be filled by increasing aid programmes.

Few consulting companies in this field resist the pressure, and rely on the merit of their proposals to win contracts. Hunting Technical Services, a British Company with a world-wide reputation, have to my knowledge lost contracts because of their refusal to offer benefits to Government officials, and dubious middlemen.

In these circumstances the real world market provides Academics and Technologists with limited clues about the real problems of developing countries requiring a remotely sensed solution, or about the most suitable methodologies.

It is therefore even more important for remote sensing specialists such are ourselves to keep our methods inexpensive and our explanations simple.

3. A GENERAL MODEL FOR MATCHING REMOTE SENSING TECHNIQUES WITH THEIR APPLICATIONS

How inexpensive is an inexpensive method, and when is the explanation simple enough? By our definition the quick answer is when the problem is solved with maximum efficiency.

The diversity of remote sensing techniques and of the problems they can be applied to would seem to defy the possibility of creating many useful generalisations. There are however three great unifying principles at work.

One of these is most frankly stated as: "There is no free lunch"; it is a biological restatement of Newton's 3rd law of thermodynamics. Because biological and human systems are essentially materialistic and competitive, it is possible to ascribe a material cost or benefit to any event or action. Modern economics is, unfortunately, somewhat obsessed with organisational (political) and artefactual (monetary) issues, and so does not offer an entirely sound means of analysing material costs and benefits, but for the moment we are stuck with it.

The second principle asserts: "There is no ground truth". Measurements, theories and truths are relative. Einstein's belief in an underlying very complex regularity in molecular, and hence all, events has fallen to a recent virtual proof of Heisenberg's Uncertainty Principle. The Steinhaus paradox can be seen as a limited methodological manifestation of the principle. Precision of measurement has to be selected. Human beings; and their science and philosophy, relate most effectively to human scales of measurement and precision.

Finally, and somewhat reassuringly following the dissmisal of altrism and truth, the third principle is: "Perception is independent of reality in any specific version of the Universe. "That is to say differences in the length of Steinhaus's empirical line measured at identical accuracy levels by different observers reflect differences in observer performance.

And so at a theoretical level there is a basis to suppose that general principles can apply in guiding the evolution and selection of remote sensing techniques.

THE PROBLEMS

The starting point is the problem; in more ways than one. The identification of real world problems depends upon a range of frail and frequently corrupt human judgements, particularly in the third world. The politically emotive issue of "two worlds", of the "north and south", of the widening gap between "haves and have nots" has its most important roots in the political systems of developing countries. A better indentification of problems may result from the monitoring of the overall costs and benefits arising from specific solutions of problems, but perfection will remain as unattainable as other goals within systems of animal behaviour.

ANALYSIS OF INFORMATION NEEDED FOR PROBLEM SOLUTION

Problem solving essentially consists of the application of energy under the guidance of information. In view of the theme of this conference we will now examine the information phase of problem solving in detail.

Most of the important real world problems remain unsolved not because of their complexity, but rather because insufficient energy and information have been applied to them.

The speakers in this conference have adequately demonstrated that academic and technological skills for information collection and analysis have developed to the almost apocryphal level of detecting the colour of the enemy soldier's nail varnish. And so, with a few "A" levels and some common sense the information needed to solve a great many important real world problems can be described in a fashion which enables an efficient approach to the collection of that information.

This description covers selection, classification, resolution, volume, precision and comprehension. In Figure 1. a model universe of information has been represented to illustrate the elements and inter-relationships of the description process.

Selection and classification of information

In selection the components and processes having strong influence in solving the problem are separated from the mass of components and processes, having smaller negligible influences. Information selected can be further grouped as more or less important implying certain degrees of precision are to be arrived at in the collection of information.

Specific resolution of information

Information about natural resources seems to be naturally classifiable as either of human scales or of global scale. Broadly speaking biotic information has spatial heterogeneities and time dynamics which match our own limitations of perception. For example a herd of animals can be seen by a human observer in its entirety; its movements can be followed by the observer by foot, and its individuals are born and die within the lifespan of the observer. On the other hand physical information generally changes in space and time on scales not easily appreciated by the technologically unassisted human being. For example individual human beings do not notice the slow erosion of mountain peaks, nor, on foot, do they realise the pattern in the distribution of mountain ranges and alluvial plains. Outside the natural resource sphere these generalisations do not hold.

Specific Resolution for any specified information element is the area or volume of the universe which a particular sensor/perceptor system requires to determine, with very high probability, the presence, absence or condition of that information in that area or volume of space.
Evolution has placed man firmly in the macro-biological sphere, with sense organs and imagination well attuned to

FIGURE 1

IMPORTANT FEATURES FOR DESCRIBING INFORMATION NECESSARY IN PROBLEM SOLVING

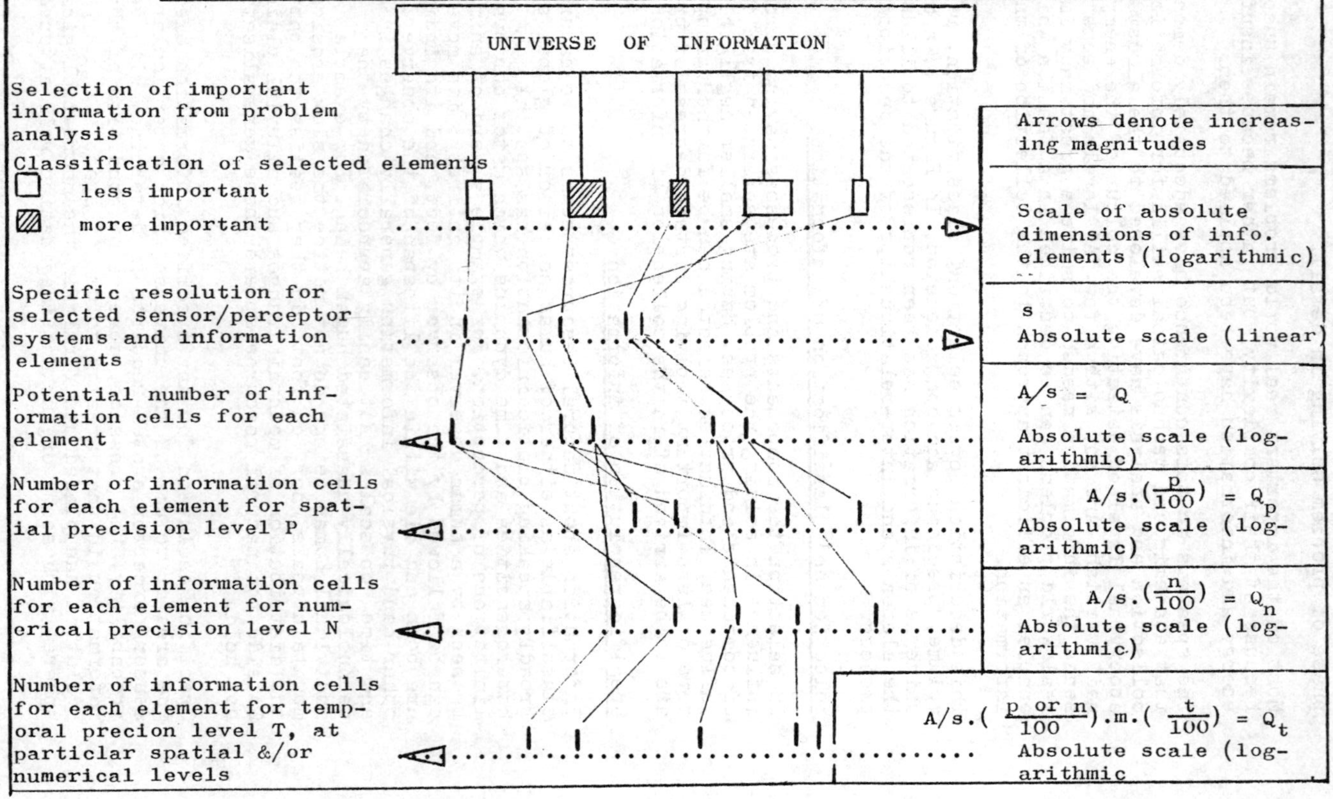

record and respond to macro-biological changes. Relatively small and relatively large phenomena present difficulties whose magnitude is roughly proportional to their scale differences.
For relatively small information elements (ie small phenomena at close quarters or large phenomena at great distances) technology has produced a large range of magnifiers and sensors, broadly termed microscopes and telescopes respectively, to enable their specific resolutions to be achieved and hence their nature studied.

However the difficulty until recently with large information elements at close quarters has not been their specific resolution, but rather their very size relative to the mobility of the sensor/perceptor system. Remote sensing platforms provide mobility over, & distance from, large information elements, thereby reducing their scale to mansized.

Volumes and precision of information

The potential quantity or volume of selected information is a function of the area or volume under study (determined by the problem), and the specific resolution of the information. The amount of information, considered as the number of cells to be filled for a complete picture, is given by the relationship:

$$Q = \frac{A}{s}$$

where Q if the potential quantity of information (ie the number of information cells)

A is the two or three dimensional size of the study
s is the two or three dimensional specific resolution for any element of information and any sensor/perceptor system.

In a general way the cost of collecting and understanding information is proportional to the number of information cells involved. Clearly then there should be sufficient selection of information to be collected, and of sensor/perceptor systems with specific resolution values matching the dimensions of the information elements. Hyde, Vesper, Goldblatt and Cobley, 1981, make this point in their most useful study of the relations between resolution, usefulness and cost.

The number of information cells involved in a study can be reduced through sampling procedures if the desired precision level, (spatial, numerical and/or temporal) can be roughly established.

The relationship between precision and sampling faction depends upon the mathamatical distribution of the information.

Jolly (1981B) has given an excellent paper on this topic, which is all the more notable for being the only paper dealing with formal sampling. The potential of sampling in the resources field in cutting costs is not at present fully exploited. For example 5% sampling fractions in extensive surveys can provide estimates for livestock with sampling errors of less than ± 5%.

It is clear from the relationship shown in Figure 1 that amounts of information required are proportional to sampling fractions. Sampling would seem therefore to hold the possibility of reducing the costs of collecting information appreciably; a 10 or 20 fold reduction would not be an unreasonable expectation.

Comprehension and utilisation of information

The usefulness of information collected for problem solution depends upon its comprehension at many levels, and its application in guiding development and management activities.

Again in general terms large numbers of information cells are less easily (ie. more expensively) converted into comprehensible forms of the ultimate users of the information. Furthermore most development and management interventions are somewhat coarse, and have no capacity to benefit from very precise guiding information.

Thus it seems likely that the problem solving returns on more precise and more resolved information not only diminish, but may become negative as more expensive interpretation and presentation systems become required. Economic returns for the consulting company however neither diminish nor become negative, at least in the short term.

Figure 2 illustrates the relationships between resolution, precision, comprehension, utilisation and costs for a single element of information in a fixed study area.
Section A in Figure 2 shows how amounts of information and the costs can be influenced by sampling fraction and resolution.
In the example proposed there are two levels of complexity requiring additional expenditure on interpretation and presentation. These levels correspond with specific quantites of information, ie. numbers of information cells. A budget limit has been added to the vertical axis. It has been assumed in this example that a 10% sampling fraction provides an estimate with precision just sufficient to satisfactorily determine the information element under consideration for the problem at issue. A further assumption is that a resolution of 0.0002 of the total study area (ie. a relative resolution of 0.0002) just resolves the information element in a way to satisfactorily answer the questions dictated by the problem.

In these circumstances the most efficient information collecting method is one with a relative resolution of just

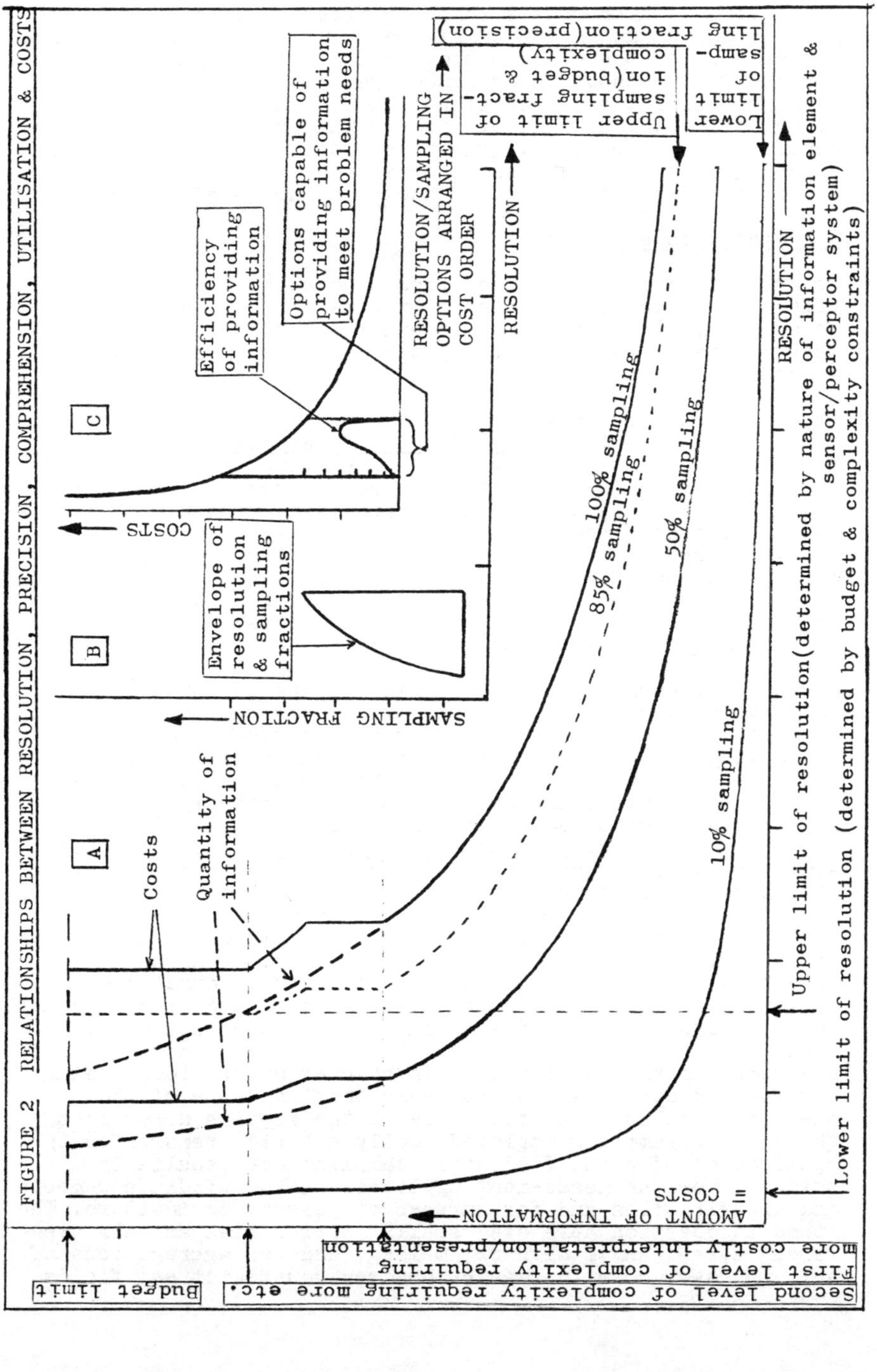

FIGURE 2 RELATIONSHIPS BETWEEN RESOLUTION, PRECISION, COMPREHENSION, UTILISATION & COSTS

over 0.0002, which uses a sampling fraction of just over 10%. This is shown in Part C of Figure 2. It follows out from the assumptions of the example that useful information cannot be derived from methods using sensor/perceptor systems with relative resolutions less than 0.000035, nor from methods using sampling fractions of greater than 85%. The envelope of resolutions and sampling fractions within the understandable and usable information can be obtained, for the given budget and complexity constraints is shown in Part B of Figure 2.

A Footnote on sampling

It appears from the analyses presented above that sampling offers appreciable benefits in information collection techniques. It is therefore somewhat surprising to see how many natural resource investigations ignore these potential advantages.

Among the factors responsible are

1. The traditions of map-making

Map-making is an ancient pastime pre-dating sampling theory. The most important remote sensing method used in map making was until recently aerial photography, and even now satellite imagery probably remains less important as a data source for mapping purposes. The wastage of information, with an associated reduction in cost efficiency, which attends traditional map making methods is impressive. Much of this waste is avoidable if the concepts of precision and sampling are formally applied to the methods.

A map is of course only a convenient means of representing information with an essentially two dimensional districution. Advances in electronics now offer alternative ways of assembling data of this sort. Some of these alternatives are undoubtedly superior for specific applications of information to problem solving, and at the same time probably more readily incorporate a sampling approach.

2. Unconscious sampling implicit in scales

A map, as well as being only a convenient representation, is also a sampling of selected information. Jolly (1981) has given an excellent paper on the role of sampling in certain types of remote sensing. In map making by traditional methods using aerial photographs, satellite images, or ground observation and plotting, the selection of information is conscious, if somewhat arbitary - when does a drainage channel become a stream?; when does a minor track become a major track?

The sampling process involved is however unconscious, being governed by common sense. It seems that the size of the phenomenon relative to the scale of the mapping determines the type of sampling employed. Jolly's (ibid) remarks on the application of stratified strip sampling are peculiarly suitable for the assessment of livestock and wildlife numbers and distributions and for surveys of peasant agriculture. The items in question here are "small" (terrestrial animals are rarely more than 4m high or 5m long, and few aggregations of animals cover more than a few hundred square metres; fields

FIGURE 3 RELATIONSHIPS BETWEEN PROBLEMS AND TECHNIQUES

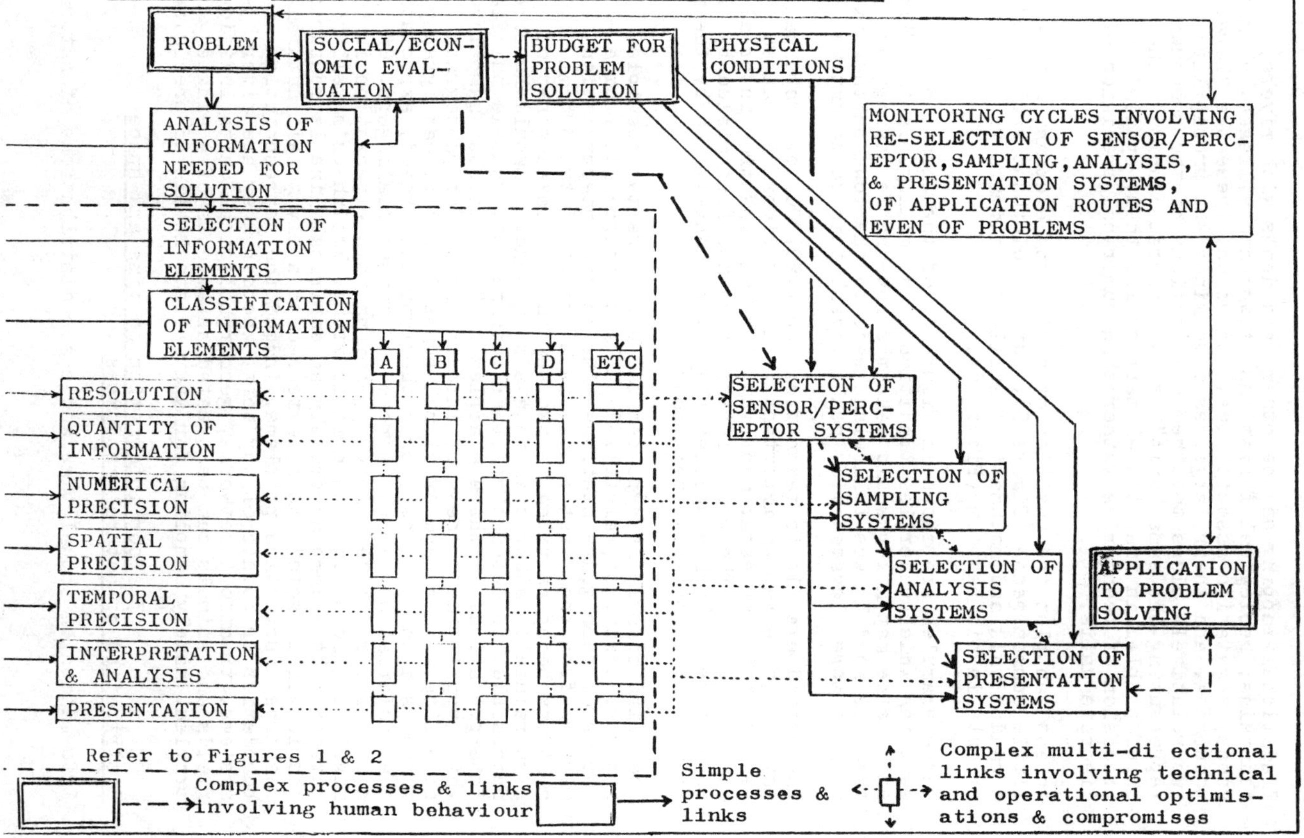

of 1 ha are large in peasant agriculture).

Traditional mapping on the other hand deals with rivers, roads, mountains, geological formations, towns and villages, climatic zones, vegatation associations, and so on. These are "large" phenomena, and their largeness determines the probability of two relatively close together observations of the phenomenon representing the same "unit" or two discrete "units". The chief problem in traditional mapping is, then, spatial precision of boundaries rather than numerical precision of discrete units.

An ant can plot the position of a sleeping elephant with a little common sense (Figure 5), but the elephant would need a random walk and mathmatical skills to estimate the number of ants in his bedroom (Figure 6).

Three examples are shown in Figure 4. of the unconcious sampling which underlies traditional map making. In the first examplethe map maker would make the reasonable assumption that the road he was following leads to the town on the next hill. A more elevated view could reveal this to be untrue.

The second example is taken from the 1:50,000 topographical mapping of Kenya. An important permanent river was mapped as descending an escarpment into the Magadi/Natron internal drainage basin. The photo-interpreters had been misled by the dense forest which concealed the river channel, and by their "expectation" of the rivers course.

The third example is hypothetical and shows two possible interpretations of rainfall data based on records from a number of rain guage stations. The second interpretation differs from the first mainly as a result of two 300 mm stations being ascribed to two separate isolated zones of high rainfall rather than to a single zone. It would be more reasonable, in absence of other information, to make the first interpretation. In these examples of "unconscious" sampling, which typify the process of interpretation (of photographs, satellite images, and of visual scenes), and of cartography, there is a common sense, or a theoretically based, expectation, that information can be "stretched" across small gaps. Of course for relatively large phenomena this expectation is right many more times than it is wrong. Steinhaus's well known paradox : "the more accurately an empirical line is measured the longer it gets" can be extended to apply to all natural resource description problems. Indeed the representation of boundaries or points of any sort, whether they are river systems, transport networks, geomorphological units individual fields and animals can be treated as aspects of a single problem, namely sampling for appropriate levels of precision.

3. <u>Poor information on time and space distribution of natural resources</u>

Information on actual time and space distributions of natural

FIGURE 4

1. PLOTTING ROADS ON THE GROUND

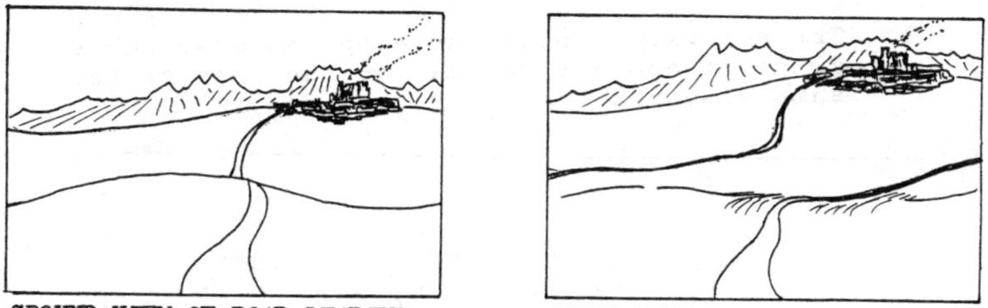

GROUND VIEW OF ROAD LEADING TO A TOWN

MORE ELEVATION REVEALS REAL POSITION

2. HYDROGRAPHIC PLOTTING FROM AERIAL PHOTOGRAPHS

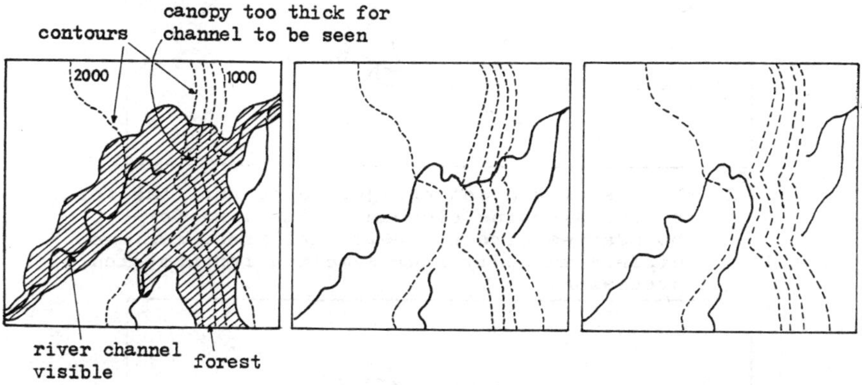

AERIAL PHOTOGRAPH

MAP DRAWN BY SURVEY OF KENYA

ACTUAL RIVER CHANNELS

3. ISOHYET PLOTTING FROM RAIN GUAGE NETWORKS

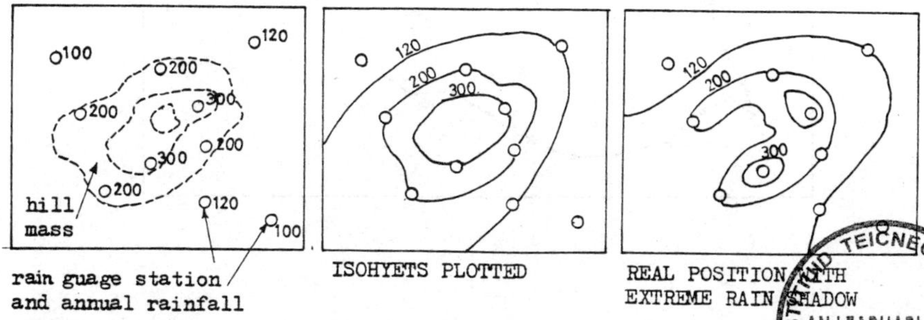

ISOHYETS PLOTTED

REAL POSITION WITH EXTREME RAIN SHADOW

FIGURE 5

AN EXAMPLE OF UNCONSCIOUS SAMPLING TYPICAL OF TRADITIONAL MAP MAKING

1. The ant walks round the elephant assuming each point of contact represents the edge of the same elephant.

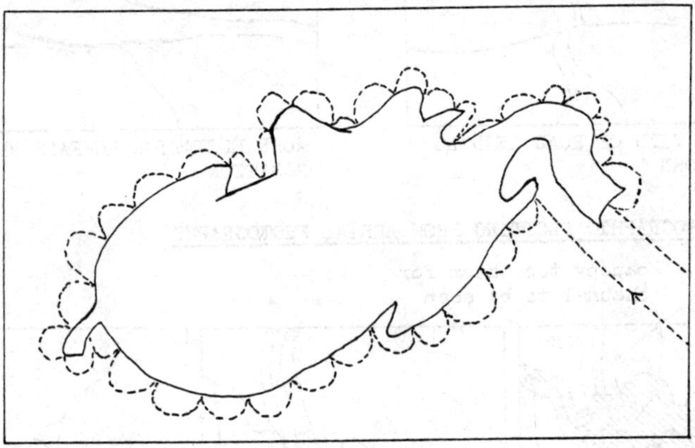

2. Points of contact are joined to show the form of the sleeping elephant. Sampling for boundaries aims at positional precision, and differs markedly from sampling for numerical precision.

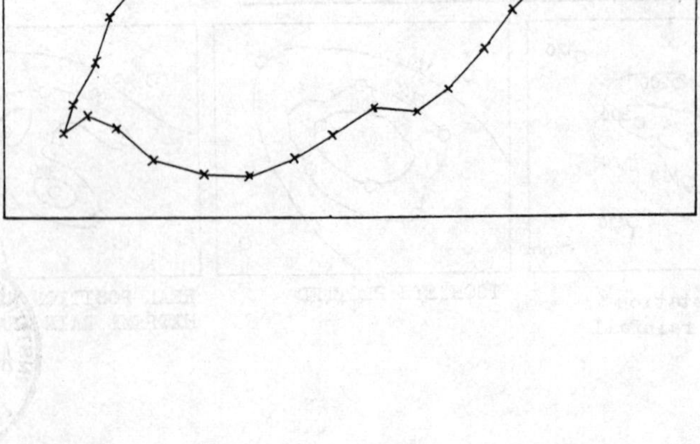

FIGURE 6

AN **EXAMPLE** OF CONSCIOUS SAMPLING SHOWING HOW SCALES INFLUENCE PERCEPTIONS AND FUNDAMENTALS IN SAMPLING

1. Dead ants left after random walk by elephant. Population of ants at the time of the walk is given by:

$$Y = \frac{\bar{y}}{n} \cdot A \quad ; \quad \bar{y}_i = \frac{c_i}{a}$$

where Y is an estimate of the ant population

$\bar{y}$ is summation of $\bar{y}_i$ from all foot prints (n in number)

A is the area of the room

$\bar{y}_i$ is the density of dead ants in each footprint

c_i is the number of dead ants in each foot print

a is the area of a foot-print

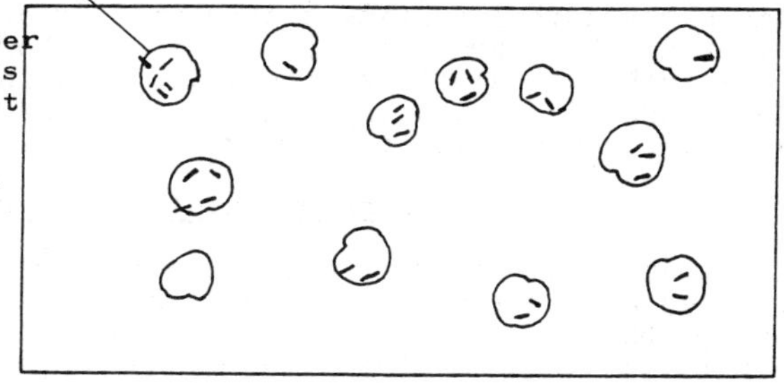

2. Ants in the room before the elephants random walk

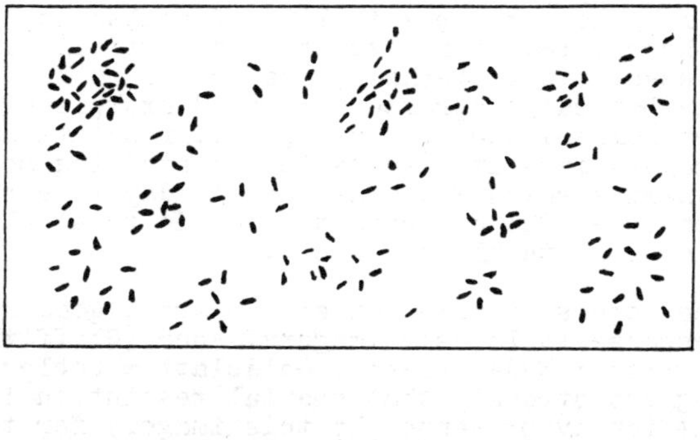

resources is difficult amd expensive to acquire. Theoretical mathematical distributions have been the basis of the development of modern sampling theory. There is little doubt that better knowledge of actual distributions will encourage the growth of sampling approaches in natural resource studies. Haggett and Chorley (1974) in their outstanding "Network Analysis in Geography" have provided geographers with their own version of D'Arcy Thompson's "On Growth and Form". It is not a large step to suggest that most natural resource and geographic features provide lines (links, edges, sides, arcs, segments, branches, routes, one cells, or boundaries) and/or points (nodes, vertices, junctions, intersections, terminals, or zero cells) suitable for the application of network analysis techniques. Such techniques, sometimes empirical, would enable the common sense expectations implicit in the process of mapping natural resources and geographical features to be refined, through some mathematical statement about the errors implicit in "stretching" information across gaps in both space and time. These mathematical statements would have influence on both the design of sampling approaches, and the precision to be expected from a specific survey.

4. Confusion of objectives

The assembly of natural resources data is frequently not directed at problem solving, but is an end in itself. Even some problem solving surveys aim at a broad range of information using a single technique. Superficially, at least, it appears to be wasteful to go to the expense of launching a satellite or taking off in an aircraft if it does not gather "all" possible information from the regions it overflys. The direct result of attempting to meet a wide range of objectives with a single information collecting system is compromise. Such compromises are often made without an aconomic evaluation. Furthermore the possibility of collecting too much information (thereby requiring expensive analytical, interpretational and presentational technologies) seems to be rarely considered. The Landsat programme is a good example of this sort of compromise. The present spatial resolutions of Landsat imagery make it unsuitable for studying most biotic space/time dynamics which would make an 18 day time resolution look sensible. The decision to provide total global cover has been almost certainly political.

The papers in this conference are a good guide to user response to Landsat imagery. Four (Griffiths & Collins; Harrison; Hyde, Vesper, Goldblatt & Cobley; and Purcell) suggest strongly that spatial resolution is the chief difficulty presented by this imagery for their specific problems. One paper (Girard) ingeniously uses April 1976 and Sept 1978 scenes to, in effect, improve spatial resolution. There are however no papers which find the temporal resolution or the overall cover of Landsat imagery a constraint.

THE WIDER ENVIROMENT INFLUENCING TECHNIQUES

Up to this point we have considered how the information required for problem solving should determine the information collecting method. The real enviroment in which information is collected introduces new features, as is suggested in Figure 3. Selection of appropriate vehicles, sensor/perceptor systems, sampling systems, analytical and presentational systems must respond to the physical, social and economic enviroment, and to the human idiosyncrasies of the method of applying the information to the problem solving. Watson & Tippett, 1981, for example, identified administrative and operational aspects as second only to sensing in providing problems in low-level aerial survey.

There is no way of systematically handling this type of issue since a significant part of its basis is irrational. Each problem has to be treated on its merits as it occurs. A few examples will show how important these factors can be.

Sampling and the consumer

There is a controversy in low-level aerial survey work between those who employ systematically positioned sampling strips and those who use randomly positioned ones. (Jolly, 1981, Norton Griffiths, 1981). The real issue here is the consumer. Gwynne, who recommends the use of systematic sampling (Gwynne & Croze, 1981) privately (pers. comm.) recognises the inherent weakness of this sampling approach but finds it easier to "sell" his Global Enviromental Monitoring System, particularly in developing countries, if common-sense, rather than statistical validity, is used to explain the sampling system.

Information and security

Many developing countries guard their maps and aerial photographs as though they were priceless military secrets. Permission to take photographs from aircraft is difficult to obtain. Patience and tact will solve most of these problems. Consulting companies in a hurry usually have neither; Fairey Surveys (in 1979) produced an interesting analysis of land resources from Landsat imagery for Northern Somalia, over an area which had already been covered by high quality 1:100,000 mapping based on 1:40,000 photography taken in 1974.

Remote sensing vehicles and legal/security difficulties

Satellites have the great advantage (at present) of freedom of the skies. Aircraft legally require permission to enter, operate in, or leave National airspace. In many parts of the developing world an aircraft is a tempting target for well armed regular troops and/or dissidents. Low over-flying of military areas, particularly along flight lines coming from the direction of hostile neighbours, should be avoided when possible.

Remote sensing vehicles and operational problems

Fuel supplies in many countries are now difficult to obtain, with AVGAS being a particularly rare commodity. The expensive solution is to use turbine engines, which confines operation to main airports; the low cost solution is to use unsophisticated piston engines which are not damaged by operation on motor petrol.

Client aptitudes and the report

It is remarkable how few advances there have been in the field of reporting. The offices of Planning Ministries in developing countries are slowly filling with reports and maps, which supposedly include all the information with which complicated development programmes are to be managed. Careful examination of these reports reveals that some are re-workings of previous reports. Some, also, cover identical or similiar ground, apparently unaware of previous studies. Few are used or even read. Paradoxically the report is usually the least effective way of conveying information to the client. It is often in a language culturally and technically foreign to both consultant (translators and report writers) and client. It is unexciting, and structurally unsuitable as a way of setting down dynamic, dimensionally complicated, inter-relating information elements.

The up market remote sensing methods of satellite imagery and image interpretation deserve an effective form of presentation. The cost of producing traditional reports (in a hundred or so copies) is no less than the cost of buying a micro-computer with printer and monitor. Resource Management and Research is working on a way of presenting reports as magnetic discs. It is intended to offer the client a report which can be read at any level, which is capable of updating, and whose underlying models can be modified as monitoring of development processes continues. Such a "communicating report" will be several times more effective than the traditional 30kg of paper.

4. SOME EXAMPLES OF DOWN MARKET REMOTE SENSING

In the examples of down-market remote sensing which follow the key down market element is the aerial platform. Tippett (1969) suggested an approach for evaluating different light aircraft platforms. Over many years and many problems I have concluded that a PA18 (Piper Super Cub) aircraft is the best all-round platform. Its characteristics are:

- fitted with long range tanks it can cruise at 90 mph for 11-12 hours without refuelling. This enables the aircraft to be positioned fairly easily at least intracontinentally, and also allows a full days sensing activity.

- it can fly at speeds between 40 & 100 mph. The slow speeds enable visual observation techniques to be applied along strip samples. The highest speeds are too low for <u>efficient</u> use of this aircraft type for hand systems mapping.

- it can land and take off from rough ground provided about 100m (less in windy conditions) is unobstructed by trees, boulders etc. This enables on the ground examination of features seen from the air, Hemmings (1981) paper on the "Flying land-rover" makes this point.

- it can operate efficiently between ground level and 18000 feet. Much sensing work is undertaken below clouds, when only radar sensors are useful in high flying aircraft and satellites. With low cost reconnaissance numeras (see later) photographs of scales up to 1:40,000 can be taken at the highest operating altitudes.

- it is mechanically and structurally simple, and therefore easy to repair and maintain.

- it has a small, virtually insignificant radar signature at low levels.

- its fabric skin can be passed through by small arms fire without structural damage.

- it is very slow and uncomfortable for the passengers (it can carry one legally). This reduces high powered requests for transport in remote areas to absolute emergency cases.

- with petrol for 11 hours cruising in the tanks the aircraft will fly with a pilot and about 100kg of baggage. This covers the weight of film, cameras, maps, tape recorders, batteries, tools, notebooks, stop watches, etc, needed for most survey contracts.

- it is cheap to buy. A good second hand PA18 in the United States would be 7-10,000 US $, (£3900 - 56000) ie appreciably cheaper than a Land-Rover.

- it is cheap to operate. It costs, at present prices, less per mile to operate a PA18 than a four wheel drive vehicle cross-country. Petrol consumption is 18 miles to the gallon at 80 mph, and overall cost, assuming 120,000 miles of self-flying annually, would be less than 0.36 US $ (20p) per mile.

Other down-market equipment carried in or fitted to the aircraft platform consists of:

- 35mm camaras with lenses (24mm, 50mm, 135mm) This equipment costs about US $ 1440 (£800)

- a five inch F24 reconnaissance camera with eight inch and five inch lenses. This equipment costs about US $180 (£100) and until recently time-expired film and paper could be found for this format at about 10% of the cost of new paper and film.

- two cassette tape recorders: these cost about S180 (£100).

- a radar altimeter for determining height of the aircraft above the ground. These range in price between US $2700 and S5500 (£1500 - £3000)

- a remote compass costing about $800 (£450)

It is not claimed that the light aircraft above can assemble all information for problem solving. Rather that down market light aircraft techniques are part of a spectrum of methods, which can be most effectively used in combination. Thalen (1981) demonstrates this in respect of surveys of arid rangelands.

One of the main strengths of the light aircraft remote sensing technique is its flexibility, which is obviously enhanced if the specialist himself flys the aircraft.

The speed and position of the platform, sampling system, type of sensor and its resolution are all under the control of the pilot, who can respond to changes in the nature of the problem more or less instantly.
If for example, low cloud and rain covers part of a survey area, the pilot can re-schedule his transects to avoid parts where sensing is temporarily impossible.

When the aircraft passes over terrain where a light resolution image is needed a photograph can be taken with suitable emulsion at a suitable scale. Over other terrain visual scanning may be sufficient.

When the pilot sees phenomena whose significance is unclear he can land or make a drogue collection from a water surface.

RANGE RESOURCE SURVEYS IN SOMALIA

A vegatation map and range resource inventory of the central and northern rangelands of Somalia has been prepared by RMR for the Somali National Range Agency (see Watson et al, 1979 and Watson and Tippett, 1979, Watson 1981)

For this study, excellent 1:200 000 enhanced colour composite LANDSAT prints and 1:100 000 topographical maps were available. Initially, these materials were annotated by systematic aerial reconnaissance flights at 305 to 3 050m, producing precise maps of range-vegatation and land-system boundaries. Except where vegatation or land-system boundaries of great complexity or poor definition were encountered, this reconnaissance work was done from a Cessna 210 aircraft looking through perspex panels. The aircraft flew in two basic patterns: a systematic series of parallel lines 20km apart at 1 525m, while the observers studied vegatation and land-system boundaries at least 10km on either side of the aircraft, and a more variable pattern over all the land-system boundaries and most of the range-vegatation boundaries.

Next, two-stage sampling quadrats were selected from the land-system units, using limited randomization. The sampling units were selected at random, but they were then moved to the nearest point at which they could be easily relocated, either from the air or on the ground. If this resulted in movement of a sampling unit to another range vegatation type or another variant of the original type, then it was rejected and another sampling unit selected. Once the sampling sites were identified, vertical or semi-vertical 35mm Ektachrome photographs were taken at 1:4 000 - 12 000. For most sites, photographs were taken at three scales. The sampling units were also visited and photographed on the ground, and land use, soils and the taxonomy and structure of the vegatation were described. Because of the importance of browse for the livestock of the area, the taxonomic composition of the vegatation was considered particularly important. A representative selection of these sampling units was visited by Hemming and Gillett, who recorded detailed botanical descriptions.

Finally a 5% cover strip-sampling census of livestock wildlife, houses, livestock enclosures, water sources and land use was carried out from the PA18 aircraft in wet and dry seasons. The aircraft flew at 350 feet (107m) above the ground, and at 60 mph (100k/hr). Sampling was stratified, using level system units identified by the process described above, and sampling units were selected with probability proportional to size.

The emphasis in this study on the relocatability of sampling sites was a response to the Somali Government's wish to establish a range vegatation monitoring programme based on fixed sampling sites. Because this survey was carried out as a preliminary to a much greater research and management effort financed through the World Bank, a wide range of materials was prepared. These included:

- a report on static range resources, including topography, climate, geomorphology, soils, drainage, erosion and land use.
- a set of maps at 1:1 000 000 accompanying the report.
- more than 16000 colour transparancies, indexed and fully documented, from about 1500 sampling sites, which will form the basis of a fixed-site range monitoring programme.
- a collection of more than 2000 identified herbarium specimens.
- a preliminary checklist of plants in the area for use by range workers.
- the enhanced colour composite LANDSAT prints, with a grid added to match the logitude/ latitude format of the 1:1 000 000 maps.
- 200 copies of a land-systems/range-vegatation map at 1:250 000 in 51 sheets of $1^o x 1^o$.
- acetate negatives and positives of 1:250 000 and 1:100 000 land system/range-vegatation maps.
- computer printouts of wet-and-dry-season census results, covering livestock, wildlife, houses by structural types, crop production patterns, surface water sources and boreholes.
- 1:1 000 000 maps of the wet-and-dry-season distribution of these features.
- a report on the survey methods used and the main implications of the information provided by the census for possible development alternatives.

The survey design is shown in simplified schematic form in Figure 7 for the Central Rangelands part of the survey.

So far these methods have been applied to almost 400,000 km^2 of Somalia at a total cost to the Somali Government of less than 800,000 US $ over a period of three years.

OIL POLLUTION STUDY IN THE NORTH SEA

Resource Management and Research took part in a remote sensing trial in August 1978 in conjunction with British Petroleum, the British Department of Trades, Warren Springs Laboratory, Aga Infra-red systems, and the Royal Signals and Radar Establishment. The RSRE flew a Viscount aircraft with radar and infra-red scanners over a "controlled" oil slick provided by the Warren Springs Laboratory vessel. RMR flew 35mm cameras and an Aga Thermovision sensor (both hand held) from a Cessna 172 aircraft (a single engine 4 seat low cost machine). At the end of the study period collections of oil were made from the sea surface using a dipped drogue.

Watson and Tippett (1978) reported from the results of these trials, that low level photography, direct visual observation and drogue sampling all offered promising possibilities for future surveillance and arrangement of oil slicks.

Surveillance and management of oil slicks can be considered to comprise of eight stages. The possible role of low-level aerial techniques in these stages is outlined below

STAGE 1. <u>Discovery</u> Large oil slicks are usually notified by the offending vessel/installation. Smaller ones can be detected from aerial surveillance. Visual observation is the most efficient sensing method for surveillance, of those tested in the trials.

2. <u>Location</u> Newly discovered oil slicks can be given a precise location by a low flying aircraft using a radar or VLF position fixing system.

3. <u>Extent</u> The extent of the oil slick can be approximately assessed if the low flying aircraft circles the area taking regular position fixes. A more precise expression of extent can be derived from the mapping described in stage 4.

4. <u>Amount of oil</u> At present the extent of a slick (so far considered to be unstructured) is thought to give no useful indication of the amount of oil in it. In assessing amounts of oil the critical feature is to determine the thickness of the slick. Visual and photographic sensing both provide a clear-cut distinction between films and thicker layers of oil (skeins and mousses). Films are considered to provide negligible contributions to the total mass of oil in a slick and so the problem becomes to measure the <u>areas</u>, <u>thickness</u>, and <u>composition</u> of the skeins and mousses (ie the non-film parts of the slick)

Figure 7 Simplified schematization of the design used for range-vegetation and land-system mapping and resource inventory of Somali central rangelands

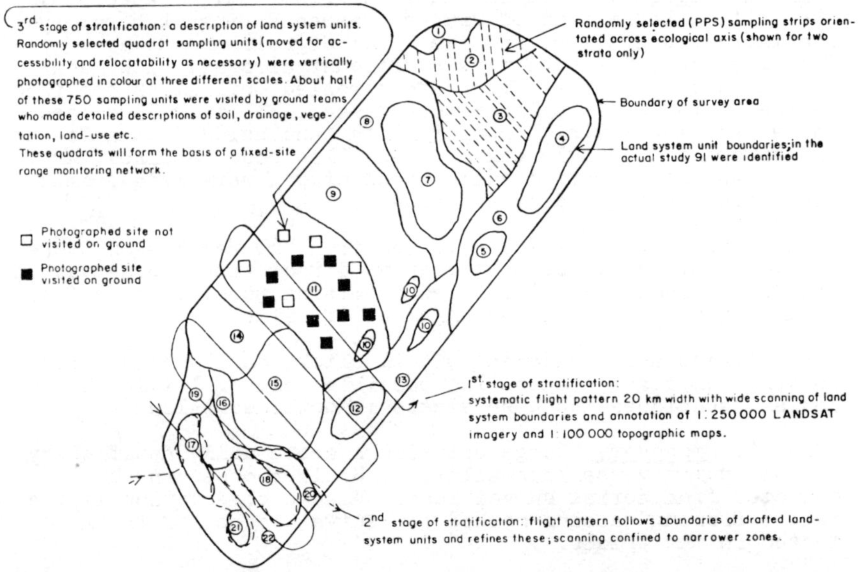

Strata

1 - 22 Strata were classified into biological classes (9) and ecological zones (26).

Strip sampling effort was proportionally greater in strata of higher livestock density and greater human use.

Quadrat sampling effort was fairly uniform over the entire survey area.

The areas of the thick oil layers can be accurately determined by photogrammetric mapping, using Ektachrome or Infra-red emulsion.

The thickness of each part of the "thick" layer is the next parameter. Visual and photographic methods seem to be able to provide information on relative thicknesses of different parts of the slick, but it is not known whether it is possible to consistently relate colour and tone, under specified conditions of lighting and sea-state, to absolute thicknesses for any specified oil. The consultants feel that colour-film records, since they provide such a wide range of information on the slick's characteristics, will be a fruitful source for further experimentation and trials, aimed at determining if a consistent set of relationships exists.

Finally, the composition of the thick layer of the slick, in terms of type of oil and its level of emulsification is necessary for the area and thickness data to be converted into an estimate of the mass of oil in the slick. Low-level aerial droguing, could be used to sample the thicker layers. Micro-analytical techniques would be needed to examine the samples.

5. <u>Type of oil</u> The type of oil involved in any slick is important information for management purposes, apart from its use, described above, for estimating masses.

Moreover information on the type of oil in a slick can provide clues as to its source, with important implications for future controls of pollution, and for meeting the costs of any management.

6. <u>Position of slick in relation to possible pollution (or control) damage</u>

The proximity of a slick to shorelines, beaches, installations, harbours, fishing grounds, bird sanctuaries, fishing fleets, etc. has bearing on the type of urgency of management (ie control) to be recommended. A light aircraft can rapidly survey the immediate surroundings of a slick and report on the features relevant for management. Aerial methods in this instance are particularly valuable for locating mobile phenomena (see Figure 2v) for example).

7. <u>Behaviour</u> The behaviour of the slick (ie. its pattern of movement, the speed with which it is fragmenting or sinking, etc) examined over a short period of time provides valuable additional information on which to generate an effective management plan. Visual observations from low flying aircraft can provide data on slick behaviour of high quality at low cost.

8. <u>Management</u> Stages 1 to 7 will provide the information necessary to determine the type of management required at any particular position and at any particular

time for an oil slick. The application of most types of management requires precise "real-time" information of high resolution about the slick.

The two principle management options are dispersal and collection.

Dispersal usually involves vessels, but in some cases could require large spray aircraft. The delivery of dispersants and the placing of booms from vessels is difficult because the vessel's crew cannot differentiate between thick layers of oil and thin films from the bridge, and therefore cannot apply dispersant or position booms efficiently. A tethered helium balloon, carrying appropriate sensors (HRV, Lowlight video, AGA Thermovision), moored to the leading vessel of the control fleet could feed all vessels with their positions relative to thick layers of oil during actual dispersing or collecting operations.

Aircraft have the same problem as vessels in staying with the thicker layers during a dispersal flight. A higher flying lead aircraft could transmit a video image to the spraying machine which shows the position of the spray-aircraft relative to the thick layers of oil. These trials cost $1200 (£670).

A FOREST INVENTORY IN SUDAN

An inventory of forest and woodlands was carried out by Watson and Tippett in southern Sudan with assistance from J. J. Beckett and F. Jolly, as a contract between RMR and the Sudan Forestry Department (see Watson et al, 1976) and (Abbas, et al, 1977). A three stage random sample was constructed, based on the identification of woodland/forest types from 1:40 000 aerial photographs which were already available. Initially, systematic aerial reconnaissance was carried out and these photographs were annotated and a 1:40 000 base map prepared. Each woodland/forest type became one stratum in the survey design.

Within each stratum, quadrat sampling units were selected at random. At the first stage, a vertical monochrome photograph at 1:3000 - 6000 was taken of all the sampling units. Within each of these quadrats (ie. photographic frames), a smaller quadrat was selected at random and marked on the photographic print. At the second stage, these smaller quadrats were photographed vertically or semi-vertically using 35mm Ektachrome film and the photographs were examined at 6 times enlargement and the species of 80% of the tree canopy identified using the key described below. At the third stage, a random selection of these sampling units was visited on the ground by trained teams of foresters. Each team had a copy of the monochrome print of the original sampling quadrat which they annotated by identifying the species of trees in the smaller second-

Figure 8 Simplified schematization of survey design used for forestry/timber inventory in southern Sudan

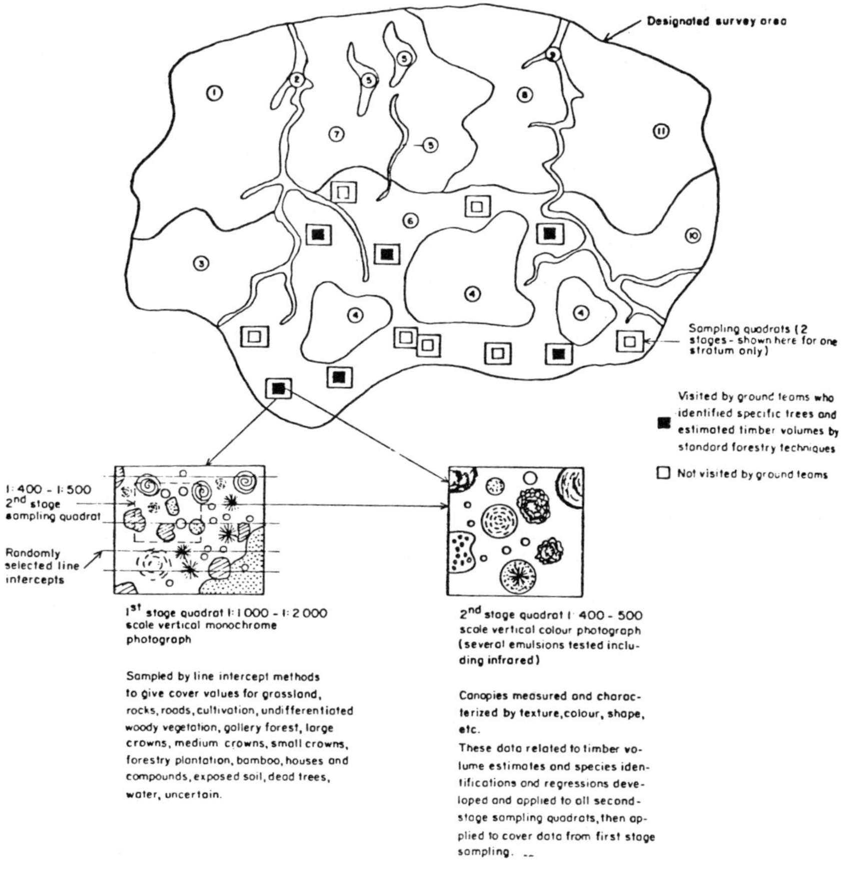

Strata

2,5,9	River valley and swampy/woody associations	Densities of timber trees not markedly different in any stratum. Sampling efforts equally distributed.
4	Highland / woody associations	
1,3,6,7,8, 10,11	Different forest / woodland types	

stage quadrat area and estimating timber volumes using standard forestry methods. The enlarged Ektachrome images were then compared with the annotated monochrome prints and a key was developed for species identification. Similarly, canopy size and timber volume relationships were constructed for each species and type of species. The key was then applied to the sampling units which were not visited on the ground, to provide estimates (of known precision) of species composition and timber volume for each forest type in the entire area. A schematic presentation of the survey design is given as Figure 8.

A slow-flying Piper Super Cub with open panels was used for this work. In addition to aerial surveying, the aircraft was used to guide ground teams to the sampling sites and to drop monochrome prints to the teams at each site, as the film was processed at some distance from the field camp.

The inventory covered about 8000 sq^2 at a cost of US $10,000 (£5600)

SODIUM FLUORIDE PROSPECTING IN KENYA

The Kenya Government has an interest in determining the location of potentially exploitable deposits and solutions of sodium fluoride. Traditional core sampling on Lake Magadi has been carried out for many years by the Magadi Soda Company, which extracts minerals from the surface trona. These show high concentrates of sodium flouride in deeper deposits of trona & silts. Resource Management and Research took a series of photographs with a vertically mounted 35mm camera between 1:5000 to 1:100 000 scales, using Ektrachrome, Ektrachrome Infra-red and Black and White Infra-red film emulsions.

These photographs have been matched with the results of a more detailed surface and core sampling of trona in the Lake to determine if colour differences could be used to detect flouride concentration.

Preliminary results suggest that the most important flouride deposits are below the surface, and that surface densities of sodium flouride have no consistent relationship with sub surface densities.

The optical colour and infra red radiations patterns photographed in the trials seem to relate to the very heterogeneuos patterns of surface and sub-surface variation of sodium flouride density. This trial cost $800(£440).

5. REFERENCES

ABBAS,B., WATSON,R.M. and BECKETT,J.J., (1977): An aerial method of forest inventory. Workshop presentation; ITC International Symposium 1976; New developments in surveys for vegetation, forestry, rural landuse and quick landuse survey. ITC, Enschede, Holland.

BARR,D.J., and BASSARAB,D., (1981): Geological applications of an interactive digital image processing system. Matching remote sensing technologies and their applications. International conference of the Remote Sensing Society, London.

COULSON,M.G., (1981): High altitude aerial photography as appropriate technology for resource studies. As above

DRUMMOND,J., and FULLER,R., (1981): Photointerpretation and computer aided cartography of Broadland vegetation. As above

GIRARD, C.M., (1981): Grassland mapping using satellite multitemporal data. As above

GRIFFITHS, G.H., and COLLINS,W.G., (1981) The development of Landsat systems for detecting and monitoring desert encroachment in N.W.Kenya. As above

GWYNNE,M.D. and CROZE,H., (1981): Light aircraft and resource monitoring. Low level aerial survey techniques; ILCA monograph no.4. Addis Ababa, Ethiopia.

HARRISON,A., (1981) : Spatial resolution of remotely sensed data and its effect on classification accuracy. Matching remote sensing technologies and their applications. International conference of the Remote Sensing Society, London.

HEMMING, C.F., (1981):Light aircraft methods and other approaches. Low level aerial survey techniques: ILCA monograph no.4. Addis Ababa, Ethiopia.

HAGGETT P., and CHORLEY,R.J., (1974): Network analysis in geography. Edward Arnold, London.

HYDE,R.F., VESPER,N.J., GOLDBLATT,I.A., et al. An evaluation of the utility of smaller Landsat ground resolution levels. Matching remote sensing technologies and their applications International conference of the Remote Sensing Society,London.

JOLLY,G.M., (1981A):A review of sampling methods used in aerial survey. Low level aerial survey techniques; ILCA monograph no.4. Addis Ababa, Ethiopia.

JOLLY,G.M., (1981B): Sampling as a cost-reducing tool. Matching remote sensing technologies and their applications International conference of the Remote Sensing Society,London.

KITCHO, C.A., (1981):Use of a GIS system for remote sensing imagery selection and coverage plotting. Matching remote sensing technologies and their applications. International conference of the Remote Sensing Society, London.

NORTON GRIFFITHS,M., (1981): Unstratified systematic sampling: rationale and method. Low level aerial survey techniques ILCA monograph no.4. Addis Ababa, Ethiopia.

THALEN,D.C.P., (1981): An approach to evaluating techniques for surveying vegetation on arid rangelands. As above

TIPPETT,C.I., (1969): A comparison of the available range of light aircraft. E.Afr.Agric.For.J. 34(special issue): 6-15.

WATSON,R.M., (1981): Somali Democratic Republic Northern Rangelands Survey: 4 vols. National Range Agency, Mogadishu.

WATSON,R.M., and TIPPETT,C.I., (1978): A report on some studies of marine oil pollution in the North Sea. Resource Management & Research, London.

WATSON,R.M., and TIPPETT, C.I., (1979): Somali Democratic Republic Central Rangeland Survey; Volume 1: Static Range Resources. National Range Agency, Mogadishu.

WATSON,R.M. and TIPPETT,C.I., (1981): Examples of low-level surveys conducted in Africa from 1969 to 1979: one firm's experience. Low level aerial survey techniques. ILCA monograph no.4. Addis Ababa, Ethiopia.

WATSON,R.M., TIPPETT,C.I., BECKETT, J.J., and JOLLY,F.,(1976) A pilot forest inventory employing aerial methods for the Loka region of southern Sudan. Forestry Department, Ministry of Agriculture, Food and Natural Resources, Khartoum.

WATSON,R.M., TIPPETT,C.I., BECKETT, J.J., and SCHOLES,V.,(1979) Somali Democratic Republic Central Rangelands Survey;Volumes 2-4: Dynamic Range Rsources. National Range Agency, Mogadishu.

WELSH,T.C., (1981): Vegetation mapping from SLAR: a question of resolution rather than wavelength. Matching remote sensing technologies and their applications. International conference of the Remote Sensing Society, London.

WILSON,M., and HENDERSON-SELLERS,A., (1981): Three dimensional histograms and quantitative image processing from low resolution sensors. As above

Part A

MATCHING REMOTE SENSING TECHNOLOGIES AND THEIR APPLICATIONS

TECHNOLOGY OR NEED - WHICH IS THE LEADING FACTOR IN THE DEVELOPMENT OF EARTH OBSERVATION SATELLITE PROGRAMMES?

C Honvault, European Space Agency, Toulouse, France

From the initial idea to the implementation of an applications space system, 10 years are necessary. From the beginning, a trade-off between what is needed and what is possible has to be performed and maintained during the development of the programme.

In the case of Meteosat, the starting point was the definition of the mission objectives by WMO and ICSU, leading to measurement requirments which were translated into technical specifications based on available technology. The introduction of a third channel in the radiometer, in the late stages of the development, is an example of the interaction Need/Technology.

For ERS-1, the original objectives had to be modified to take into account the technical limitations. Experts were involved from the beginning of the definition process. However, as stated in one of the keynote addresses, the identification of end users is more difficult than in the case of meteorology; at present the measurement capabilities have been defined and the Agency is starting the process of identifying the main end users of these parameters who can intervene, if necessary, before the system specifications are frozen.

GLOBAL SATELLITE REMOTE SENSING INTERNATIONALIZATION AND INDUSTRIALIZATION FOR ENERGY, MINERALS AND OTHER RESOURCE DEVELOPMENT DURING THE 1980s.

F B Henderson, The Geosat Committee, San Francisco

During the 1970s, the development of civilian satellite remote sensing began on a global scale primarily with the introduction of the US series of Landsat satellites in 1972. These satellites were designed principally for agriculture, hydrology and land use planning. Applications have demonstrated the value of satellite synoptic perspective, global coverage and increased efficiency of reconnaisance geology in mapping from space to the energy and mineral resource industries.

The international industrial geological community represented by the Geosat Committee has worked with NASA, JPL and others to demonstrate the potential benefits of the Landsat system and additional satellite capabilities to be realized in the 1980s to the energy and mineral exploration community. Capabilities to be added to current Landsat systems should include fixed and pointable stereoscopic coverage, increased resolution (to 10 meters IFOV), additional rock/soil sensitive bands, the synthetic aperture radar and corresponding ground segment systems for digital data processing and applications. The exploration and engineering value of these capabilities have been demonstrated in the joint Geosat-NASA/JPL Test Case Program for oil/gas, uranium and copper deposits.

Several international satellite systems in various stages of proposal/development may include the US LANDSAT D and D', with both the 4-band-multi-spectral scanner and the 7-band Thematic Mapper scanner. In France, there is the SPOT system (1985); in Japan, MOS(Marine Observation Satellite), mid-80s, and the 1987 proposed launch ERS (Earth Resources Satellite); and ESA's proposed marine/land observations systems late in the 1980s. All these developing satellite systems recognise the necessity of international cooperation, primarily because of the commonly held view of preserving the "Open Skies" policy (mutual accessibility and availability of data) in civilian satellite remote sensing. To the globally oriented geological community, this is of the utmost importance.

The success of the various national developing civilian satellite remote sensing systems will in large measure depend on international cooperation and exchange of data, which will inherently lead to the eventual industrialisation of these operational systems. Such industrialization will occur in those areas where industry can more efficiently supply the user community with the required data than can government agencies.

Because of the high cost of obtaining, processing, distributing and preserving the data from these future civilian satellite remote sensing systems and because of the high potential value in future exploration and development of non-renewable resources utilizing the satellite data, all national systems are encouraged to move towards internationalization and industrialization of complementary and compatible systems in the 1980s for benefit of all mankind.

SAMPLING AS A COST-REDUCING TOOL

G.M. Jolly
Agricultural Research Council Unit of Statistics
University of Edinburgh
Mayfield Road
EDINBURGH EH9 3JZ

Abstract: In resource surveys on the ground or from aircraft, the required information is usually adequately provided by a sample. The method of strip-sampling from light aircraft is described. Essential to the method are calibrating techniques to eliminate bias in estimation of the area of a sample strip and from faulty observation; to maintain overall cost-effectiveness, sampling for these should not materially increase the total sampling error. Examples of other double sampling procedures are cited in which a correlated, low-cost variable replaces the observation of interest in the main survey and calibration is performed on a relatively small sample. Wider application of double sampling is suggested, especially with use of LANDSAT. Improved methods of surveying large objects such as farms are discussed. Fully valid sampling methods are preferred to those involving additional, dubious assumptions such as random distribution of the population.

1. INTRODUCTION

The role of sampling in a resource survey is varied, and it may be useful to consider some of the principles involved with particular reference to the use of aircraft as one source of information. These will be discussed under separate headings, although most are closely interrelated.

Paper presented at a conference on Matching remote sensing technologies and their applictions convened in London by the Remote Sensing Society, 16-18 December, 1981

2. SAMPLE OR COMPLETE COVER?

This question applies to aerial or ground surveys where cost is invariably related to sample size.

If the requirement were a complete enumeration or mapping of, for example, all water sources in a region, then clearly 100 percent cover of the region is necessary. More often a resource may need to be evaluated only as a single total for the region, or as sub-totals for separate sub-regions. In such instances the sample size will depend on the level of precision required in the estimates. There can be no merit in a 100 percent sample as such. When a survey is undertaken on a "one-off" basis and the results are to be used in decisions for the future, there is likely to be a substantial and unknown component of variability associated with time, and this sets a limit to the precision necessary. Table 1 from Scottish Agricultural Colleges Technical Notes (1975, 1976) exemplifies an extreme case of high variability in a characteristic which depends heavily on climate; clearly, the numbers of samples examined were more than sufficient to establish the mean and general pattern of failure rate.

Table 1

Percentage of barley samples failing to reach Intervention standards in three regions of Scotland (number of samples examined in brackets)

Year	Region			All Scotland
	East	North	West	
1973	68.0(25)	40.0(30)	75.0(24)	59.5(79)
1974	71.4(56)	74.5(51)	88.9(54)	78.3(161)
1975	26.3(57)	60.0(60)	74.2(62)	54.2(179)

3. DETERMINING AREAS OF SAMPLING UNITS IN AERIAL STRIP SAMPLING

Some of what follows refers to the methods of sampling from light aircraft described yesterday by Dr. R.M. Watson. Observations are made by observer or pilot/observer from 100-120 m on sample strips approximately 250 m in width and defined by parallel marker rods attached to the port wing strut. Selection of a sample of strips is by conventional methods of random sampling from the total of contiguous parallel strips within a previously defined stratum. The direction of strips is chosen to be the direction of greatest expected variability, thus minimising variability between strips. Further details are given in Jolly and Watson (1979), or in the references to particular surveys cited by Watson.

As the width of strip viewed at any moment is highly variable on account of movement of the aircraft and of the observer's head, an essential ingredient of the technique is an accurate calibration procedure to obtain an estimate of the mean strip width seen. This is achieved by accumulating a large body of data from repeated flights over markers placed at known spacings on the ground and simultaneously recording strip width and height of aircraft. The mean strip width as seen during a survey is then estimated as

$$\hat{\bar{w}}_s = \frac{\bar{w}_c}{\bar{h}_c} . \bar{h}_s$$

where $\bar{w}$ and $\bar{h}$ denote mean width and height, c refers to data obtained during calibration and s to survey data. The h_s are observations of height made from radio altimeter at approximately 5 min intervals while the aircraft is surveying from an intended fixed height. In subsequent calculations $\hat{\bar{w}}_s$ is used in calculating the area of each sampling unit (strip). Four sources of random sampling error are associated with the procedure. Two arise directly from the calibration, these being estimation errors of the ratio $\bar{w}_c/\bar{h}_c$ and from variation of the h_s about their mean $\bar{h}_s$. However, by obtaining sufficient calibration data and observations h_s , it is easily shown that those errors become entirely negligible. If it were otherwise, then the surveyor's top priority would be to supplement these data. The two remaining components of error are associated with fluctuations in aircraft height occurring between the observed h_s , and movement of the observer's head and general instability of the aircraft. Both of these affect directly what the observer sees so are automatically incorporated in the formula for the general sampling error, on which they can be shown to have a negligible effect.

The application of sampling in calibration generally would be regarded not so much as a cost-reducing tool but as an essential to eliminate what could otherwise give rise to serious biases. In this sense it is merely desirable to ensure that associated random components of error are suitably small.

4. CORRECTING FOR BIAS IN OBSERVATIONS

Sources of potential bias occur in the observations themselves, for example, through misidentification of crop species, or in counts of animals concealed by vegetation or other animals. Watson has described his procedure of observing groups of animals, independently of the survey itself, first in the normal way and then from lower altitude when the group has moved out of cover and when by camera or eye a count can be made that is confidently considered the "true" count. A correction factor equal to the ratio of the accurate to the initial counts is thus established by which the survey estimate for that particular type of cover will be multiplied. This also is a form of calibration and again the effect on the overall sampling error can be made extremely small by having adequate data for estimation of the correction factors.

Assuming bias in animal counts to be unacceptable, then the above could be regarded as a means of reducing the cost from what would otherwise be prohibitive if every animal group in a survey had to be examined closely.

5. DOUBLE SAMPLING

The procedure just described for obtaining correction factors is an example of the general method known as double sampling in which both the variate of interest, y , and a supplementary variate, x , are observed on a relatively small "sub-sample" from which the relationship between x and y is established. Only x need be observed in the main survey sample and $\bar{y}$ is then estimated from the x-y relationship. Suppose that x and y are related by a linear regression. Whether the "sub-sample" is a true sub-sample from the main sample or an additional independent sample, Cochran (1977) shows that for double sampling to give increased precision over sampling of y only, when total cost is fixed, the correlation, ρ , between x and y , must exceed a critical value of approximately

$$|\rho| = \frac{2\sqrt{c_n c_{n'}}}{c_n + c_{n'}},$$

where $|\rho|$ is the value of ρ taken positively, c_n and $c_{n'}$ are the respective costs per sampling unit for the sub-sample and main sample and n and n' are the corresponding sample sizes.

Watson, Tippett et al (1977) have used the double sampling principle in other ways, for example, to estimate numbers of sheep and goats separately from mixed herds where the species are difficult to distinguish from the air, and to estimate the human population in rural areas from aerial estimates of numbers of different types of dwelling. In both instances ground samples provide the relationship required, that is, the ratio of sheep to goats in mixed herds, and average numbers of persons associated with particular types of dwelling.

LANDSAT imagery must surely provide unlimited opportunities for the application of double sampling to estimate characteristics not distinguishable from satellites but correlated with other characteristics which <u>are</u> distinguishable. The calibration sample might be from aerial or ground samples or possibly treble sampling could make use of all three levels. One difficulty liable to be encountered is psychological reluctance to accept one measure made indirectly through another. Provided that the sub-sample is always representative of the region from which the main sample is taken, the only requirement for double sampling to be more efficient is that the correlation should exceed a critical value. More sophisticated applications might include, for example, replacement of x by one or more derived variates from a principal component analysis.

6. OTHER APPLICATIONS OF SAMPLING

The function of stratification and the multi-stage sample to reduce cost is well established and needs no further comment.

One problem that has received scant attention in the past is how best to record objects whose size is not negligible compared with the dimensions of the sampling unit. The customary approach to sampling of farms from maps has been to include the farm only if some pre-determined point such as the farmhouse falls within a selected sampling unit. Depending on the pattern of costs in relation to the sampling method, this rule is not necessarily the most efficient. If, in aerial strip sampling, observations are to be related to specific farms, information on farm boundaries has to be obtained on the ground. An efficient procedure then utilises all data recorded from the air, but, to ensure that every farm in the population has the same probability of being included in the sample, an observation is weighted by the reciprocal of the total number of sampling units containing part of the farm concerned. For this particular example a more readily available weighting might be the reciprocal of the proportion of the farm area contained in the observed sampling unit, that is, the weight might be replaced by an estimate of it.

The theory of this and other similar methods is given in Jolly (1979). Watson, Tippett et al (1977) use the above technique for arable farms under irrigation and also to improve efficiency in counting of animals occurring in groups or herds. By simple marking devices on the aircraft the observer can quickly assess the number, w, of sample strips which would include the herd. The complete herd is then recorded but weighted by a factor of $1/w$. A count of numbers of herds must be similarly weighted. Thus, for three herds covering respectively 2, 1 and 3 sample strips the number of herds recorded would be

$$1 \times \frac{1}{2} + 1 \times 1 + 1 \times \frac{1}{3} = \frac{11}{6}$$

The resulting, unbiased information on distribution of herd size is a bonus additional to the improved efficiency of the total count of animals. The maximum increase in efficiency possible for a herd dispersed over w sampling units is w.

7. CONCLUSIONS

The first requirement of any sampling procedure is undoubtedly validity - at least to a degree well within the bounds of sampling error. It may be noted that all techniques referred to above, if properly applied, are valid in this sense. Other techniques are often used which depend on special assumptions such as random distribution of the population. Unless no sensible alternative exists, these are best avoided. A sample can more appropriately furnish information on distribution rather than rely on a distribution property such as randomness, a rare characteristic in nature.

Given validity, introduction of an additional sampling procedure or change in existing procedures should always aim at increasing efficiency, never the reverse. Since efficiency is measured as the precision of an estimate for fixed total cost, C, and precision is inversely proportional to the variance, V, of the estimate, this implies minimisation of VC.

The exploration into new or modified procedures founded on basic sampling principles may be rewarding. Our approach to sampling may be akin to that of most animal species to potential danger. At an early age innocence prevails and mistakes are made; in maturity we settle down to the known, tried methods which we have learned to trust; finally, if we survive long enough, a new confidence develops to try something which hitherto may have seemed more than a little peculiar.

8. ACKNOWLEDGEMENTS

Dr. R.M. Watson from whose initiative and cooperation several of these ideas developed.

9. REFERENCES

COCHRAN, W.G. (1977): Sampling techniques. Wiley, New York. Third edition, 428 pp.

JOLLY, G.M. (1979): Sampling of large objects. In Sampling Biological Populations. R.M. Cormack, G.P. Patil and D.S. Robson, eds., Satellite Program in Statistical Ecology, International Co-operative Publishing House, Fairland, Maryland, U.S.A.

JOLLY, G.M. and WATSON, R.M. (1979): Aerial sample survey methods in the quantitative assessment of ecological resources. In Sampling Biological Populations. R.M. Cormack, G.P. Patil and D.S. Robson, eds., Satellite Program in Statistical Ecology, International Co-operative Publishing House, Fairland, Maryland, U.S.A.

SCOTTISH Agricultural Colleges Technical Notes: Nos.5 (1975), 9 (1975) and 13 (1976).

WATSON, R.M., TIPPETT, C.I. et al (1977): Sudan National Livestock Census and Resource Inventory. Ministry of Agriculture, Food and Natural Resources, Khartoum, Sudan (33 vols. mimeo).

THE CANADIAN RADAR-SAT PROGRAMME

Ed Shaw

Programme Manager, Radar-Sat Planning
Office, Department of Energy & Mines
Ottawa KIA OE4 Canada

ABSTRACT

Phase A study has begun on a synthetic aperture radar satellite to provide information over the frozen and open oceans bordering Canada. Secondary applications, forestry. geology, hydrology and agriculture will also be investigated. The base-line parameters for the radar are a swath-width of 150 kilometres with a ground resolution of 25-30 metres and an incidence angle of 35-40 degrees. Preliminary orbit studies indicate that with this swath-width it is possible to obtain twice daily coverage over the North-West Passage, the primary Arctic transportation corridor.

An important requirement will be for direct telemetry of radar data to be sent to three ground stations equipped with SAR processors with sufficient through-put to maintain zero backlog for the area requiring near real-time response. An interdisciplinary team has completed a preliminary study of mission requirements for many applications which are discussed.

THE EUROPEAN SAR-580 CAMPAIGN AS A PREPARATION FOR THE ERS-1 SATELLITE SAR

A Haskell, ESA, Toulouse

This paper provides an account of the European SAR-580 Campaign conducted in June and July 1981. The relevance of the campaign to the preparation for the synthetic aperture radar sensor for the ERS-1 satellite is emphasised.

MATCHING REMOTE SENSING TECHNOLOGY TO OPERATIONAL RENEWABLE LAND RESOURCE MANAGEMENT REQUIREMENTS SYSTEMS IN CANADA

J Cihlar, E J Langham and W M Strome, CCRS, Ottowa, Canada

Experimental applications of multispectral and other advanced forms of remotely sensed data have demonstrated that remote sensing can make a valuable contribution to the monitoring and management of Canada's land resources. More frequent coverage and additional spectral bands on satellies planned for the mid 1980s and beyond will increase the opportunities for regular use of remotely sensed data. To effectively utilize these data in resource management, the remote sensing input must be matched with the resource management systems existing at that time. Thus, it is essential to anticipate the form of resource management systems of late 1980s and 1990s, to determine the appropriate role for remotely sensed data and to develop and implement a plan which will yield the remote sensing systems and methodologies necessary to meet the operational resource management requirements.

Previous studies of resource information needs indicate that there will be a need for geocoded remotely sensed data, improved image analysis techniques and better information integration concepts for future resource management systems. To develop a plan for meeting the anticipated requirements one must first consider the flow from the recording of the remotely sensed data to the end use of the derived information. The timeliness and accuracy requirements of different users, the diverse data types and forms required for individual applications, the analysis methods/decision models needed, and the implication of these factors on the configuration(s) of remote sensing input into the future resource management systems must be examined. From this analysis, areas requiring further work (research, development, demonstration, transfer) can be identified, and a plan of action can be suggested.

THE ROLE OF AIRBORNE DIGITAL MULTISPECTRAL SCANNING - A REVIEW OF RECENT OPERATIONAL EXPERIENCE

J L Van Genderen: G De Man, Belfotop-Walphot, Wemmel, Belgium

In 1979, the first digital Daedalus multispectral scanner arrived in Europe. This paper reviews the experience gained in various European countries by this twelve channel digital airborne MSS (Model 1260) over the past two years.

Various practical operational problems are discussed, including those of data acquisition, data processing and data analysis.

Data acquisition is carried out by using the ERSAF-platform (ERSAF = European Remote Sensing Aircraft Facility), equipped with three camera bays, real-time on-board imaging in two spectral bands, and coupled to a digital navigation system.

The digital MSS, which has four detectors, records the spectral reflectance data onto a HDDT (High Density Digital Tape Recorder), from where, by means of a specially designed interface, it can be transcribed to CCT or direct to disc, for immediate processing on a digital image analysis system.

The paper will give examples of research projects of acquiring multispectral reflectance data at altitudes ranging from 1.000 feet to 20,000 feet. By flying at greatly different flying heights over the same test area on the same day, many of the factors of prime importance for spectral signature analysis have been able to be studied, such as effect of pixel size on spectral reflectance, atmospheric effects, look angle effects, etc.

Especially for spectral signature analysis, the digital MSS system, now operationally available, will allow European workers not only to simulate spectral data from Landsat D, SPOT and other satellite MSS data, but also to carry out controlled, basic, airborne spectral reflectance measurements.

The use of digital MSS technology will be illustrated by means of applications in the field of oil pollution monitoring, soil studies, agriculture and vegetation analysis.

LANDSAT IMAGERY FOR MAPPING AND MONITORING VEGETATION IN NORTHERN KENYA

G H Griffiths and W G Collins
Remote Sensing Unit,
Civil Engineering Department,
University of Aston,
Birmingham,
B4 7ET,
England.

Abstract: False colour composite imagery in combination with panchromatic aerial photographs and ground-data, were used to map semi-arid vegetation in Marsabit District, Northern Kenya. Results showed that a high reflectance from a dry soil surface dominates a combined spectral response from vegetation and soils, in an area where reflectance from a sparse cover of dry vegetation is low. Thus only two distinct vegetation classes, covering a small part of the study area could be accurately mapped from the imagery, making it difficult to apply the method to other areas and to monitor the impact of desertification on the vegetation cover. Nevertheless, the study identified the problems involved in mapping the semi-arid vegetation of this region, and provided essential information about the spectral characteristics of soil and vegetation types.

1. INTRODUCTION

More than half of the land surface of Kenya, with an annual rainfall of less than 600mm, is classified as semi-arid and supports only a sparse cover of bush with grassland vegetation. Traditionally these rangelands have been used by nomadic pastoralists to support their herds of livestock. Recurrent drought, notably between 1968-1976, combined with pressure upon resources from an expanding population, has in many areas resulted in a loss of rangeland productivity leading to desertification.

Paper presented at a conference on Matching remote sensing technologies and their applications convened in London by the Remote Sensing Society, 16 - 18 December, 1981.

Figure 1: Location of study areas A & B

Source : IPAL Technical Report (1977)

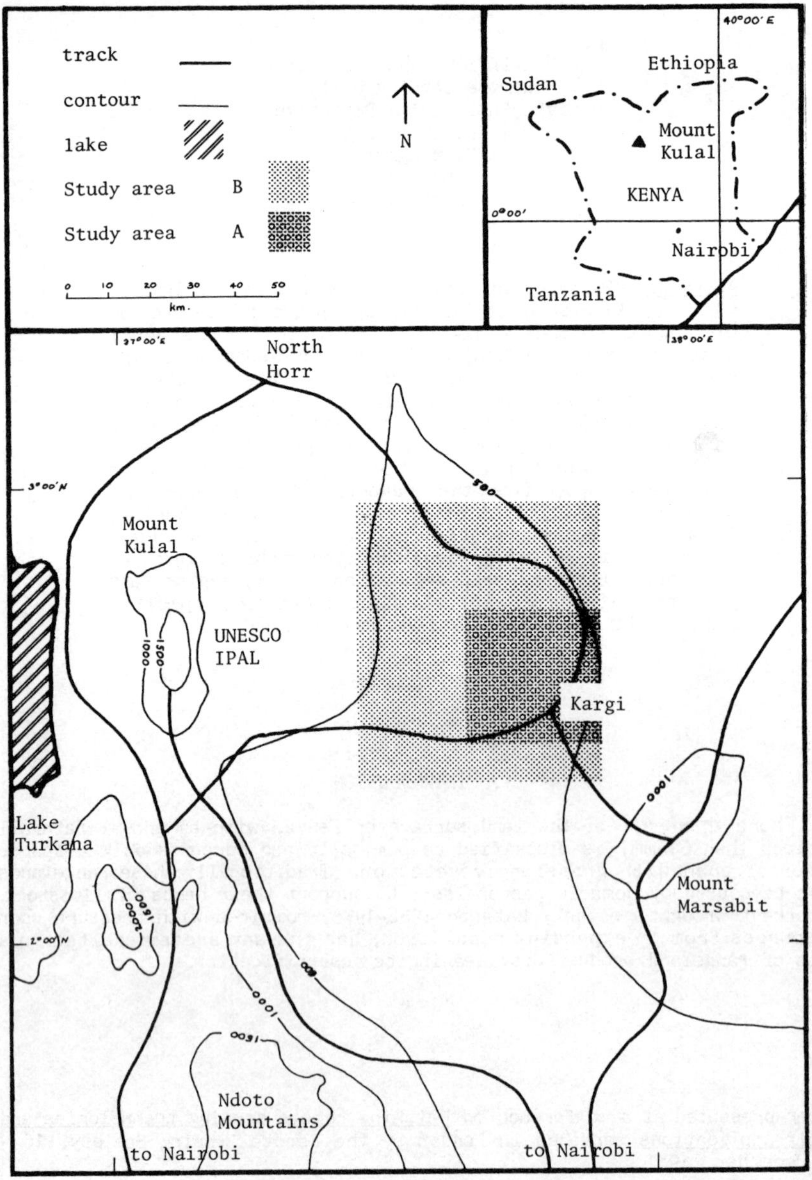

The primary objective of this paper is to identify the problems involved in mapping semi-arid vegetation using LANDSAT false colour composite imagery and in particular to determine whether LANDSAT imagery can be used to detect changes in the composition and canopy cover of the vegetation resulting from desertification. The area covered by this investigation is part of a larger region in Northern Kenya at present being studied by a United Nations Integrated Project for Arid Lands (IPAL), the aim of which is to find,

> "direct solutions to the most urgent environmental problems associated with desert encroachment and ecological degradation of arid lands". (Lamprey, 1978).

In particular, base-line data are being gathered to serve as a reference point for a programme of continuous monitoring capable of measuring ecological change. It was felt that the repetitive and sequential nature of LANDSAT data could provide useful information, and this paper describes some of the problems, and points to the likely potential, of applying LANDSAT data to a project of this kind.

2. THE STUDY AREA

Figure 1 shows the position of the area under study which is situated to the east of Lake Turkana in Marsabit District, Northern Kenya, and is divided into two parts:

1) area A of 20 x 30km, with its centre at Kargi (Long. 37`35'E, Lat. 2`30'N),

2) area B of 50 x 60km, which mostly lies within the basin defined by the Chalbi desert to the north and the Kulal and Marsabit lavas to the west and east.

3. SOILS AND GEOLOGY

There is very little published information about the soils and geology of the area on a scale suitable for a study of this kind, (Herlocker, 1979). Two major geological lithologies dominate the area; Pleistocene basaltic lavas flowing from Mount Kulal (2335m) in the west, and Mount Marsabit (1865m) in the east. Quaternary sediments derived from the Pre-Cambrian basement massif to the south form basin deposits which partly overlie the basaltic lavas to east and west. These two lithologies can be accurately mapped from the LANDSAT imagery; the highly reflective sediments appearing in sharp contrast to the basaltic lavas. Soil types are closely related to the geology, with soils derived from lavas being generally loamey to clayey in texture, whereas soils derived from basement material are sandy and more permeable, (Hemming, 1972). Both study area A and area B are situated within the extensive, flat, sedimentary basin at an elevation of 530 - 760m. For the purposes of the present study, areas of basalt lava have been excluded.

4. CLIMATE

Rainfall is low (<150mm/annum) and is combined with very high evaporation rates (2600mm), consequently little water remains for plant growth (Edwards et al., 1979). The rains generally arrive in the two wet-season periods of April/May and November, though rainfall is unreliable and variable in time and space, resulting in long periods of drought. The 1968-76 drought has been followed by a series of relatively wet years. In particular the heavy rains of April/May 1981 were unusually high and prolonged. This fact had to be taken into account during fieldwork in the summer of 1981.

5. VEGETATION

The rangeland vegetation classification of Pratt et al., (1966), assigns this vegetation to ecological zone VI,

> "Rangeland of low potential, the vegetation being dwarf shrub grassland or a very dry form of bushed grassland, in which Acacia reficiens is a characteristic species".

This classification, modified and adopted by Herlocker, (1979) for the compilation of an IPAL vegetation map, has been used for this study.

Pure stands of annual grassland dominated by Aristida adscensionis and A. mutabilis, cover only a small part of the study area. The occurence of these two grassland species is normally associated with dwarf shrubland, the dominant species Indigofera spinosa protecting the grass from severe grazing pressure. The largest unit in the area is deciduous shrubland, dominated by Acacia reficiens with an understorey cover of mostly I. spinosa dwarf shrubs*. Finally, a small part of the study area is covered by Acacia tortilis woodland, associated in this case with Salvadora Persica evergreen species, and with an understorey of Duosperma eremophilum and I. spinosa dwarf shrubs.

6. SETTLEMENT

Traditionally, the nomadic Rendille tribe have moved across an area stretching from the Chalbi desert in the north, between Lake Turkana (Rudolph) in the west and Mount Marsabit in the east, and southwards to Samburu country at the base of the mountains in the south. This traditional pattern of movement, whereby family groups travel across the rangeland in search of good grazing, began to break down during the colonial period, but especially following independence in 1963, (Walther and Herlocker, 1980). Settlement became more sedentary as Missions and Schools were built, and a system of pumps, wells, roads and "dukas" established. These social and economic changes are particularly marked at Kargi, (Figure 1), where large concentrations of people and stock are creating a degraded environment. Local people cut woody vegetation for fuel and building materials, whilst intensive grazing by livestock has denuded the vegetation cover up to 6km from Kargi, with restricted plant growth occuring over a much larger area, (Griffiths, 1979, unpublished).

The relative uniformity and structural simplicity of the area in terms of vegetation and soil, combined with social change and the onset of desertification, suggested that this would be a good test area in which to apply LANDSAT imagery and to evaluate its usefulness for mapping and monitoring semi-arid vegetation.

7. METHODOLOGY AND RESULTS

This has so far involved a comparison of satellite false colour composite imagery with other data bases, including:

1) a vegetation map at a scale of 1:500,000
2) R.A.F panchromatic aerial photographs at a scale of 1:50,000 for 1972

* The canopy cover of a composite vegetation type is the sum of its separate components, e.g. the percentage of a unit area covered by shrubland and dwarf shrubland.

3) ground-data on vegetation composition and canopy cover, and soil type, collected during summer 1981
4) vertical, low-altitude, full colour aerial photographs taken on a sample basis from a height of 500m during summer 1981
5) 1:100,000 UTM projection topographical maps.

In the early stages much of the work was done with an I2S Colour Additive Viewer using sections of 1:1m black and white positives in MSS (multispectral scanner system) bands 4, 5, 6 and 7 (0.5 - 1.1 um), and enlarged to a scale of 1:150,000 on the viewer screen. Although this method is flexible and allows the researcher to combine different bands with different colour combinations for better discrimination of classes, and to produce ratio and multi-temporal images from negative and positive transparencies, the problems of producing cheap, high-quality "hard-copy" are considerable. At a later date, very high quality, cloud-free NASA false colour composite images were obtained at a scale of 1:250,000 in the dry season month of January for 1973, 1976 and 1979*. Dry season images were obtained, partly because of cloud-free skies, and partly because atmospheric and surface conditions are likely to be more uniform from year to year during the dry season, thus enabling comparisons between images to be made.

Interpretations of soil and vegetation units from the colour composite imagery were transferred to a topographical UTM projection map-base using a Bausch and Lomb Zoom Transferscope. Although NASA images are "system corrected" to compensate for earth rotation, non-systematic changes in the altitude , attitude and velocity of the satellite can result in pixel position displacement of up to 400m in an extreme case, (Colvocorsces and McEwan, 1973). This distortion can be crudely but adequately compensated for with a Zoom Transferscope.

A crucial problem in mapping semi-arid vegetation is that reflected energy sensed by the satellite from each pixel is a composite of reflectance from both the vegetation and the soil, (Hielkema, 1977).

A second problem is that dwarf shrubland and shrubland, the vegetation types which cover most of the area, are xerophytic, drawing water for growth only during the wet-season, when the appearance of the rangleland from the air is overwhelmingly green (H. Lamprey - pers. comm). During the dry season vegetation is dry and dormant, the deciduous shrub species losing their leaves and the grassland drying back, so that the peak of reflectance in MSS 7 (0.8 - 1.1um), characteristic of green vegetation in temperate regions, is absent, (Siegal and Goetz, 1977).

A third problem is that the dry soils of the area (<5% moisture content), are highly reflective, the reflectance increasing steadily with wavelength (Otterman, 1980). Again, this situation contrasts with temperate regions where reflectance from moist soil peaks in the visible wavelengths (MSS 5: 0.6 - 0.7um) before decreasing sharply in the near infrared (MSS 6 and 7: 0.7 - 1.1um).

These facts suggest that it may be difficult, given the spectral and spatial resolution of the present generation of earth resource satellites, to separate spectrally the low reflectance of sparse, dry, non-green vegetation from the high reflectance of dry soils.

* Imagery for the area (Path 181/Row 58) is not available after 1979 due to satellite technical problems.

Figure 2: Comparison of interpretation from panchromatic aerial photographs of soil and vegetation boundaries for study area A

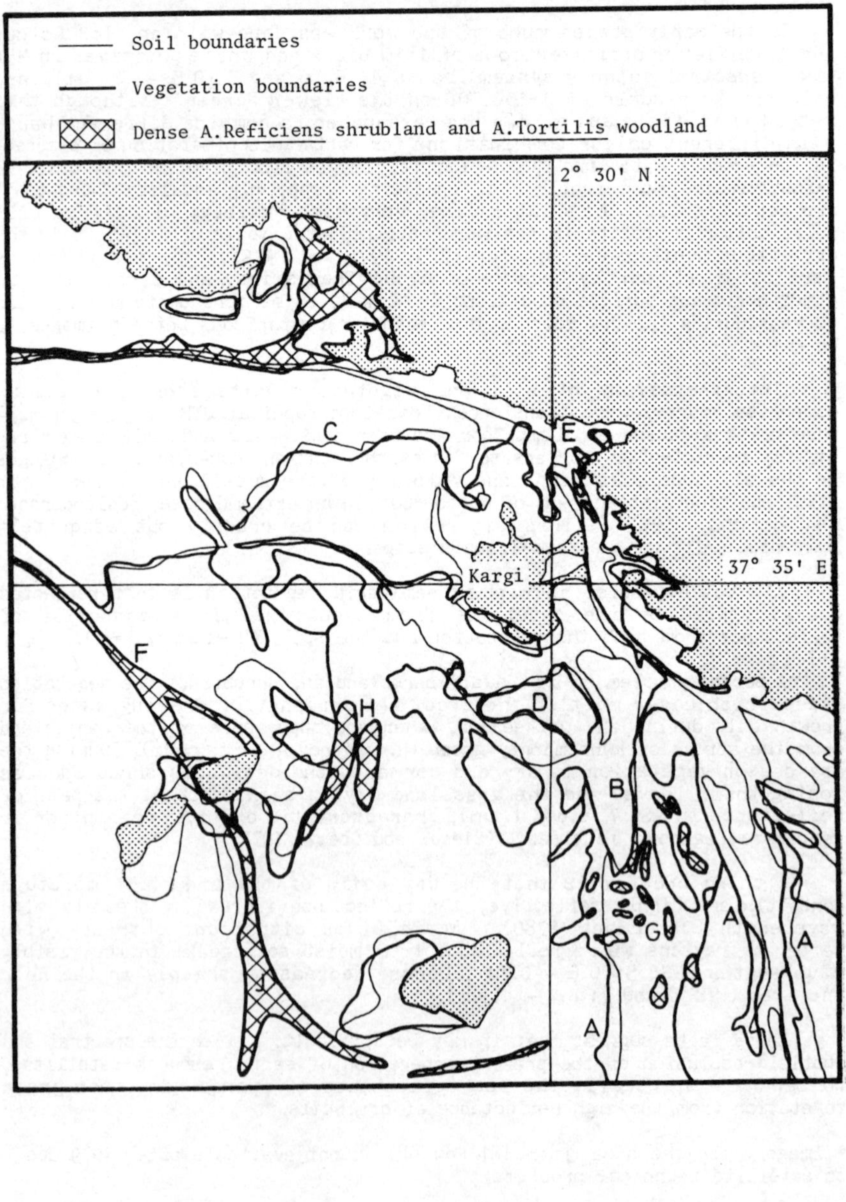

Figure 2 is an interpretation of soil and vegetation boundaries based on the R.A.F. panchromatic aerial photographs for study area A (figure 1), transferred from the photographs at a scale of 1:50,000 to a topographical base map at a scale of 1:100,000, and reduced to A4 size for this paper. Criteria for interpretation were based upon canopy cover and type for vegetation, and tonal and textural variations for soil.

The interpretations were confirmed with reference to full-colour, low-altitude aerial photographs and ground-data. The identified boundary lines for vegetation and soil types indicate that there is a close correspendence between soil and vegetation boundaries. For example, the boundaries shown as A on figure 2 mark a transition from a sandy loam soil which supports dense thickets of A. reficiens shrubland with a canopy cover of 40 - 60%, to a silt loam supporting a more open cover of A. reficiens shrubland with a canopy cover of 10 - 20%. The boundary shown as B marks a transition from dense A. reficiens shrubland to an open cover of I. spinosa dwarf shrubland, the difference again reflecting a change from a sandy to a silty loam. The boundaries marked as C show a transition from an open cover of A. reficiens shrubland growing on a silt loam to a sparse dwarf shrubland cover growing on old, stabilised sand dune formations. Finally, D marks a transition from a moderate cover of A. reficiens shrubland (20 - 40%) to an open cover of I. spinosa dwarf shrubland.

In an area where water for plant growth is scarce, the water retentive properties of different soil types strongly influence the distribution of vegetation types, (Herlocker, 1979). It is predictable, therefore, that there is a close correspondence between soil and vegetation boundaries as interpreted from aerial photographs. However, in view of the low reflectance of dry, semi-arid vegetation in all MSS wavelength bands (0.5 - 1.1 um), the sparse nature of the cover and a high reflectance from dry soils, the finding above suggests that the satellite colour composite imagery shows mostly variation in soil-type and not changes in the canopy cover and composition of the vegetation.

Neverthless, it can still be argued that the very close correspondence between vegetation and soil boundaries allows for the accurate mapping of vegetation classes. Bentley, (1976), in a study of ephemeral and perennial rangeland vegetation in Arizona with a canopy cover ranging from 2 - 20%, showed that vegetation classes could be accurately mapped only when associated with homogenous soil and landform units, making it very difficult to extrapolate results and methods to other areas without the need for extensive ground-data collection. If the composition and canopy cover of a vegetation type in association with a particular soil or landform unit is known, this method is adequate for broad, small-scale mapping. A comparison between the IPAL vegetation map, (Herlocker, 1979) and boundaries interpreted from the LANDSAT imagery, shows a close correspondence between the two. For example, areas of basaltic lava covered with a semi-permeable silt-clay soil support a sparse annual grassland vegetation and although these same areas can be identified on the LANDSAT imagery,(Figure 2), further information about the characteristics of the canopy cover is not obtained. For a study of desertification, where changes to the vegetation cover in time and space are small, accurate large scale mapping of vegetation types is required.

Figure 2 also shows that certain vegetation boundaries do not coincide with soil boundaries. The shaded areas marked E, F, G, H, I, J show patches of vegetation that either mask underlying soil patterns or which are unrelated to changes in soil type. Ground-data and aerial photographs

Figure 3: Comparison of vegetation boundaries interpreted from panchromatic aerial photographs with an interpretation from the LANDSAT false colour composite image (January 1979), for area A.

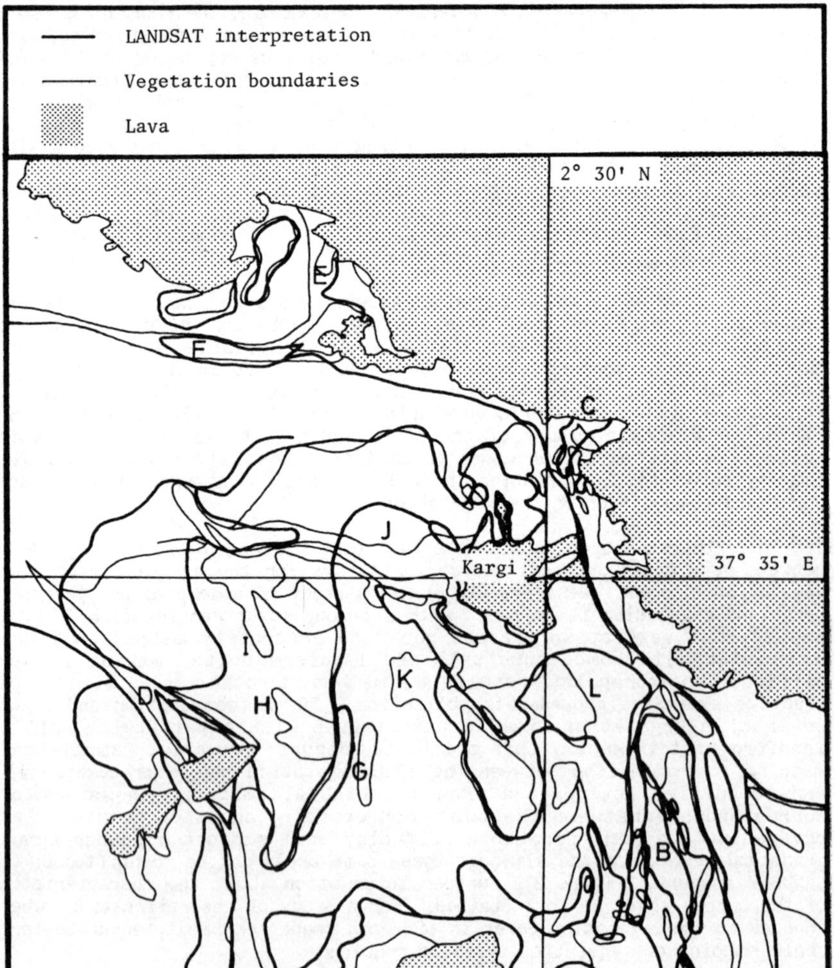

identify these areas as patches of vegetation which are either very dense thickets of A. reficiens shrubland which have a low reflectance and appear dark on the image in contrast to the surrounding, sparsely vegetated soil, or are more open groves of A. tortilis woodland which appear deep red and are highly reflective in the near infrared (MSS 6/7: 0.7 - 1.1um).

In figure 3 the vegetation boundaries shown in figure 2 are compared with an interpretation from the LANDSAT colour composite image for January 1979. This has been transferred to a topographical base map at a scale of 1:100,000 and reduced to A4 size for this paper. The shaded areas marked A, B, C, D, E and F are those vegetation units which were interpreted from the aerial photographs as either dense A. reficiens shrubland or A. tortilis woodland. They show a correspondence with the boundaries interpreted from the LANDSAT image. However, certain areas which were identified as homogenous vegetation units on the aerial photographs cannot be detected on the LANDSAT colour composite. These shaded areas are marked G, H, I, J, K and L on figure 3. From ground-data and full colour low-altitude aerial photographs these units are identifiable as areas of either I. spinosa dwarf shrubland, with a canopy cover of <20%, or A. refieciens shrubland, with an open canopy cover of 10- 20%. In addition, broad scale shrubland/dwarf shrubland boundaries shown on the IPAL vegetation map, (Herlocker, 1979), in particular, the boundary between these two vegetation classes to the south of the area which runs for 50km east to west, cannot be detected on the satellite imagery.

Figure 4, which has been transferred from the LANDSAT imagery onto a topographical map base at a scale of 1:100,000 and reduced to A4 size for this paper, shows the extent and distribution of Acacia woodland and A. reficiens shrubland within the wider study area B, that can be accurately mapped from the LANDSAT colour composite image for 1979. From an extrapolation of the results obtained in the area A to the wider study area B, the figure shows that the two vegetation classes cover only a small part of the total study area.

Apart from the possibility of mapping vegetation on the basis of its local correspondence with soil type, it appears that the two factors, the greeness and canopy cover of the vegetation determine the ease with which vegetation can be mapped from colour composite imagery in this region of Northern Kenya. Green vegetation is highly reflective in the near infrared (MSS 6/7: 0.7 - 1.1um), in this case swamping a lower reflectance from the soil surface. Furthermore, it is likely that higher soil moisture levels associated with Acacia woodland depress the infrared reflectance of the soil, and encourage the growth of a vigorous shrub understorey which further reduces soil reflectance. In the case of a dense A. reficiens shrubland soil reflectance is also masked by the high combined canopy cover of shrub and dwarf shrub species, and by a significant shadow component which is added to and further lowers the composite reflectance from soil and vegetation, (Hielkema, 1977). The lack of green leaves and the cell structure of stem material gives a very low reflectance in all wavelengths, where the absence of chlorophyll lowers the reflectance peak characteristic of green vegetation in the visible wavelength (MSS 5: 0.6 - 0.7 um), (Carneggie et al., 1974). This vegetation type can be very clearly seen on the colour composite image for January 1973, when the contrast between the dense shrubland and surrounding dry, sparsely vegetated and highly reflective soil is accentuated under the drought conditions prevailing at that time. Only a small percentage of study area B is covered by vegetation classes which display these two characteristics - either a vegetation type which remains green to some degree at all times of the

Figure 4: Interpretation for area B from LANDSAT false colour composite (January 1979) showing distribution of shrubland and woodland that can be accurately mapped from the imagery

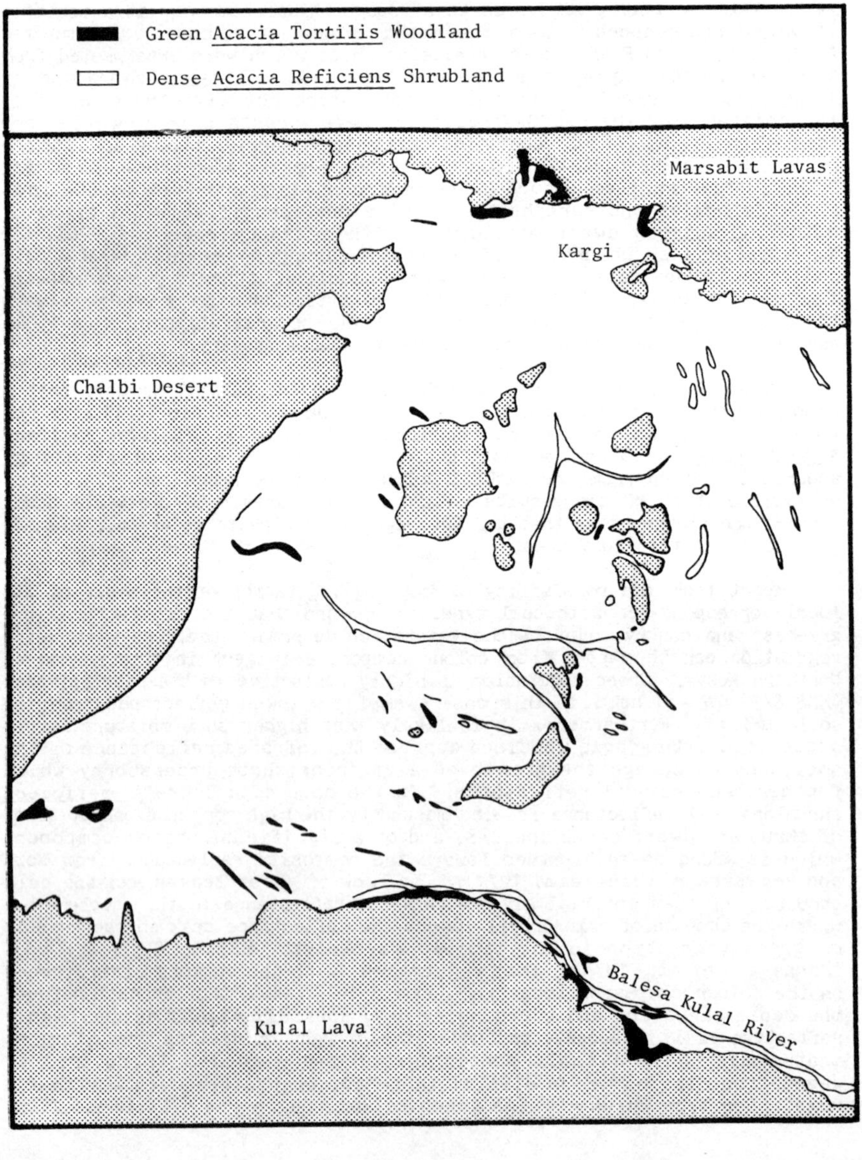

year, or a dry vegetation type that has a canopy cover exceeding approximately 40%. The remainder and larger part of the study area is covered by shrub/dwarf shrub and annual grassland vegetation types, all of which are important for livestock grazing, but cannot be classified with colour composite imagery.

8. DISCUSSION

Although the applicability of the method used in the present study is limited, the results obtained point the way forward to the development of an alternative approach. Some qualitative insights into the relative differences between the spectral curves for vegetation and soils can be obtained from the colour composite imagery, but precise knowledge of the combined spectral response of vegetation types of given composition and canopy cover growing on particular soils is lacking. Without this knowledge of the positions of the combined soil/vegetation classes in the feature space, it is unlikely that discrimination between vegetation types can be improved upon, even if conventional digital enhancement and processing techniques are used, e.g. histogram-equalisation, density-slicing, psuedo-colour, ratio images and classification algorithms.

A preliminary comparison of histogram equalised images in MSS bands 5 and 7 (0.6 - 0.7um/0.8 - 1.1um) for the study area A, with the colour composite image of January 1979, has shown that although boundaries are more clearly defined, further information about the characteristics of the vegetation cover and an improved discrimination between classes is not obtained. The initial analysis of the colour composite imagery described in this paper has provided the insights into the relative spectral characteristics of soil and vegetation types necessary for the formulation of an alternative approach based upon digital techniques.

Siegal and Goetz (1977), argue that the dry, non-green sage bush vegetation of Arizona with a canopy cover of as little as 10%, can be detected using LANDSAT digital data. The sage bush cover modifies the amplitude of the spectral soil curve without significantly changing its shape.

It would appear therefore, that for the study area, further progress may be made by careful characterisation of the spectral response of the various soil types, together with an analysis of this modification by vegetation types of varying canopy cover.

9. ACKNOWLEDGEMENTS

The authors wish to thank IPAL for their support in the field, and Hunting Technical Services Limited for the loan of the I2S Colour Additive Viewer. Much valuable assistance was given by Miss M. Biddle and Miss J. Corbett with the collection of field data, by Miss G. Thomas with drawing the diagrams and by Mrs G. Jones with typing the script.

10. REFERENCES

BENTLEY, R.G., et al, (1976): A LANDSAT study of Ephemeral and Perennial Rangeland Vegetation & Soils. NASA Report YA-300-1700-1012, Bureau of Land Management, U.S. Department of Interior, 234 pp.

ÇARNEGGIE, D.M., et al, (1975): Range Resources; Inventory, Evaluation and Monitoring in Manual of Remote Sensing, American Society Photo. 1460-1466.

COLVOCORSCES, A.P., and McEWAN, R.B., (1973): Eros Cartographical Processes. Photogramm Engng Remote Sensing, 39 No 10, 1303 - 130.

EDWARDS, K.A., FIELD, C.R., and HOGG, I.G.G, (1979): "A Preliminary Analysis of Climatological Data from the Marsabit District of Northern Kenya. IPAL Technical Paper B - 1, (UNEP-MAB Integrated Project in Arid Lands), 44 pp.

GRIFFITHS, G.H., (1979): The Detection of Semi-Arid Overgrazed Vegetation surrounding the water-hole at Kargi in Northern Kenya using LANDSAT Satellite Images. Undergraduate Dissertation (Unpublished), University of Cambridge Library.

HEMMING, C.F., (1972): The South Turkana expedition, Scientific Paper VIII, The Ecoloy of South Turkana: a reconaissance classification, Geographical J, 138, 15 - 40.

HERLOCKER, D.J., (1979): Vegetation of Southwestern Marsabit District, Kenya. IPAL Technical Paper D - 1, (UNEP-MAB Integrated Project in Arid Lands), 65 pp.

HIELKEMA, Jelle U., (1977): Application of LANDSAT Data in Desert Locust Survey and Control. FAO, AGP:LCC/77/11. AGD(RS)5/77, 28 pp.

LAMPREY, H., (1978): The Integrated Project on Arid Lands (IPAL). Nature and Resources, XIV 2 - 11.

OTTERMAN, J., (1980): Satellite and Field Studies of man's impact on the surface in arid regions. Tellus 33, 68 - 77.

PRATT, D.J., GREENWAY, P.J., and GWYNNE, M.D., (1966): A classification of East African rangeland with an appendix on terminology. J. Appli. Ecol.3 369 - 382.

SIEGAL B.S., and GOETZ, A.R.H., (1977): Effect of Vegetation on Rock and Soil Type Discrimination. Photogramm Engng Remote Sensing, 43 No 2, 191 - 196.

WALTHER, D., and HERLOCKER, D.J., (1981): Indicators of Environmental Degredation and Range Conditions in the Korr area of Southwestern Marsabit District, Kenya. Paper prepared for the International Workshop on strategies for developing the resources for arid and semi-arid areas of Kenya, Nairobi. (In the press) 10 pp.

VEGETATION MAPPING FROM SLAR: A QUESTION OF RESOLUTION RATHER THAN WAVELENGTH

Thomas C. Welsh
EASAMS Ltd.
Lyon Way
Camberley
Surrey

Abstract: If all the variables which contribute to backscattering from the vegetation canopy are investigated, the task of selecting the critical parameters is extremely difficult. However, within the resolution cell there may be more than one type of scattering surface, and it is the strongest of these which governs the backscatter variations. The composition of the canopy, in terms of the different types of surface present, their distribution and extent, and the properties of the boundaries between them (such as the effect of edges), may be just as important to vegetation discrimination as the backscattering process. During research with real-aperture, Q-band SLAR imagery of upland vegetation in the Derbyshire Peak District, it was observed that the variation in cover characteristics within the cell was important. This paper puts forward the case for vegetation interpretation with reference to these broader characteristics rather than the micro-relief at the order of magnitude of wavelength.

1. Introduction

The advantages of all-weather, night and day, synoptic imaging have long stimulated enthusiasm for SLAR as a solution to various mapping tasks where conventional methods are defeated or hindered. This optimism persists although, after more than a decade, the application of SLAR has been the preserve of a few with the resources to procure the imagery and the technology to analyse it.

Paper presented at a conference on Matching remote sensing technologies and their applications convened in London by the Remote Sensing Society, 16-18 December 1981

Unfortunately, a long standing pedigree in accurate positioning and identification of objects against their surroundings, and in relief mapping, has not been balanced by an equal pedigree in interpretation experience of extended surface areas, such as is necessary for application to vegetation. The largest contribution in this sphere has been in the mapping of vegetation in tropical rainforest areas which has relied on extrapolation from known associations between vegetation and terrain types.

The application of SLAR to vegetation began about fifteen years ago, when radar sensitivity and image quality improved sufficiently to allow extended surfaces to be distinguished. However, from very early on, civil users were able to take advantage of dual polarisation and dual frequency imagery. This changed the outlook of vegetation interpretation from one of careful decision making through detailed examination of the image to a much more flexible approach which relied on comparison of two different images. More than this simple distinction, comparison of any one characteristic, such as image density, on two or more simultaneously recorded images, has been much more adaptable to sophisticated methods of analysis.

Over more than a decade the literature on vegetation interpretation from SLAR has assumed that either multipolarised or multifrequency imagery is available. In consequence, single parameter imagery, (obtained at one frequency and one polarisation), has been looked upon as something less than adequate for civil applications, and except by means of contextual evidence, is impractical for use for vegetation mapping. There are very few guidelines available on how to make the best use of the information available on the image.

However, an even greater obstacle to wider and freer use of SLAR imagery has been the quantity and diversity of ground data which must be collated, in a real time situation, for correlation with the image data. Since a number of factors other than simply the roughness of the vegetation canopy can affect returns, the data collected may include the depth of cover, leaf size and shape, plant moisture content, proportion of plant material to free air within the canopy, spatial characteristics of cover, soil moisture content, soil type and properties. Moreover, measurements of the spatial characteristics of the plant cover would require to take into account the likely incidence angles at which energy would be received. Such a quantity of detailed inventory implies a small test area and accurate knowledge of the time and coverage of the radar flight. Without large manpower resources it is difficult to undertake adequate provisions for real time data collection to support subsequent interpretation of radar imagery.

Given both the level of processing and analysis necessary to make the optimum use of multiple radar images, and the outlay and precision required for real time ground data, it is unlikely that the availability of side-looking radar imagery for training in vegetation interpretation will extend beyond a few well-equipped centres, and occasional restrictive and short term research projects. Whereas other sensors have benefitted from a period of widespread availability during which ideas have developed, experience and understanding have been disseminated, and support for the application of the sensors has been established, SLAR is likely to remain the preserve of specialists, without the confidence in its capabilities, or the awareness of its limitations, to render it viable and competitive as a tool.

When the opportunity arose to work with single image SLAR coverage, one of the main objectives, in developing a methodology for the interpretation of vegetation, was to devise readily available techniques to make SLAR more accessible as a sensor. The main areas investigated were an understanding of the image information so that at least some direct interpretation of the image was possible, before resorting to contextual and physiographic evidence, and simplifying ground data collection. However, this paper is concerned with an alternative approach to the categorisation of vegetation in relation to the sensor capability which has implications both for interpretation and ground measurement. This alternative emerged during the research, when it was observed that certain cover types which should have produced different image responses, were instead confused with one another. The reason this should occur is given as a matter of conjecture, but from it may derive a simpler and more practical approach to image interpretation.

The conventional approach to differences in returns from vegetation is to consider the many variables which affect the amount of absorption, backscatter and depolarisation of the incident radar energy, not only by the surface, but from within the canopy, and from the underlying terrain. This means that time sensitive parameters such as moisture levels and windspeed at the instant of sensing can have a marked effect on returns. Surface roughness at the micro-level, related to the wavelength of the transmitted energy, becomes the key factor in determining which types of cover will appear different on the image, and which will appear similar. This paper proposes that, while processes related to the wavelength of the radar energy undoubtedly determine the amount of energy returned to the receiver, discrimination depends ultimately on the broad roughness characteristics approaching the magnitude of the resolution cell.

It is these gross features of the canopy which should be used to determine which types of cover will be distinguished, and which types of cover are likely to be confused.

2. Two Approaches to Surface Evaluation.

In a research project at the Department of Geography, Sheffield University, from 1975 to 1978, real-aperture Q-band imagery, (wavelength approximately 0.8cm), with horizontal transmitted and received polarisation, was used to develop a methodology for the interpretation of vegetation from SLAR. The imagery was obtained from a modified EMI 391 radar flown by the Royal Signals and Radar Establishment, aboard a Viscount flying at 1,830 metres above ground. The resolution of the system was 15 metres approximately in azimuth, varying in range from about 7.5 metres at 4 kilometres to about 15 metres at 8 kilometres. The coverage was an area of the Derbyshire Peak District westward from the fringes of Sheffield. There was no user control over the coverage and timing of flights, which were at the discretion of RSRE. Imagery was obtained on the 18th of June 1976 and the 16th September 1977. As only single imagery was available, the research was directed to evaluation of the full range of information of the image.

A problem which frequently affected accuracy levels during the research, and recurred during post-mortem assessment of interpretation errors, was the enigmatic confusion of surfaces which, both in visual terms, and in terms of anticipated radar detection differences, were consistently confused. Two examples serve here as illustration. Conifer plantations were often misinterpreted as cultivated grass or permanent pasture. Calluna vulgaris in moorland controlled by seasonal burning was often misinterpreted as permanent pasture. Two factors were involved: a break down in the usual assumption about boundaries used as contextual information, and a close similarity in tone and texture characteristics. Conifer plantations were usually clearly defined on the image because of the strong returns along the edges facing the radar, and the corresponding radar shadow, which gave the appearance of bas-relief. On facing slopes with steep incidence angles, or in areas where intervening ground obscured the perimeter of the plantation, these edge effects were subdued or absent, and the regular shapes of plantations were indistinguishable from adjacent fields. In other situations, clearings within the plantations were not identified. In the second example, areas of young regrowth of Calluna vulgaris were delimited by rectilinear outlines, marking the paths followed during annual muir burning, yielding a pattern not dissimilar from the larger, less regular enclosures of adjacent permanent pasture on hill farms.

With this failure of contextual evidence, much more weight is transferred to whatever tone or texture information is available. Three forms of image information were considered: tone, local texture produced by tonal variations, and microtexture. It was found during the research that certain significant and useful discriminations were possible using microtexture. However, for all three parameters, returns were found to be similar in the cited cases of conifer plantation and agricultural grassland, and heather moor and permanent pasture. Both during interpretation, and in the subsequent re-examination of the imagery to establish the source of errors, both direct and contextual evidence failed to distinguish these cover types.

In putting forward an explanation for this problem, there are two approaches. Either, the failure to distinguish indicates an inherent limitation in the sensor capability for mapping vegetation. Or, the failure tells us something about our expectations for radar discrimination, which are related to our needs, rather than sensor capability. Examples of the former case may be used to justify arguments against using SLAR for vegetation mapping, except by the very indirect method of known associations with physiographic regions. However, both the examples given are related to peculiar situations which could be anticipated in the interpretation methodology. In order to provide the necessary understanding to meet this anticipation, the reasons for the confusion should be examined.

One way of explaining the confusion of dissimilar surfaces is to investigate the probable relative contributions of the many variables which control the amount of absorption, surface and subsurface scattering, and depolarisation of the incident energy by the surface cover. Unfortunately, the complexity of the variables at the surface is multiplied by the relationship between these and the system parameters, particularly incidence angle, and as a result it is doubtful whether a single explanation can be offered to account for all occurrences of confusion between two cover types over the whole image. This uncertainty can only serve to undermine confidence in SLAR as a sensor for vegetation mapping. During the research it was felt that this approach to SLAR was self-defeating, as it was clear that even with the necessary resources for investigative analysis of sensor/vegetation relationships at this level, it would be difficult to provide a firm foundation for interpretation. Instead, attempts were made to explain surface differences in terms of broad characteristics, such as the continuity and gross configuration of the cover.

Initial experiments were centred on a method for approximating surface roughness, which was described in a paper at an earlier annual conference of the Remote Sensing Society, in December 1976. This used a chain line arranged over the surface, to copy as closely as possible the breaks and undulation of the canopy. A rudimentary approximation of the surface roughness was calculated from the difference between the distance across the end points of the laid chain, and its fully extended length. Originally, this had been intended to register the micro-roughness of the surface, given the constraints of the chosen size and weight of the chain links. However it was found that roughness on the meso-scale, which ommitted the variations in stem and leaf height where these did not support the chain, as particularly (and predictably) occurred with tall grasses, could be used to differentiate cover types. Together with shape, size and height of canopy features, it became evident that measurement of the gross configuration of the canopy could be used to categorise cover within the context of ground data. In addition, an indication of the variability of roughness over each site was provided by calculating the variance for a number of line measurements. It was found that the variance value was very effective at discriminating types of vegetation. The question which arose from this experience was whether radar was sensitive to the same characteristics. If the chain line technique, however crude it might appear as a method for quantifying surface roughness, was able to distinguish surfaces by both gross configuration and variation in configuration, could this principle be applied to image interpretation?

As has already been mentioned, one of the resources on the image considered during the research was microtecture, otherwise known as speckle or pepper and salt. Unfortunately, the observed relationships between microtexture and vegetation run against the grain of accepted theory. Microtexture is considered to be a wholly random, system induced phenomenon (Estes & Simonett 1975), and thus it is controversial, at the least, to incorporate this image noise in a methodology, without attributing to it a valuable role in interpretation decisions. It is hoped to produce a paper in the near future to explain and justify its use in image interpretation, but in the meantime the observed relationships are offered as illustration of a further development of the ideas discussed in this paper. Two broad distinctions could be made using microtexture. There were distinct differences in microtexture patterns between very smooth and very rough surfaces, although between these extremes the evidence was not so reliable. Similarly differences were observed in the pattern between uniform, homogeneous surfaces, and interrupted, varied or otherwise heterogeneous surfaces.

In other words, a complimentary tool on the image had been found which appeared to achieve the same distinctions as the chain line roughness measurement.

One implication of the findings in surface measurement and microtexture is the possibility of determining whether or not a surface can be described in terms of its gross characteristics rather than its dielectric and microroughness properties, for the purpose of distinguishing vegetation types of SLAR imagery. Before discussing this further, several more examples of interpretation problems where both tone and microtexture were considered, serve here as illustration.

During a project which included vegetation on the limestone plateau south-west of Castleton, Derbyshire, several areas were assigned to a category of rough moorland associated elsewhere in the project area with Ericaea species on the gritstone uplands. The returns were brighter than adjacent areas of permanent pasture and moorland dominated by tussock forming species, and it was assumed that these areas represented residual moorland with sufficient acidity to support Calluna vulgaris or a similarly rough cover type. When the ground check was carried out, the observed classification was marginal pasture, which comprised very closely grazed grass, and a uniform microrelief. However, at intervals over the ground there were small, shallow hollows, between 1 and 3 metres across, and about 0. 5 metre deep, each with a scatter of stones containing traces of galena (lead sulphide). The area had been affected by surface lead prospecting, and had subsequently returned to pasture, but the ground had not been levelled. It was clear that the radar returns from each cell were dominated by discrete reflectors from the hollows and mineral scatter, giving the appearance over an extended area of a much rougher vegetation cover.

The contributions of ground surface relections in areas where the vegetation cover is not continuous is also relevant to the discussion. Ericacaea moorland sites were compared to evaluate the different types of response. Brighter, and from the microtexture point of view, rougher, heterogeneous responses were recorded both for low, carpet-like patches and for mature bushes of Calluna vulgaris Erica tetralix and Vaccinium myrtillus, when the cover was interrupted by bare stony ground compared to areas of continuous cover. This occurred even where patches of Ericacaea species no more than 5 cm high, and uniform, were interrupted only locally by bare ground with stones and moss. In another example differences were noted for Juncus squarrosus tussocks where these occured on bare, stony ground, or on a shallow regolith with mosses. Clearly, in these examples, the ground surface returns are more important than those from the vegetation.

Once again there are two approaches to the explanation of these examples. The first one, and the conventional approach, is to conclude that these are further instances where radar fails as a sensor for the interpretation of vegetation. The backscatter returned to the radar was predominatly due to non-vegetation factors. Therefore, in these situations, the vegetation cover was not identified. It is quite justifiable, on this basis, to dismiss SLAR as unsuitable for vegetation mapping in any situation where the vegetation canopy does not dominate the returns. But this is self defeating. There are likely to be many situations, which cannot be anticipated, where other factors contribute to the returns. It may be that we are asking too much of the backscatter process, when we expect it to provide all the answers to vegetation discrimination.

The alternative approach is the one which re-examines our expectations. Instead of concentrating on the micro-roughness and dielectric properties of the canopy, which are related to the wavelength of the transmitted energy, to determine the amount of backscatter, it may be possible to describe vegetation in terms of its gross characteristics, and use this information to determine which surfaces can be distinguished from one another, and which cannot. This may mean a rethink of vegetation categories. It does not mean that the process responsible for scattering energy is disputed; only its usefulness for explaining the ability to interpret vegetation from SLAR.

The simplest example of discrimination from gross characteristics is the difference between continuous and broken cover. In the former case, the backscatter characterisitics of the surface will decide the returns. However, as has already been suggested, these characteristics can vary widely with conditions. In the case of broken cover, the dominant returns will either be from bare ground or edge effects within the canopy, which will have stronger backscatter than the canopy itself, and will boost the returns. Instead of worrying about the process-related variations in back-scatter between surfaces, the comparison is between two distinct morphological types. If two types of vegetation can be found in a project area, perhaps even different habits of the same species, one continuous and one interrupted, there should be sufficient evidence on the image to discriminate them, even where sensor-terrain relationships vary the amount of backscatter.

Returning to the earlier example of confusion between two outwardly contrasting surfaces, the explanation considered in the research was that they were similar because, although the surface roughness differed, the cover was uniform. Because of such factors as absorption and specular reflection, surface scattering was reduced, and similar returns were recorded for both sites in each case.

However, the decisive factor was perhaps the regular spacing, or uniform configuration of the canopies. Conifer plantations often comprise close stands of uniform height and maturity of growth, so that there are very few changes in the canopy. Both grass and young, continuous Calluna vulgaris produced a flat surface to the radar, even if the micro relief differed.

3. Conclusion

The return from each resolution cell is a product of one or several scattering processes. If the former, one species in the canopy is the dominant influence, either in areal extent, or ability to re reflect energy. The cover may be composed entirely of a single species, or a discrete plant or object, surrounded by vegetation which absorbs and dissipates the incident energy, and which will determine the return for an area many times its size. If returns are determined by several characteristics, whether species or different growth or structure, the response may be the average of all or only some of the characteristics present. Alternatively the response may be generated by the interaction of several levels of scattering, if such an occurrence can be supported by theory. Or again, edge effects, exposed ground, or other discrete reflectors may make the larger contribution. The interpreter has no way of knowing the relative contributions of surface characteristics to each return. If his approach to the problem anticipates differences arising from different levels of scattering and depolarisation of the incident energy due to the many sensor-terrain variables involved his attempts to determine different types of vegetation from the image are likely to be frustrated. The return only fulfils his expectations optimally, if the cover is uniform over the whole resolved area. In any other habit of vegetation cover, an average or some other aggregate of the surface properties, both vegetation and non-vegetation, may give rise to deceptive levels of return. Even for the uniform cover, other factors may be involved.

If. however, the interpreter accepts these variations in gross configuration as natural attributes of vegetation, and can identify such characteristics with certain cover types he wishes to distinguish on the SLAR image, and these characteristics adequately differentiate the required cover types, he may be better able to use the information on the image to advantage. It may be neccessary to narrow the specification or to use a description which differs from the classification ideally required. At the same time, using these terms, the interpreter will also be able to anticipate situations where vegetation types cannot be distinguished.

Where this occurs, other information, such as contextual evidence, must be sought.

SUMMARY

Past researches on the interpretation of vegetation from SLAR have concentrated on the arrangement and electical properties of the individual facets of the canopy. This paper proposes that the arrangement of the individual plants, and the broader variations in height, breadth and shape of the canopy should form the basis for describing vegetation in radar terms. Surface scattering is in effect a mechanism which controls the strength of radar returns from surfaces. However, discrimination of vegetation on the image may depend, not on the spatial distribution of individual scatterers relative to wavelength, but the spatial arrangement of different types or groupings of scatterers, and the relationships between them, over the area which makes up each resolution cell. This approach may simplify the application of radar imagery to vegetation mapping, and enable it to be more freely used at different levels of training and research, in order to build up confidence in SLAR as a sensor.

ESTES & SIMONETT D.S. (1975): Interpretation and Applications, Ch. 14 in Manual of Remote Sensing (Reeves R.G. et al editors), American Society of Photogrammetry, Virginia USA Vol. II

WELSH T.C. (1976): Interpretation of Vegetation from SLAR.A method for obtain roughness values for ground data. Unpublished Paper given at annual conference of the Remote Sensing Society in 1976.

WELSH T.C. (1981): Unpublished PhD Thesis.

HIGH ALTITUDE AERIAL PHOTOGRAPHY (HAAP) AS APPROPRIATE TECHNOLOGY FOR RESOURCE STUDIES AND CARTOGRAPHIC INFORMATION

M G Coulson
Topographic Sciences
Department of Geography
University College of Swansea

Improvements in aerial cameras, and the availability of small commercial jet aircraft have meant that high quality aerial photography can be acquired at scales between 1/80,000 and 1/130,000.

More emphasis should be given to the potential of this 'Intermediate Technology' particularly for applications in developing countries. HAAP draws upon the proven techniques of photogrammetry and uses existing human skills in photointerpretation. Its use may not necessarily demand further costly investment in equipment or manpower by the user.

Problems of image acquisition and costs are discussed, and comparisons made with the proposed developments in Landsat, SPOT and space borne cameras. The economics of operation show that although this is a data source which will mostly be paid for by governamental agencies, it may not be economically viable for individual governments to purchase aircraft.

The cartographic accuracy and applications of HAAP are illustrated, in particular the role it can have in rapid mapping. Certain problems of interpretation of small scale imagery are discussed. It is concluded that the relative inflexibility of photography may not be a serious drawback for general resource studies.

SPOT LINKS WITH REMOTE SENSING USERS

M. TRAIZET - C.N.E.S. - Division des Programmes d'Application

-=-=-

Space based Earth Observation Systems offer particularly powerful means to collect information on our res ources and our environment. The French Government decided in February 1978 to undertake the development of the Système Probatoire d'Observation de la Terre, or SPOT. Sweden and Belgium are now associated with France in this programme.

1. OUTLINE OF THE SPOT SYSTEM

The SPOT satellite is designed for a sun synchronous orbit at 822 km. It is composed of a multipurpose platform and two identical HRV instruments (HRV stands for High Resolution Visible). The first satellite will be launched by ARIANE in early 1984. Details of the satellite are given in the appended tables.

The SPOT ground segment is designed to satisfy the requests of the users on a world wide basis with high quality products in a short time delay. A central receiving station is being implemented near TOULOUSE for a direct reception of images over Europe and North Africa, and world wide data acquisition via the on-board tape recorders. A special company (SPOT IMAGE) will be set up to disseminate the images and provide services to users. Foreign receiving stations are foreseen,generally in conjunction with LANDSAT station, to handle regional requests.

2. FEATURES OF THE SPOT SYSTEM

SPOT is a new generation earth observation satellite which includes salient features to improve its remote sensing capability :

1) high resolution imaging

Each HRV instrument can be operated in one of two modes, either 20 m sampling or 10 m sampling, or both simultaneously. The user is free then to select a mode well adapted to his needs.

2) refined spectral bands

In the 20 m mode, three spectral bands have been selected which match precisely the significant features of the vegetation spectral signatures. In addition they give access to free water quality assessment and to bare soil mapping ; water vapour absorption bands are carefully avoided.

In the 10 m mode, which provides 4 times more pixels than the 20 m mode, only one band has been implemented, adapted to the needs of the cartography of vegetated areas as well as urban or natural areas.

SPOT HRV INSTRUMENT (each)

	Color Mode	B/W Mode
Ground sampling step (m)	20	10
Array sampling period (m.sec.)	3	1,5
Spectral bands (μ m)	XS1 0,50-0,59 XS2 0,61-0,68 XS3 0,79-0,89	P 0,51-0,73
Grey levels	256 (8 bits)	either 64 (6 bits) or 256 compressed to 6 bits
Data rate (M bits/sec)	25	25
Swath width (km)		60 (Nadir)
Central viewing direction		+0,163 and -0,163 deg.
Field steering direction		0 ± 45 steps 0,6° apart
Outermost observable direction (deg)		± 29.23
Corresponding distance to track (km)		± 475

ORBITAL PARAMETERS OF SPOT

Revolutions/day	14 + 5/26
Nodal Period	101.46 min
Mean altitude (45°Nothern Lat.)	832 km
Inclination	98.70°
Orbital (repeat)period	26 days
Number of tracks	369
Intertrack distance (equatorial)	108,4 km
Accessibility pattern at 45° latitude	1,4,1,4,1,4,1,4,1,4,1 days
Mean local solar time at descending node	10h30m

3) pointing of each instrument

The optical axis of each HRV can be steered to any of 91 positions sidewise, from -27° to +27°. This corresponds on the ground to a band of access of 950 km centered on the satellite ground track, independantly for each instrument. Steering of any of the mirrors is executed upon commands stored in the on board computer.

The two HRVs can be operated jointly, covering a swath of 117km at nadir or a little more when off nadir. This mode includes a 5% overlap between the two fields of view (3km).

The consequences of cross-track pointability are :

- higher accessibility to a given point
- stereoscopic capability
- off nadir reflectance measurements

4) various image products

All image products will be radiometrically corrected (calibration, equalization of detectors sensitivity).

As regards the geometry, three different types of products will be available :

level 1b : System corrected geometry
level 2 : Fine geometric corrections for digital cartography
level S : Image to image registration for change detection

These products will be offered either on computer compatible tapes or on 241 mm films or prints, or both.
In addition, images at level 1a with raw geometry can be made available on request.

3. HOW TO MAKE USE OF THE SPOT SYSTEM

The SPOT IMAGE Company and the foreign regional stations will be the users' points of contact. The dialogue could be much more elaborated than experienced with past satellites.

The user will
- search the catalog file for archived images.
- order some archived images, or decide to ask for a special programming of the satellite.
- specify the type of products, standard or custom made, depending the processing capability of the producer.

To ask for a special programming of the satellite, the user should be assisted by the station operator, to :

- Select the resolution (10 m or 20 m) and the associated spectral bands.
- Require the large field of view of the joint HRVs, or the independant pointing capability of each instrument.
- Adopt the nominal gain setting or select specific values for different sensitivity.
- Specify the geometry of the observation , by giving the acceptable domain of incidence angle for vertical viewing, or for slant viewing ; or the range of base over height ratios adequat for the stereoscopic pairs required.
- Select the dates of observation : period, repetitivity ;

Such requirements can be established for each image ordered. More conveniently, global requests can cover the needs of users on a permanent basis, say once a week or once a month for given periods. We are considering to build up some "standard modes" to specify most of the requests in a simple way, still offering the complete selection process for specific needs.

The regional station operators will combine the various requests to elaborate the "programme of work" of SPOT inside their station coverage. These programmes are then integrated by CNES to operate the satellite itself from the control and command station in TOULOUSE.

4 - CONCLUSION

After its launch in mid 84, SPOT should provide users with valuable information ; a significant advantage could be derived from the simultaneous use of the complementary capabilities of SPOT and LANDSAT.

A thorough dialogue will link the users with the serving regional station or SPOT IMAGE to program the SPOT system and get the most significant information.

Simulations are going on to help users to assess the value of SPOT data for their needs. We encourage you to look at the simulation, and already to think about your SPOT observation plan ; on request, we can check your preliminary plans.

For the future, the continuity of the SPOT system will be ensured by a serie of satellites; the SPOT n° 2 has been decided in October this year, for a launch in 1986.

UNDERSTANDING THE LANDSAT MARKET IN DEVELOPING COUNTRIES

M Willard
Menlo Park
California

The National Oceanic and Atmospheric Administration (NOAA) of the United States Department of Commerce recently headed an interagency task force to determine the technical parameters and institutional arrangements for a US operational earth resource sensing system. One of the most important and least understood inputs confronting the task force was the market for Landsat products and ground processing equipment. While the US government represents somewhat over half of the 1979 Landsat market (about 52%), the foreign segment of the market is substantial (about 36%) and is expected to grow rapidly. In particular, the developing nations of the world represent a large potential market for Landsat data and products. However, this "potential market" is not very well understood by US policy makers. This paper is an effort to understand the Landsat market in developing countries, and the constraints on the growth of that market which stem from the process of development itself and from a country's technical, political and institutional attributes.

AN UPDATED CATALOG SYSTEM FOR REMOTE SENSING DATA

R S Singh
Civil Engineering
Lamar University
Beaumont
Texas

Indexing and Retreival of Remote Sensing Data is not an easy task. In 1971 the author accepted a challenging proposition to develop a catalog system for many different formats of Remote Sensing Data at the University of Wisconsin, Madison, USA. The project was funded by NASA, and took two years to complete.

The objective in developing the system was to encompass the many different formats of Remote Sensing Data; to leave spaces for future additions in formats; and to make it practical for small as well as large agencies, such as NASA to adopt the system.

The system is divided into three parts as follows:

1. A card catalog
2. A site-index-map, and
3. Site-index-file; Industry-index-file; Project-index-file; and Catalog-cards

In the summer of 1974, the author was employed at NASA to demonstrate the workability of the system and again in the summer of 1977 to catalog the audio tapes of the conversations between the astronauts in orbit or on the moon or earth with the Mission Control in Houston and other tracking stations around the world.

Since 1973 many changes have taken place in format size and codes. In this updated article, these changes have been incorporated.

POLITICAL PERSPECTIVES ON REMOTE SENSING

N Garnder, School of Geography, University of Oxford
and
A R Harrison, Department of Geography, University of Reading

Remote sensing technology offers new methodologies for land resource appraisal and management planning. But to what extent is remote sensing politically neutral? The present paper evaluates some of the potential political overtones of remote sensing applications.

In the developing world, the very real political problems associated with technology transfer have received scant attention in the remote sensing literature. Technological sophistication necessarily limits widespread comprehension and therefore concentrates decision making in the hands of educated elites in society. It also enhances the dependence of the developing world on a handful of technologically advanced nations. Furthermore, remote sensing from space leads to an emphasis on physical attributes of the land, yet, at the synoptic level at which remote sensing is most commonly applied, social and political factors may be the prime causes of land mismanagement. We may therefore be exporting an inappropriate technology.

In advanced countries, civilian remote sensing is heralded as a peaceful application of satellite technology in an era of increasing militarisation of inner and outer space. But there remains considerable overlap between military and cilivian applications of remote sensing. Theoretical and practical advances in remote sensing technology are often pioneered for military purposes. The political implications of this interdependence between civilian and defence spheres are explored.

GEOLOGICAL APPLICATION OF AN INTERACTIVE DIGITAL IMAGE PROCESSING SYSTEM

D J Barr: D Bassarab
University of Missouri
Rolla

A state-of-the-art image processing and data base system consisting of a minicomputer driven Comtal color graphics display and NASA Earth Resources Laboratory software has been adapted for mineral and petroleum exploration studies and geological engineering analysis.

System application modules provide for a variety of LANDSAT and aircraft multispectral scanner data analysis techniques including contrast manipulation, principal components, filtering, ratioing and pseudo-color composite displays as well as multispectral signature analysis. In addition, natural resource data bases containing digitized thematic parameter maps can be created, processed and merged with remotely sensed data sets. This allows interaction and visible display of such features as elevation, slope, gravity and stratigraphic isopachs.

The results of three exploration projects are illustrated and described. A potential oil and gas site in Eastern Colorado was studied with LANDSAT ratios and principal components to yield enhanced joint sets, lineaments and a number of elliptical shaped anomalies. Enhanced LANDSAT data covering a portion of the Laramide thrust fault of southwestern Montana clearly displayed major faults in a highly deformed rock mass. Spatial filtering and principal components analysis provided for an effective display of localised shear zones. A data base containing isopachs of subsurfaces geologic parameters was created of the Missouri New Lead Belt and analysed with respect to locations of potential new mining sites.

Studies of the LANDSAT data were undertaken in order to determine the types of techniques that would be most appropriate for geological feature definition in a number of differing applications. Results have been positive and encouraging. We believe that digital image processing and data base development hold the key for major advances in mineral exploration.

AN EVALUATION OF LANDSAT-3 RBV IMAGERY FOR OBTAINING ENVIRONMENTAL INFORMATION IN TANZANIA

R B King
Land Resources Development Centre
Tolworth Tower
Surbiton
Surrey
KT6 7DY

Abstract: Landsat-3 RBV imagery of south-west Tanzania obtained during the 1978 six month dry season, is compared with MSS imagery, and checked with ground information. The higher resolution and greater stereoscopic coverage of RBV imagery renders it particularly suitable for reconnaissance terrain mapping involving landform, soil and drainage surveys. The superior ground resolution of the RBV imagery was generally found to be more cost-effective than the superior spectral resolution of MSS; although both types were generally used together. In particular, ground position could usually be easily located during fieldwork on 1:125 000 scale enlargements of the RBV imagery, and subsequently determined on the MSS imagery, thus overcoming the frequent problem of pixel location in the field. The 40 m RBV resolution was found to be especially suitable for the East African Central Plateau Land Province with its covering of *Brachystegia-Julbernardia* woodland; and it is doubtful whether any further improvements in resolution would be worthwhile, because of the generally coarse terrain parameters of this land province. The investigation also found that it is usually best to obtain Landsat-3 RBV imagery from as early in the dry season as cloud cover will permit, because there is less atmospheric haze, less burning and more distinctive vegetation reflectances early in the dry season.

1. INTRODUCTION

There appear to be two main reasons why the development of remote sensing technology has been mainly influenced by instrument designers rather than those who apply them: (1) instrument designers are in a more powerful position; (2) singular irresolution by environmental scientists to state their needs. We find ourselves in the strange situation where decades of experience with interpretation of panchromatic aerial photography are frequently ignored in the blind belief that it can all be done by computers, which usually only consider one of the four basic elements of interpretation, namely reflectance, generally ignoring the other three: shape, size and position.

Paper presented at a conference on *Matching remote sensing technologies and their applications* convened in London by the Remote Sensing Society, 16-18 December, 1981

Environmental scientists are however beginning to make their own demands, witness the United States 'geologists' request for stereoscopic sensors. Lack of stereoscopy and low resolution have been the two main technical disadvantages of satellite imagery; but the main conceptual weakness in much of current remote sensing development is the over-concern with technological solutions. More attention should be given to aiding interpretative skills than to computer application. More usable information will most often be gained by visual interpretation of a stereoscopic image and/or multitemporal imagery, than any amount of automated interpretation of the narrowest of spectral bands.

Futher, remote sensing is most effective in those parts of the world where data are sparse, i.e. most of the least developed countries, but by their very nature, these same countries are least able to afford expensive sophisticated equipment; and even if such equipment is donated, they often do not have the infrastructure to maintain them. Often their only alternatives are either to rely on analysis by foreign experts, who are sometimes little aware of ground conditions; or to take the less expensive but usually more productive option of visual and intelligent interpretation.

With the help of the International Development Research Centre (IDRC) and the Overseas Development Administration (ODA), Tanzania's Bureau of Resource Assessment and Land Use Planning (BRALUP) was given the opportunity to assess the developmental value of not only Landsat imagery perse, but also some of the more sophisticated analyses. The investigation (King et al., 1979), which covered the Rukwa Region in south-west Tanzania (Figure 1) - a sparsely inhabited semiarid to subhumid region straddling the western branch of the East African Rift Valley, confirmed the value of Landsat imagery, but only if supported by airphoto interpretation and fieldwork. The statistical classification techniques were found to have little value, and were certainly not cost-effective (King and Kikula, 1979). The investigation also demonstrated that a resolution of 80 m was generally too low for this kind of environment. Accordingly when Landsat-3 was due for launching, BRALUP requested the Goddard Space Flight Center to activate the 40 m resolution RBV sensor over south-west Tanzania during the 1978 dry season, partly to be used for a land resource survey of Mbeya Region (Figure 1), and partly to compare with the investigation of MSS imagery already completed in Rukwa Region.

Mbeya Region is even more varied than Rukwa Region, straddling as it does the intersection of three branches of the East African Rift Valley: the Rukwa Valley containing Lake Rukwa at 800 m altitude to the north-west, the Usangu Plain at an altitude of 1 000 m to the north-east, and the Malawi Rift to the South-east with Lake Malawi at an altitutde of 500 m but also containing the Poroto Mountains volcanic massif rising to over 2 800 m between Mbeya town and the lake. The intervening plateaux between the rifts generally lie at altitudes of 1 200 to 1 500 m. The wide altitudinal range creates broad climatic differences such that the Rukwa and Usangu Valleys are semiarid while the Poroto Mountains are temperate. The rest of the region is semiarid to subhumid. The vegetation varies from grassland and *Acacia* savanna in the Rukwa and Usangu valleys to upland forest on the Poroto Mountains, with *Acacia* savanna or *Brachystegia-Julbernardia* woodland in the rest of the region.

Figure 1 Location of Rukwa and Mbeya regions, showing area covered by Figures 2, 3 and 4.

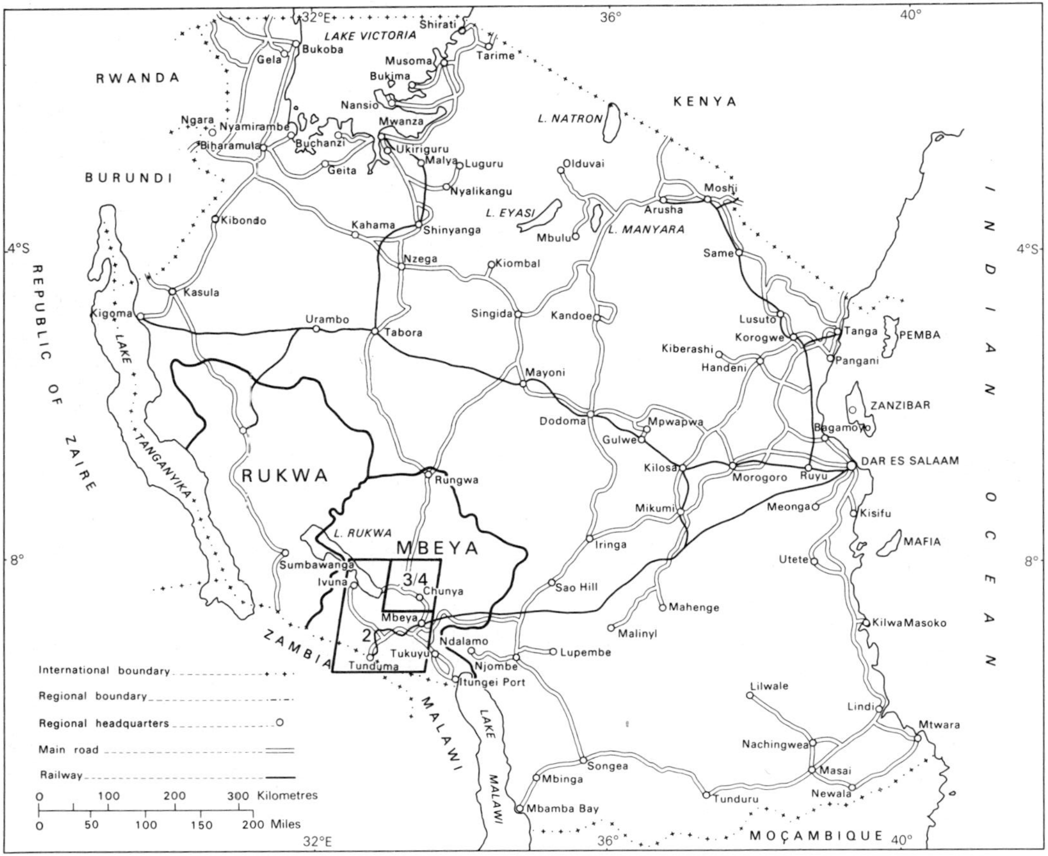

2. DETAIL

Figure 2 is the optimum quality MSS (band 5) image covering the area shown. It was taken on the 12th July 1975 in the middle of the dry season which lasts from May to October inclusive. Lake Rukwa is seen in the north-west. The north-eastern shore of the lake is bounded by a fresh fault scarp which continues further to the south-east, past the small Flamingo Lake to meet the older, somewhat retreated, NNE trending Chunya Escarpment bordering the Usangu Graben. Mbeya town can be seen at the junction of the two escarpments. The escarpment forming the south-western edge of the Rukwa Graben can be seen, as well as the Mbozi Horst between this escarpment and Mbeya town. The Tanzania-Zambia border town of Tunduma can also be seen on the crest of the south-western Rukwa Graben escarpment. The Poroto Mountains are hidden by the clouds in the south-eastern corner of the scene. Figure 3 is the RBV B subscene of the Figure 2 image taken on the 12th May 1978.

Significantly greater detail can usually be seen on the RBV image than on the MSS. Relief, roads and villages are usually much clearer on the RBV. For example, it is possible to trace on the RBV image (Figure 3) the Mbeya-Rungwa road (see Figure 1) from where it emerges from the forest plantation in the south-east of the scene along the top of the Chunya Escarpment through Chunya town and other villages to the Lupatingitingi settlement on the northern edge of the scene. Only occasional traces of the road can be seen on the MSS image (Figure 2). The dense drainage north-west of the forest plantation is also much clearer on the RBV image.

The RBV image shows a distinct landscape change between the southern triangular shaped Chunya Dissected Plateau and the Kipembawe Inselberg Plain to the north. The two areas are separated by a zone of 'mbugas' (seasonally waterlogged grassland displaying a light grey tone on Figure 3) and inselbergs. The boundary is not so distinct on the MSS image, but this is largely due to the time of the year, which will be discussed later. The inselbergs in the Kipembawe Inselberg Plain are more easily recognisable on the RBV image than on the MSS, e.g. those lying just to the north of the mbuga zone and just east of the main road.

The darker tone of the Kipembawe Inselberg Plain is caused by low reflectance from the *Brachystegia-Julbernardia* woodland, compared to the lighter more mixed woodland, savanna and subsistence cultivation of the Chunya Dissected Plateau. The drainage in the Kipembawe Inselberg Plain is revealed by the relatively higher reflectance of the valley bottom mbugas, which are again clearer in Figure 3 than 2, although again this is partly a seasonal factor.

Another important example of the greater RBV detail is village location. In Figure 2, Lupatingitingi on the northern edge of the scene and Chunya (for location reference, see Figure 1) are easily recognised; but Makongolosi (where the road changes direction) is hardly recognisable on the MSS image, whereas the RBV image shows not only Makongolosi in some detail but also another small village just north of the mbuga zone. Village location is particularly important to remote sensing, because whateven conclusions are made from environmental analysis, they ultimately should be related to people. It is important to know precisely in what environmental landscape a particular village lies. Difficulty of village location from MSS imagery confounded the effectiveness of the Rukwa Region survey.

Figure 2 MSS band 5 image of part of Mbeya Region.

Figure 3 RBV May image of subscene B of Figure 2.

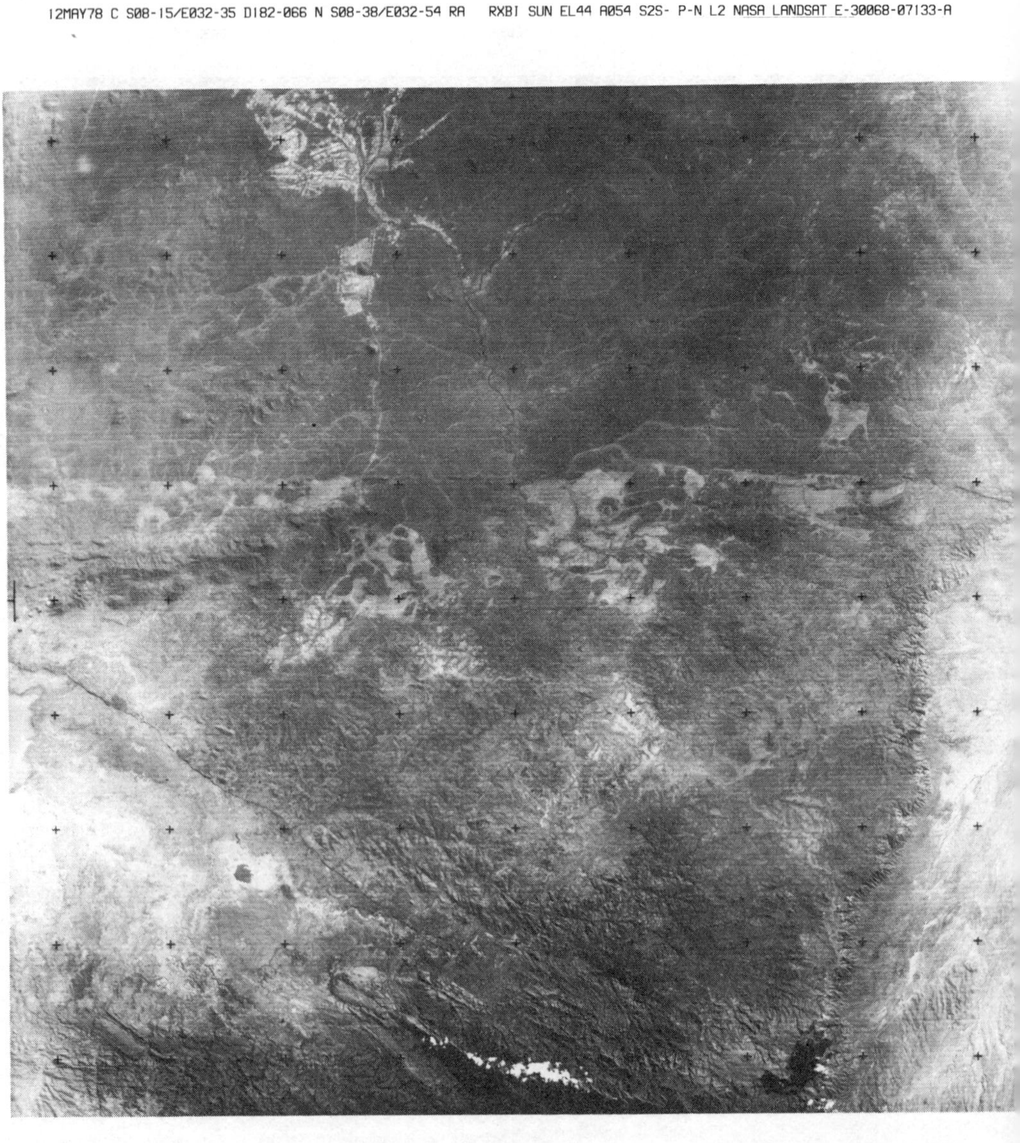

Figure 4 RBV August image of subscene B of Figure 2.

10AUG78 C S08-12/E032-35 D182-066 N S08-38/E032-54 RA XAOR SUN EL43 A058 S2S- P-N L2 NASA LANDSAT E-30158-07143-A

10AUG78 C S08-18/E033-20 D182-066 N S08-38/E032-54 R B XAOR SUN EL43 A058 S2S- P-N L2 NASA LANDSAT E-30158-07143-B

Although these examples have been specific to the scenes depicted in this paper, the general conclusion that detail - particularly relief, roads and villages - is significantly clearer on RBV imagery than MSS was gleaned from a study of all the imagery covering Rukwa and Mbeya regions.

3. STEREOSCOPY

Not only can relief be inferred by visual interpretation more easily on Landsat-3 RBV imagery than MSS, but there is also more stereoscopic coverage. MSS imagery only has stereoscopic coverage along the sidelap with imagery from the neighbouring orbit. The scanning mode prevents forelap stereoscopy, i.e. in the direction of satellite motion; but the photographic mode of RBV provides forelap stereoscopy for both subsequent scenes and subscenes. (There is no subscene sidelap stereoscopy.) The average MSS sidelap in Tanzania is 17%, so that for a project area covered by two orbits, the stereoscopic coverage is 25%. RBV sidelap stereoscopy is the same as for MSS, but to this should be added 17% forelap stereoscopy, which gives a total of 40% stereoscopic coverage for a project area covered by two subsequent scenes in each of two neighbouring orbits. Furthermore, although sidelap coverage cannot be penecontemporaneous, RBV forelap coverage is separated by an interval of only a minute or less, so that for example it is possible to measure cloud height with a parallax bar if the cloud falls within the forelap. Penecontemparaneous stereoscopy is more satisfactory because of the confusion than can be produced by not only seasonal and cloud cover changes, but also burning and atmospheric haze which can even change on imagery from the following day.

Linescan imagery stereoscopy cannot be considered at all photogrammetrically accurate; but comparison with 1:50 000 scale contour maps indicated a stereoscopic height resolution of 200 m. The RBV stereoscopic coverage was a significant aid to the land resource survey of Mbeya Region, particularly the extra sample strip provided by subscenes.

4. SEASONALITY

Investigation showed that it is generally best to obtain Landsat-3 RBV imagery from as early in the dry season as possible. During the wet season and shortly thereafter, all woodland vegetation has a dark tone, whereas grassland has a noticeably lighter one, as can be seen by the contrast between the seasonably waterlogged grassland mbugas and woodland in the northern part of Figure 3. As the dry season progresses, some trees lose their leaves, the grass becomes drier and more and more of it is burnt. Woodland reflectances vary and it becomes more difficult to distinguish grassland from woodland. Partly because of atmospheric haze later in the dry season, linear features such as roads, railways and even strand lines can be seen more clearly on early dry season imagery. Similarly, relief contrast is usually stronger early in the dry season because of the atmospheric haze which obscures late dry season imagery. Later in the dry season, grassland becomes distinguishable because it is burnt more often and burning is more visible than under woodland. Nevertheless burning late in the dry season is a less reliable indicator of grassland than the tonal differences at the beginning of the dry season.

These differences between early and late dry season imagery can be seen by comparing Figures 3 and 4. The main Mbeya-Rungwa road is significantly less perceptible on the late dry season (10th August) image (Figure 4), particularly at the southern Mbeya end. Although relief difference are not so marked in this example, the dense drainage dissection in the southern part of the Chunya Dissected Plateau can be seen more clearly on Figure 3 than 4. Grassland recognition is the most noticeable difference, particularly the mbuga zone separating the northern Kipembawe Inselberg Plain from the Chunya Dissected Plateau to the south. This zone would not have been recognised at all on the August image (except for the lineament) nor on the MSS one (Figure 2). Detailed comparison of Figures 3 and 4 shows that by August (and as can be seen from Figure 2 even by early July), the mbugas have been partly and variably burnt; and the reflectance differences between grassland and woodland are not so significant as early in the dry season. It can be further seen from Figure 4 that the drainage in the Kipembawe Inselberg Plain is largely revealed by valley bottom mbuga burning, but that this is usually a less reliable drainage indicator than the early dry season grassland-woodland reflectance differences illustrated in Figure 3.

The reflectance differences between early and late dry season imagery can sometimes be put to advantage by comparing imagery from the two periods under a stereoscope. Terrain features which produce the most change as the dry season progresses, e.g. floodplains - both due to changes in water level and grassland burning, catch the eye during stereoscopic vision.

Generally, it is recommended here for those parts of the tropics where there is a marked wet season, Landsat-3 RBV imagery should be chosen either from the wet season or early in the dry season provided, of course, there is no obscuring cloud cover. (However MSS colour composites should be chosen from the middle of the dry season because the colour at that time reveals the greatest vegetation density constrasts, which are often associated with relief.)

5. MSS/RBV COMBINATION

The main constraint of Landsat-3 RBV imagery is its poor spectral resolution. MSS imagery is preferable for any project specifically requiring spectral rather than ground resolution. Most projects require good ground and spectral resolution; and both MSS and Landsat-3 RBV imagery should be obtained if they are available. Theoretically the Thematic Mapper (TM) will in future provide both good ground and spectral resolution; but its initial worldwide availability seems very unlikely in view of the very low scheduled processing rate. Furthermore, there seems to be considerable uncertainty as to what TM products will be made available, and whether users will need their own computers to analyse them.

For the purpose of the BRALUP investigation, MSS colour composites were combined with RBV images under a mirror stereoscope to provide both spectral and ground resolution. The technique had varying success; and it must be admitted that it often produced eye strain. MSS and RBV images were also viewed together under a pocket stereoscope in the field for the purpose of locating MSS pixels. During the Rukwa Project, it was often difficult and frequently impossible to locate MSS pixels in the field, and even on aerial photography - a problem also realised by other investigators (e.g. Townshend and Justice, 1980). However, during the Mbeya Project, the field position could nearly always be located on 1:125 000 scale enlargements of the RBV imagery. The located position could then be related to an MSS pixel by viewing 1:125 000 scale enlargements of the RBV and MSS imagery together under a stereoscope.

6. CONCLUSIONS

As indicated above, it is clearly best to obtain both MSS and Landsat-3 RBV imagery of any particular project area; but although both types of imagery were obtained for all of Mbeya Region for the purposes of the general land resource survey, the RBV imagery was used to a much greater extent than the MSS mainly because of the better relief representation provided by the former.

If remote sensing users are forced to state the kind of imagery they require, they usually ask for stereoscopic imagery with high resolution, good synoptics, large spectral range and as multitemporal as possible, as well as a minimal time delay in acquiring the imagery! Most of these requests are either antipathetic or very expensive. It behoves users to indicate to sensor designers the kind of trade-off they prefer. Considering that the purpose of satellite imagery is mainly synoptic, and the limited availability and/or excessive data relay generated by high ground resolution imagery, 40 m would seem to be a reasonable satellite ground resolution, especially for areas with generally coarse terrain parameters, like the East African Central Plateau, of which the Kipembawe Inselberg Plain is a part. Higher ground resolution should be provided by aerial photography.

Landsat-3 RBV imagery would certainly be more effective if it was multispectral, but the panchromatic option for this sensor was presumably in order to diminish the data flow. Inexpensive readily available 40 m resolution panchromatic imagery is preferable for many purposes to expensive multispectral imagery, which may be difficult for ordinary users to obtain if combined with ground resolution of the order of 40 m or less.

A standard Earth Satellite Corporation (Earth Sat) enhancement, incorporating distriping, radiometric and geometric restoration and contrast and edge enhancement, of a Rukwa Region scene to the north-west of that shown in Figure 2 was compared with Landsat-3 RBV images of the same area. The Landsat-3 RBV imagery revealed detail generally better than the MSS enhancement. Roads, villages, drainage, cultivation, relief, water bodies, small and large mbugas, deltas and strand lines were all clearer on the RBV imagery. Only lacustrine bars and state farms were clearer on the enhanced MSS image. (It is often forgotten that no enhancement can improve on the resolution of the original sensor.) Spectral information was of course much superior on the enhanced MSS image compared with the RBV image, but if the $1 500 cost of a 1:250 000 enlargement of the enhanced MSS image is compared with the total of $80 for the four RBV images at a scale of 1:125 000 covering the same area, then the RBV imagery is found to be definitely more cost-effective.

The BRALUP investigation of south-west Tanzania (and recently supported by a field workshop on the slopes of Mount Meru in the northern part of the country), an area with similar environmental and economic characteristics to much of the developing world, demonstrated that for general land resource surveys the current most cost-effective remote sensing imagery (including aerial photography) is Landsat-3 RBV. The investigation further showed that the most significant improvement of satellite imagery would be the provision of complete stereoscopy, which although probably expensive, would cut the time taken for land resource survey by up to 90%.

7. ACKNOWLEDGEMENTS

The investigation would not have been possible without the support and encouragement provided by IDRC, ODA, Uyole Agricultural Centre - particularly Mr C W Rombulow-Pearse, and various members of the BRALUP staff, particularly the Director, Professor A C Mascarenhas. In addition, I would also like to thank the many local authorities in both Rukwa and Mbeya regions who where always very cooperative and helpful.

8. REFERENCES

KING R B and KIKULA I S, (1979): Assessment of computer processed Landsat data in areas of subsistence or no cultivation. Remote Sensing and National Mapping, Remote Sensing Society, Reading, 123-146.

KING R B, ROMBULOW-PEARSE C W, KIKULA I S, LEY G and KAMASHO J A, (1979): Land resources of the Rukwa Region: A reconnaissance assessment, BRALUP, Dar es Salaam, 522 pp.

TOWNSHEND J R G and JUSTICE C O (1980): Unsupervised classification of MSS Landsat data for mapping spatially complex vegetation. International Journal of Remote Sensing 1 No. 2, 105-120.

UTILITY OF AVHRR OF NOAA 6 AND 7 FOR VEGETATION MAPPING

John R G Townshend
Department of Geography
University of Reading
Reading
Berkshire, UK
and
Compton J Tucker III
Goddard Space Flight Center
Greenbelt
Maryland, USA

Abstract: The first two channels of the advanced very high resolution radiometer (AVHRR) on board NOAA satellites 6 and 7 have similar spectral responses to channels 5 and 7 of the multispectral sensor system (MSS) of the Landsat series. The latter have been shown by numerous studies to be very useful for vegetation mapping, notably when combined to produce the so-called "vegetation index". Although having similar spectral responses, the sensors have very different spatial and temporal resolutions. The spatial resolution of the AVHRR imagery is much coarser than that of the MSS of Landsat, namely 1.1 km. compared with 79 m. Such comparatively poor resolution may well suggest that the imagery has little to offer for vegetation mapping. However, AVHRR imagery is available every one or two days thereby much improving the chances of obtaining cloud-free imagery compared with the Landsat system, hence monitoring of vegetation change is much more feasible. Moreover, data-handling problems are much reduced due to the lower resolution of the AVHRR. Evaluation of AVHRR images is made through qualitative examination of images from several parts of the world and through a quantitative comparison of registered AVHRR and MSS digital images. It is concluded that at very broad scales the AVHRR may well provide a more appropriate technology for vegetation monitoring than the Landsat MSS.

1. INTRODUCTION

Multispectral scanner data from the Landsat series of satellites have been extensively used for vegetation mapping for diverse purposes. These

Paper presented at a conference on Matching remote sensing technologies and their applications convened in London by the Remote Sensing Society, 16-18 December, 1981

have included crop survey (e.g. Swain and Davis, 1978; Tokerud and Quirein, 1978), forest survey (e.g. Hoffer et al, 1978), identifying likely sites of locust breeding (Hielkima, 1980), and identifying geobotanical anomalies for mineral exploration (e.g. Cole et al, 1974) along with many others. Although these and many other studies have clearly demonstrated the benefits of satellite-derived data, much of the early promise of Landsat imagery as an operational tool especially for monitoring has not been realised. Amongst the reasons for this, is the low frequency of suitable images for many parts of the world. In part this arises from the fact that the Landsat series of satellites do not form a fully operational system as yet; especially for those areas out of sight of ground receiving stations there is reliance on onboard recording. Even where Landsat data are regularly collected by ground receiving stations, the frequency of suitable cloud-free images may be very low. At best with two satellites collecting data, the frequency of imaging has been once every 9 days, and at most times the frequency has been once every 18 days. In many parts of the world this may be insufficient even to make one annual cloud-free image highly likely.

A second disadvantage of the current Landsat system for some applications, is that the quantity of data that has to be processed is enormous. For example, comprehensive monitoring of the Sahel zone would involve analysis of 150 Landsat frames, each with more than 200 million bits of data, and even if Landsat coverage could regularly be obtained, any form of computer-processing would be extremely time-consuming and costly using currently available computer technology. Whereas many would argue that Landsat MSS is inappropriate because its resolution is too coarse, we would argue conversely that for some applications it is inappropriate because its spatial resolution is too fine.

Until recently the only regularly available satellite data suitable for vegetation mapping came from the Landsat multispectral scanner system (MSS). With the launching of the recent NOAA (National Oceanographic and Atmospheric Administration) satellites we have an additional source from the Advanced Very High Resolution Radiometer (AVHRR). Two of its bands are similar to those Landsat MSS bands which have proven most useful in vegetation studies, namely numbers 5 and 7 (Justice, 1978; Tucker, 1978). However, its spatial resolution at 1.1 km. is much coarser than that of the Landsat MSS which is 79 m. In the following paper we describe the AVHRR system, followed by a qualitative evaluation of its usefulness and then finally a quantitative assessment is presented using registered MSS and AVHRR data planes.

2. SENSOR-SATELLITE CHARACTERISTICS

NOAA satellites 6 and 7 and TIROS-N form the beginning of the third generation of atmospheric and oceanographic operational polar orbiting spacecraft. Their new capabilities include higher resolution day and night time cloud cover observations on local and global scales, and improved sounding observation of temperature and water vapour profiles on a global scale (Hussey, 1979). The characteristics of the payloads are outlined in Table 1. In the current paper we are exclusively concerned with the AVHRR. As Table 1 shows in outline and as described in more detail in Table 2, three types of AVHRR data are available. Within direct view of ground receiving stations at Wallops Island, Virginia, Gilmore Creek, Alaska, Lannion, France and Dundee, Scotland, full spatial resolution images are available (HRPT). For the remainder of the globe more limited

Table 1: NOAA 6 and 7 and TIROS-N payloads
Source: Kidwell (1981)

AVHRR - Advanced Very High Resolution Radiometer

a. HRPT (High Resolution Picture Transmission) - High resolution direct read out AVHRR data

b. LAC (Local Area Coverage) - High resolution recorded AVHRR data

c. GAC (Global Area Coverage) - Reduced resolution recorded AVHRR data

TOVS - TIROS Operational Vertical Sounder

a. MSU (Microwave Sounding Unit)

b. SSU (Stratospheric Sounding Unit)

c. HIRS/2 (High Resolution Infra-red Radiation Sounder 2)

Table 2: Availability of AVHRR data
Source: Kidwell (1981)

Direct readout:

HRPT - High resolution data within approximately 2500 kms. of ground receiving station: nominal resolution 1.1 km., 10 bit precision

Recorded readout:

1) GAC - Global Area Coverage: real time AVHRR data sampled along scan lines to produce 4.0 km. resolution, 10 bit precision retained

2) LAC - Local Area Coverage: full resolution data. Maximum of 10 minutes of LAC data recorded per orbit (a strip of imagery approximately 4000 kms. long)

images only can be obtained. Global coverage, but at approximately one-quarter resolution is recorded on board the satellite (GAC). Full resolution images are also recorded on board the satellite for part of the globe (LAC). Table 3 summarises the spectral bands sensed by the five channels of the AVHRR. On NOAA-6 the fifth channel is not available and channel 4 is simply repeated on the CCT's. Channels 1 and 2 are for cloud mapping, delineation of land and water boundaries and snow and ice extent. Channels 3, 4 and 5 are for determination of sea surface temperatures and 3 and 4 are used for cloud mapping (Hussey, 1979).

Initially on TIROS-N AVHRR channels 1 and 2 had very significant overlap, channel 1 extending from green parts of the visible spectrum well into the near infra-red (Table 3). However, subsequently the AVHRR channel 1 was restricted to 0.58 - 0.68 μm in the yellow/red part of the spectrum. Comparison with the Landsat MSS shows that channel 1 is very similar to MSS channel 5 and channel 2 is roughly equivalent to MSS 6 and MSS 7 combined. For vegetation studies channel 1 is spectrally reasonably satisfactory lying approximately in the chlorophyll absorption part of the spectrum, though ideally its lower limit could be moved to a slightly higher value, to make it comparable to Landsat D's thematic mapper channel 3 (Tucker, 1978). Channel 2 of the AVHRR is potentially less satisfactory since although it includes the highly reflective plateau in the near infra-red beyond 0.8 μm associated with healthy green vegetation, it also includes part of the chlorophyll absorption band as well between 0.7 μm and 0.8 μm. Nevertheless, the majority of the received signal will be from the former region (Figure 1).

It has commonly been found that a ratio of a near infra-red band and a visible band in the chlorophyll absorption region is a good indicator of vegetation status (e.g. Tucker, 1979; Curran, 1980). In particular, a normalized difference vegetation index defined as follows has often been found to show a strong linear relationship with vegetation characteristics such as green biomass:

$$\frac{\text{MSS 7} - \text{MSS 5}}{\text{MSS 5} + \text{MSS 7}}$$

It is reasonable to suppose, therefore, that a similar ratio using the AVHRR bands might also be used, namely:

$$\frac{\text{AVHRR 2} - \text{AVHRR 1}}{\text{AVHRR 2} + \text{AVHRR 2}}$$

Comparison of the NOAA and Landsat systems reveals some interesting similarities and differences (Table 4). Their orbital inclination, height above the Earth's surface and number of orbits per day are very similar. Since the NOAA satellites do not have an integral number of orbits each day, the sub-orbital tracks do not repeat on a daily basis, and the longitude of the satellite nadir will vary.

The times of coverage are similar though in the winter the early morning images may have a significant part of the scene in shadow. The principal differences between the two systems lie in the much bigger overall field of view of the AVHRR, compared with the MSS resulting in an overall width of imaging of approximately 3000 kms. compared with 185 kms Consequently HRPT and GAC images (Table 2) are available on a daily basis rather than every 18 days for Landsat. The angular instantaneous field

Table 3: Spectral bandwidths of AVHRR and the Landsat MSS (50% response limits in μms)

TIROS-N AVHRR		NOAA-6 and 7 AVHRR		Landsat MSS	
Channel 1	0.55 - 0.90	Channel 1	0.58 - 0.68	Channel 4	0.5-0.6
Channel 2	0.725- 1.10	Channel 2	0.725- 1.10	Channel 5	0.6-0.7
Channel 3	3.55 - 3.93	Channel 3	3.55 - 3.93	Channel 6	0.7-0.8
Channel 4	10.5 -11.5	Channel 4	10.5 -11.5	Channel 7	0.8-1.0*
		Channel 5	(11.5 -12.5**)		

* This is a more accurate representation of the spectral bandwidths than the value of 0.8 - 1.1 μm which is normally quoted

** On NOAA-7 only

Table 4: Comparison of the NOAA AVHRR and Landsat MSS

Sources: Kidwell (1981); NASA (1976) and General Electric (undated)

Characteristic	Landsat/MSS	NOAA/AVHRR
Inclination of orbit	98.91^{o} - 99.1^{o}	99.092^{o}
Height above surface	916.6 kms.	833 kms.
Number of orbits/day	14	14.2
Time of coverage at equator	09.30	07.30 descending) 19.30 ascending) NOAA-6 02.30 descending) 14.30 ascending) NOAA-7
Orbital period	103.3 minutes	102 minutes
Latitudinal coverage	82°N - 82°S	90°N - 90°S
Cycle duration	18 days	c. 1 day
Ground coverage	185 kms.	c. 3000 kms.
< FOV	± 5.78^{o}	± 56°
< IFOV	0.086 milliradians	1.39-1.51 milliradians
Ground resolution (nadir)	79 m.	1.1 km.
Ground resolution (maximum off-nadir)	79.5 m. along track 80 m. across track	2.4 km. along track 6.9 km. across track
Samples per IFOV	1.411	1.362
Number of channels (see Table 3 for further details)	2 visible 2 near infra-red	1 visible 1 near infra-red 2 or 3 thermal infra-red
Data precision	6 or 7 bit	10 bit

Figure 1: Comparison of the spectral responses of channels 5 and 7 of the Landsat MSS with channels 1 and 2 of NOAA-6 AVHRR

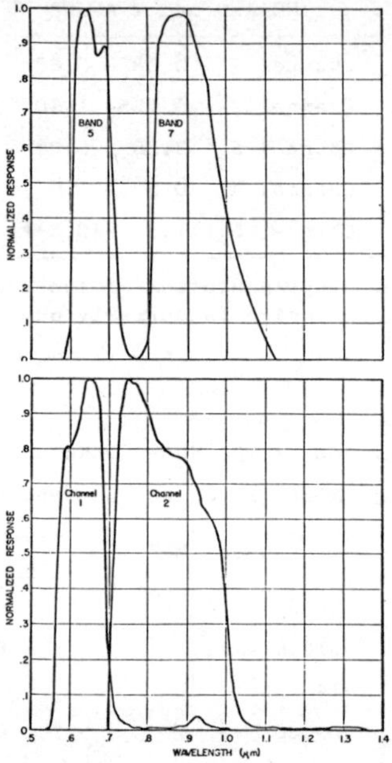

(From Schneider et al., 1981)

Figure 2: Demonstration of atmospheric effects due to marked off-nadir pointing of the AVHRR. Channel 1 image of the eastern Mediterranean. Note the grouping of the density isolines over the Mediterranean Sea at the edges of the images.

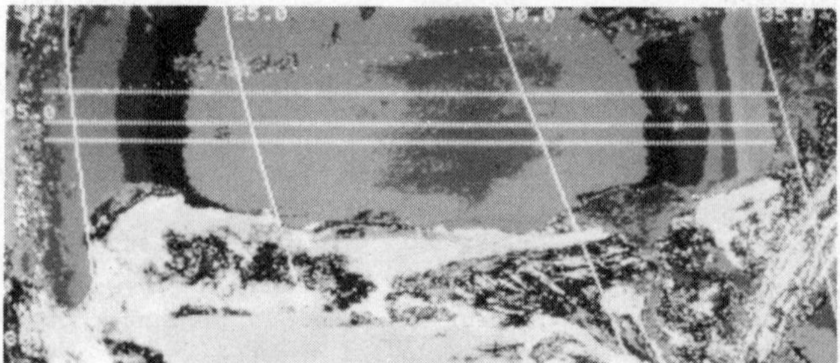

of view is similarly much greater for the AVHRR, resulting in a corresponding ground pixel size of 1.1 kms. compared with 79 m. for the Landsat MSS. These values are accurate only for pixels at the nadir. For the Landsat MSS maximum pixel size at the maximum off-nadir angle is little different from the nadir value (Table 4), but for the AVHRR the large field of view results in a pixel size at the edge of the images over three times that at nadir. An additional consequence of the high off-nadir look angles of the AVHRR is that atmospheric effects vary considerably across the scene and are very marked at the margins (Figure 2). Consequently it is necessary only to use the central 600 kms. approximately in studies of vegetation.

In summary, the AVHRR system has the possibilities of providing much more frequent images of the Earth's surface and hence raises the chance of obtaining cloud-free images. The spatial resolution of the system is much coarser than that of the Landsat MSS. This has the obvious disadvantage of substantially increasing the minimum size of objects which can be distinguished and accurately classified, but has the advantage of significantly reducing computer processing time per unit area of ground surface. Channel 4 has been found useful for accurate identification of areas covered by thin clouds difficult to detect using the visible channel.

Although we do not consider geological applications in this paper, it is worth noting that NOAA 6 and TIROS-N have two thermal channels and that NOAA 7 has three thermal bands (Table 3). For regional geological mapping such channels may well provide useful information when combined with the visible and near infra-red channels.

3. POTENTIAL UTILITY OF DATA

Visual inspection of standard black and white AVHRR images tends to be rather disappointing since relatively little variation is apparent on the land areas. Much information, however, is hidden in this form of data display, since the black and white images have to show a 1024 (10 bit) range of values. Simple contrast stretching of the data on the CCT's reveals a considerable wealth of detail as shown in Figure 3 of southern Italy. It is also of value to produce digital images of the normalized difference vegetation index described above. Figure 4 of part of the Imperial Valley allows us readily to distinguish the main terrain components of irrigated land, lake surface and desert surfaces, using such an image.

Channels 1 and 2, and in particular images of the vegetation index, have been examined for several other parts of the world. These include the Nile Delta where it has proved possible to observe the progressive increase of the vegetation index through this year's growing season. Further details may be found in Tucker <u>et al</u> (1982). In west Africa considerable contrasts in the vegetation index have been observed. Interestingly, at the Senegal-Mauritania border there is a considerable drop on the Mauritanian side, possibly relating to different management of the pastures. In the Mato Grosso, between the Rio das Mortes and Xingu River at the margins of the Amazonian rain forest, Landsat MSS images obtained in 1973 reveal extensive areas of relatively undisturbed cerrado and forest, in a condition similar to that observed by the first author of this paper in 1968. AVHRR images obtained this year show large areas of both cerrado and forest which have been cleared since 1973.

These observations do provide some <u>prima facie</u> evidence that AVHRR imagery is indeed of value in vegetation mapping. A fundamental objection to the use of such images, of course, lies in their low resolution. Specifically, there is clearly the risk that each large AVHRR pixel integrates so much significant variation at ground level, that the digital values obtained bear only a weak relationship to the actual radiance reflected (or emitted) from the surface. Thus, although visually the images may be seen to contain recognisable features, more refined quantitative analysis might be inappropriate. The following section attempts to assess whether or not this is true.

4. QUANTITATIVE ASSESSMENT OF AVHRR IMAGES

Although possibly desirable, it is not easy to collect ground data appropriate for assessing AVHRR images, because of their large pixel size. For example, homogeneous sample areas as large as 2 to 3 times the pixel size would be needed (Justice and Townshend, 1981a) and thus would be between 2.2 and 3.3 kms. in size. Consequently it was decided to use an indirect method of assessment, namely by making comparisons between AVHRR images and those from a higher resolution sensor, namely the Landsat MSS. Given the demonstrated value of the latter system, high correlations between the two data sets would clearly suggest that the AVHRR data are of potential value and that the pixels are not simply averaging out most of the significant variation at ground level. Low correlations would indicate the lack of potential value of the imagery.

Correlations between the two data sets were derived as follows. Using the E.S.L. IDIMS system at the Eastern Regional Remote Sensing Centre at NASA's Goddard Space Flight Center, common points were found on the two sets of images. Third order equations to transform the AVHRR images to the same geometry as the MSS images were calculated and then the predicted position of the common points was calculated. Those with a large deviation from their actual position were either eliminated or more accurately positioned by re-examining the images. When accuracies of approximately half an AVHRR pixel were achieved this was deemed satisfactory, since significantly higher accuracies are unlikely to be achieved. Using these equations interpolation was then performed using a nearest neighbour algorithm.

Registered data sets were produced for part of southern Italy, including the provinces of Basilicata and Puglia (Figure 3), and for part of the southern Imperial Valley, California (Figure 4). For the latter area we obtained a cloud-free MSS image of the 10th May 1980 and a cloud-free AVHRR image of the 17th June 1980, so that direct comparison should be possible. In the case of southern Italy, images with such similar dates were not possible. The AVHRR image was obtained on the 26th July 1981 and the Landsat image on the 20th September 1980. Sample areas selected avoided areas of intensive agriculture, the predominant land cover types being either quasi-natural vegetation, or grazed pastures of varying quality. Moreover arable areas are almost entirely harvested by the end of July, and sowing does not commence until the end of October. Consequently the contrast in spectral responses of those arable areas which were included should not be very great between the two years.

Nevertheless, the difference in dates will inevitably reduce the correlation between the data sets. The correlation coefficients will also be lowered due to imperfect registration between the data sets, different sun elevations, and the spectral responses of the two sensors being different.

From both of the two sets of registered images, six sample areas of approximately 100 sq. kms. were selected, each containing 20,000 Landsat pixels and approximately 103 of the larger AVHRR pixels. Product moment

Figure 3: AVHRR Channel 2 (near infra-red) image of southern Italy

a. Unstretched image

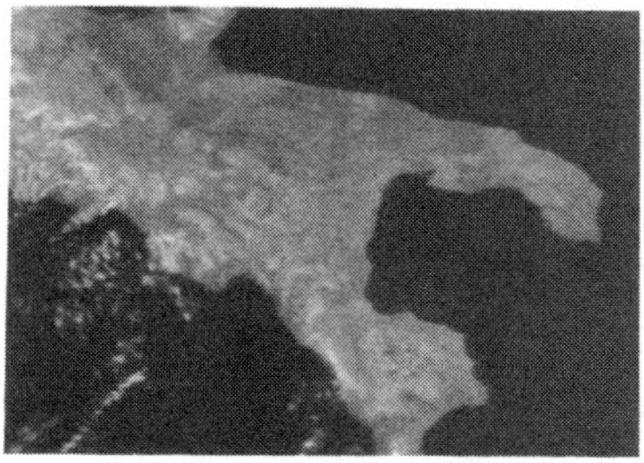

b. Contrast stretched enlargement

Figure 4: AVHRR "vegetation index" image of the Imperial Valley, Salton Sea and surrounding areas

correlation coefficients were calculated between AVHRR channel 1 and MSS channel 5, between AVHRR channel 2 and MSS channel 7, and between the vegetation indices derived from the two AVHRR channels and the MSS channels.

Results from the twelve sample areas are presented in Table 5. For the southern Italian sample areas, the correlations are surprisingly high, particularly for the vegetation index. Only for sample area 1 is the correlation low. For the individual channels, correlations are in general somewhat weaker, especially for channel 2.

In the Imperial Valley sites we find that in general the correlations between the vegetation indices are moderate to high for four of the areas, but are weak for the other two. In contradistinction to the southern Italian sites the correlations for the vegetation indices are rather worse than those for the individual channels. The rather better vegetation index correlations for the southern Italian sites compared with those for the Imperial Valley are surprising for two reasons. The sets of images in the latter area were from dates much closer together and apparently the southern Italian areas are very much more spatially complex than those in the Imperial Valley. One possible explanation of this apparent paradox may lie in the much larger areas of relatively uniform ground conditions in the Imperial Valley. Hence within a single sample area only a relatively small proportion of the total variation may be found and correlation coefficients may therefore be low. This is supported by the satisfactory overall correlation coefficients obtained when the sample areas were all combined. Gray and McCrary (1981) obtained a correlation of 0.86 from an area on the Brazil-Argentina border between AVHRR channels 2 - 1 and MSS channels 7 - 5. This lends further weight to the wider applicability of the AVHRR data.

5. CONCLUSIONS

1) Evidence has been presented that AVHRR images from NOAA satellites have a significant potential for the mapping and monitoring of vegetation. Visual examination of images initially suggested that much useful information could be extracted, especially when the vegetation index images were examined.

2) Quantitative comparisons between AVHRR images and Landsat MSS images revealed surprisingly strong correlations despite the very great differences in spatial resolution of the two systems and despite the differences in dates between the sets of images.

3) These results suggest that where broad scale vegetation surveys of large areas are required, AVHRR data may provide a satisfactory data source. Moreover such data are available at a much higher frequency than Landsat MSS data, improving the chances of cloud-free imaging. Also data processing will be reduced by a factor of approximately 200.

4) Small changes in the spectral response of future sensors would probably improve the utility of the data. The visible channel should operate at slightly longer wavelengths, from approximately 0.6 to 0.7 μm and the near infra-red channels should have a lower 50% response limit of about 0.8 μm rather than 0.7 μm.

6. ACKNOWLEDGEMENTS

The authors are grateful to Jim Gatlin and colleagues in Code 942 at the Goddard Space Flight Center for assistance in the processing of the AVHRR

Table 5: Correlations between AVHRR and MSS sensors

SOUTHERN ITALY

Sample Area	1	2	3	4	5	6
AVHRR1 and MSS5	0.53	0.76	0.72	0.68	0.34	0.76
AVHRR2 and MSS7	0.46	0.79	0.47	0.34	-0.26	0.56
VI_{AV} and VI_{MSS}	0.42	0.70	0.81	0.66	0.84	0.84

IMPERIAL VALLEY, SOUTHERN CALIFORNIA

Sample Area	1	2	3	4	5	6	Whole area
AVHRR1 and MSS5	0.26	0.69	0.85	0.32	0.89	0.88	0.85
AVHRR2 and MSS7	0.25	0.71	0.90	0.36	0.81	0.89	0.85
VI_{AV} and VI_{MSS}	0.36	0.27	0.44	0.59	0.78	0.68	0.73

AVHRR1 and AVHRR2 — Channels 1 and 2 respectively of the Advanced Very High Resolution Radiometer

MSS5 and MSS7 — Channels 5 and 7 respectively of the Multispectral Scanner System

VI_{AV} — Vegetation index $= \dfrac{\text{AVHRR2} - \text{AVHRR1}}{\text{AVHRR2} + \text{AVHRR1}}$

VI_{MSS} — Vegetation index $= \dfrac{\text{MSS7} - \text{MSS5}}{\text{MSS7} + \text{MSS5}}$

All correlation coefficients are population values for sample areas each containing 20,000 MSS pixels data. Stan Schneider of NOAA/NESS and Mike Craig of USDA were most helpful in obtaining AVHRR and Landsat data respectively. John Hardy made helpful suggestions concerning an earlier draft of the paper, especially concerning off-nadir resolution estimates. Other useful comments were made by Tim Munday.

7. REFERENCES

COLE, M.M., E.S. OWEN-JONES, N.D.E. CUSTANCE, and T.E. BEAUMONT, 1974: Recognition and interpretation of spectral signatures of vegetation and associated environmental parameters from aircraft and satellite imagery in W. Queensland, Australia. In: Battrick, B.T. and Duc, N.T. (Eds) European Earth Resources Satellite Experiments, Frascati, Italy, ESRO SP-100, 243-87.

CURRAN, P.J., 1980: Multispectral remote sensing of vegetation amount. Progress in Physical Geography, 4, 315-41.

GENERAL ELECTRIC (undated): Landsat 3 Reference Manual, Valley Forge Space Center, Philadelphia, General Electric.

GRAY, T.I. and D.G. McCRARY, 1981: Meteorological satellite data - A tool to describe the health of the world's agriculture. AGRISTARS Report EW-N1-04042, Johnson Space Flight Center, Houston, Texas, 7pp.

HIELKEMA J.U., 1977: Application of Landsat data in desert locust survey and control, FAO/AGP: LCC/77/11 AGD(RS)5/77.

HOFFER, R.M., S.C. NOYER, and R.P. MROCZYNSKI, 1978: A comparison of Landsat and forest survey estimates of forest cover. Proc. Am. Soc. Photogramm. Fall Meeting, Albuquerque, New Mexico, 220-31.

HUSSEY, W.J., 1979: The TIROS-N/NOAA Operational Satellite System, National Oceanic and Atmospheric Administration, Washington, D.C.

JUSTICE, C.O., 1978: An examination of the relationships between selected ground properties and Landsat MSS data in an area of complex terrain in southern Italy. Proc. Am. Soc. Photogramm., Fall Meeting, New Mexico, 407-21.

JUSTICE, C.O. and J.R.G. TOWNSHEND, 1981a: Integrating ground data with remote sensing. In: Townshend, J.R.G. (Ed) Terrain analysis and remote sensing, London, George Allen and Unwin, 38-58.

JUSTICE, C.O. and J.R.G. TOWNSHEND, 1981b: The use of Landsat data for land cover inventories of Mediterranean lands. In: Townshend, J.R.G. (Ed) Terrain analysis and remote sensing, London, George Allen and Unwin, 133-53.

KIDWELL, K.B., 1981: NOAA Polar Orbiter Data (TIROS-N and NOAA-6) Users Guide, National Oceanic and Atmospheric Administration, Washington, D.C.

LAURITSON, L., G.J. Nelson, and F.W. PORTO, 1979: Data extraction and calibration of TIROS-N/NOAA radiometers. NOAA Technical Memorandum NESS 107, United States Department of Commerce, Washington, D.C.

NASA, 1976: Data Users Handbook, NASA/Goddard Space Flight Center, Greenbelt, Md.

SCHNEIDER, S.R., D.F. McGINNIS, and J.A. GATLIN, 1981: Use of NOAA/AVHRR visible and new infra-red data for land remote sensing. NOAA Technical Report NESS 84, Washington, D.C., Dept. of Commerce, NOAA/NESS.

SWAIN, P.H. and S.M. DAVIS, 1978: Remote sensing: the quantitative approach, New York, McGraw-Hill.

TOKERUD, R.E. and J.A. QUIREIN, 1978: An assessment of LACIE and related methodologies for conducting crop inventories. In: Collins, W.G. and Van Genderen, J.L. (Eds), Remote sensing applications in developing countries, 91-101.

TUCKER, C.J., 1978: A comparison between the first four thematic mapper reflective bands and other satellite sensor systems for vegetational monitoring. Photogramm. Engng. and Remote Sensing, 44, 1369-80.

TUCKER, C.J., 1979: Red and photographic infra-red linear combinations for monitoring vegetation. Remote Sensing of Environment, 8, 127-50.

TUCKER, C.J., J. GATLIN, R. SCHNEIDER, and M.A. KUCHINOS, 1982: Monitoring large scale vegetation dynamics in the Nile Delta and River Valley from NOAA AVHRR data. Proceedings of Conference on Remote Sensing of arid and semi-arid lands, Cairo, Egypt., Jan. (in press).

GRASSLAND MAPPING USING SATELLITE MULTITEMPORAL DATA.

C-M. GIRARD

Laboratoire de Botanique et Ecologie Végétale
Institut National Agronomique Paris-Grignon
78850 THIVERVAL - GRIGNON – France

Abstract :

This experiment done under contract with Centre National d'Etudes Spatiales (C.N.E.S.), aims to inventory and monitor grasslands.

Two ground surveys (Spring 1980 and Spring 1981), helped to collect information concerning botany and agronomy at some test-sites.

Field data are used to improve satellite data classification. Two different Landsat scenes of the test-area (April 1976 and September 1977) have been classified on the data handling unit from Institut Geographique National (I.G.N.). Different classification programs have been used on each scene separately.

One program using thresholds on one channel of a Landsat scene enabled six land use units to be separated in the test area, five of them concern grasslands.

The classifications for the two scenes have been compared. They show similarities which assist in a better classification on the basis of both sets of satellite data.

A map of grasslands has been prepared with units related to agro ecological parameters.

It is proposed to determine how such a methodology may be extended to other grassland production areas.

1. INTRODUCTION

France, like other european countries, possess few maps concerning the spatial distribution of its main agronomic productions. The only available information consists of figures obtained by the Service Central des Enquêtes et des Etudes Statistiques (S.C.E.E.S.). For these statistics, land is divided in cultivated and non-cultivated areas. The latter concern areas covered by grasslands (Surfaces toujours en herbe : STH).

For the whole country, STH represents 23.5 % of the surface against 31.7 % for cultivated land-grasslands with an annual production of 577, 254, 333 quintaux represent a valuable agronomical production. This production consists of grass directly grazed or eaten as hay. Although grasslands may have very different productivity and thus agronomical value, statistics give few information on their nature. STH is just divided into : mown meadows, grasslands and rangelands.

In order to obtain some information about the spatial distribution of grasslands and more detail about their productivity, a research contract has been made between I.N.A. Paris-Grignon and Centre National d'Etudes Spatiales (CNES) on grassland mapping and survey with remote-sensing techniques.

France is not the only country interested by such a topic, countries with large grassland areas such as USA, Canada, or Netherlands have developped methods using remote-sensing techniques for estimating aerial biomass production from, in the field, spectral reflectance measurements (Tucker, 1980 ; Pearson and al., 1976 ; Thalen, 1980 ...) or for mapping different units from spatial or aerial multispectral data (Pearson and al., 1972).

The research begining in France concerns both topics, but only results concerning grassland mapping will be published here, as aerial biomass estimation has to be tested on different campaigns before being proposed as a reliable method. This estimation is here more complicated than for monospecific crops and needs more time and attention as species have different growing rates and show differnet phenological stages at the same time.

2. THE TEST - SITE AND FIELD RESULTS

The selected test-site is a rectangle of 223 sq km in Côte d'Or (France) centered on the city of VITTEAUX (47°24'N - 4°33'E). (Figure 1). It is the heart of a cattle breeding area, with the following characteristics :

Canton de Vitteaux	: 21,627 ha
Agricultural land (SAU)	: 18,333 ha
Grasslands (STH)	: 11,768 ha

STH = 54 % of the county's area (which is a very high figure)

The area consists of parallel N.-S. oriented valleys separated by narrow strips of plateau. The cultivated fields are grouped on top of these plateaux, the grasslands occupying the bottom and both sides of the valleys. The geological sketch (Figure 2) shows that plateaux consists mostly of limestone from Jurassic and Bajocian, while slopes and valleys correspond to liasic marl and alluvial deposits. The strong geological differences stressed by geomorphology are illustrated by soils and ecological factors which are derived from them.

The ecological conditions are very different from top to bottom of the slopes according mostly to the soil depth and moisture variations. Botanical composition and agronomic value of grasslands vary in function of these ecological conditions.

Sixty botanical stations chosen in different grasslands have been studied. They were described according to the classical phytosociological manner (Delpech, 1975). Size of stations varied from 25 to 50 sq m, the environmental data were noted as well as the complete list of species affected with the Braun-Blanquet cover-abundance value.

Descriptions were analyzed and compared, they show that grasslands may be linked to the following types corresponding to phytosociological associations (Table 1).

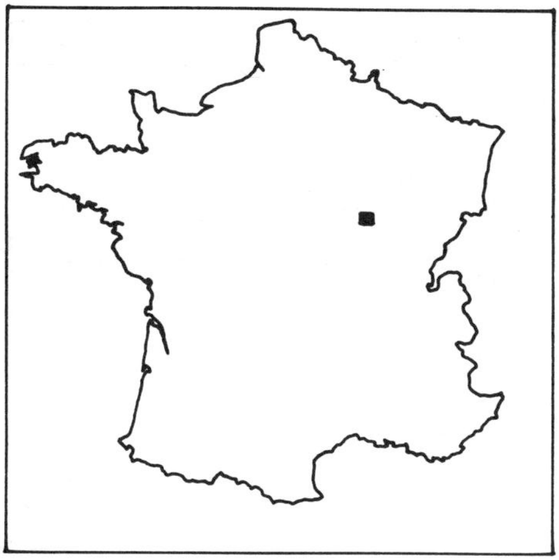

Figure 1 - Geographical situation of test-site.

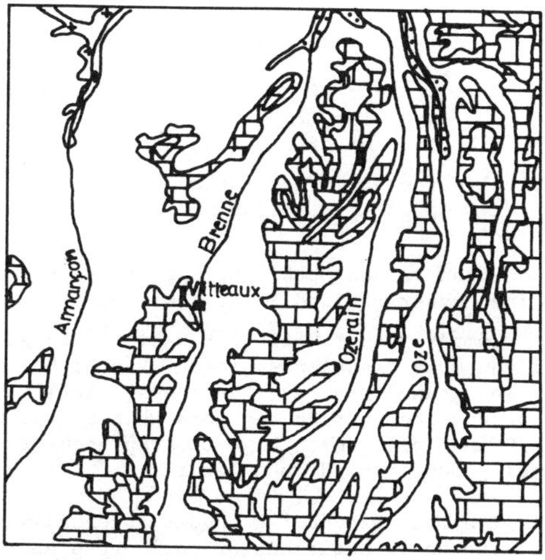

alluvial deposits

Lias

limestome from Jurassic

Granitic rocks

Bajocian

Figure 2 - Geological sketch at 1/320 000.

Grassland type	Phytosociological association
Calcicolous, heliophilous grazed grasslands on rendzina -	Mesobrometum
Moist grazed grasslands	Lolio - Cynosuretum
Moist mown grasslands	Dauco - Arrhenatheretum
Wet grazed grasslands	Association with Oenanthe peuceda-nifolia
Damp grazed grasslands	Sub-association with Caltha palustris

An agronomical interpretation of the results have been tempted. In order to compare different stations, corresponding to different fields receiving different treatments, a method has been used. Species observed in the different relevés are clustered in different groups.

All fodder plants are noted f, the most productive species being noted fp. Poison species such as : Cardamine pratensis L, Ranunculus acris L, Primula veris L are noted t. Non fodder, and non poison species are not noted.

For each relevé the following values are calculated :

- S : sum of all cover-abundance values (+ = 0.1).
- F : sum of cover-abundance values for fodder species (f and fp).
- Fp : sum of cover-abundance values for productive fodder species (fp).
- T : sum of cover-abundance values for poison species (t)

and are also the following ratios :

F/S - Fp/S - T/S.

These values enable to appreciate the agronomical value of a grassland. The higher F/S and Fp/S, the lower T/S, the better a grassland is.

Comparison of different ratios with grassland types show that moist grazed and mown grasslands present the highest F/S and Fp/S ratios while dry or wet grasslands have the lowest. Ratios give a good idea of grasslands production, but do not work when grasslands are over grazed. In that case F/S and Fp/S remain high and represent the grassland potential, while the real production is much lower

The most productive fodder species, with the highest food value grow in the best conditions of soil depth and moisture. In this test-site, ecological parameters are so important that they condition the presence in grasslands of the main fodder species. These parameters being strongly dependent on the soil types, themselves directly linked with geology and geomorphology, it may be concluded, in this particular test-site, that spatial distribution of grassland types is linked with spatial distribution of geological and geomorphological units.

3. LANDSAT DATA HANDLING

Since there were no data available corresponding to the periods of ground data acquisition (April till July 1980 and April till July 1981), two dates were selected corresponding respectively to the begining and to the decreasing of photosynthetical activity in grasslands. These periods seemed interesting to study as they correspond to periods where differences between favourable and unfavourable ecological conditions are stressed. Grasslands are green and photosynthetically longer active in favourable ecological conditions, they are already yellow and nearly unactive in the other cases. The dates of April, 29th, 1976 and September, 6 th, 1977 of scene 213-027 were selected and studied.

Different classifications have been done on these data on data handling system (TRIAS) of Institut Géographique National (IGN). (Denègre, 1980).

Three methods were successively employed.

1) punctual radiance measurements

On a combination of channels 5, 6, 7, pixels covering stations are selected and their radiance values in the different channels noted.

2) Test - zones

This method relies qualitatively to the knowledge obtained on the ground. It consists in delineating on visualization screen small surfaces corresponding to homogeneous grassland areas studied in the field (botanical stations). The automatic handling is only possible when the test-zones thus selected have homogeneous luminance values for each Landsat channel.

Every pixel corresponding to the whole-test-site is explored and linked to one or the other of the test-zones.

Such a classification has been applied to the April 1976 scene. It distinguishes three grasslands units in relation with soil moisture : dry, moist, moist and wet.

3) Thresholds

Such a method, in opposition ot the two former ones, is based on the study of only one Landsat channel, here channel 7 (800 - 1 100 nm).

All the luminance values have been clustered into classes, all the pixels have been assigned to one of these classes. The difficulty of such a method lies in the choice of the luminance values defining the classes. The thresholds may be quite arbitrary, they are satisfying only when chosen through a good knoledge of the fields.

At a first glance such a classification, done on one channel, should give less detailed results than the test-zones classification, done on the four channels. It is not the case for this April 1976 scene on which five grassland units were differenciated. The good result of such a classification may be due to a very good contrast between grassland and not grassland areas (bare soil and bare deciduous forests in April), and a good contrast also inside grasslands themselves due to the different ecological conditions stressed by the particularly dry 1976 weather.

The resulting threshold classification (table 2) was visualized and fixed on a cromalin support (Cromalin).

Affected colour	Land use or grassland unit
Green	Bare soils and forests
Red	Dry grassland
Salmon	Moist grassland
Blue	Wet grassland
White and cream	Damp and mown grassland

Table 2

Results of threshold classification.

The same type of threshold classification was realized on September 1977 data. The handling was made independently from 1976 classification. The thresholds have of course different luminance values in 1976 and 1977 as the luminance ranges are different.

4. COMPARISON OF LANDSAT DATA. INTERPRETATION

Automatic classification do not really give maps. They only give images with coloured dots which may be an aid to the thematician but which still need an interpretation to be transformed into maps.

Such a visual interpretation was made on the April 1976 classified data. Surfaces have been delineated according to the different colours showing on the visualization but also to geological and geomorphological units and of course according to the knowledge acquired in the field.

The result is a sketch at a scale of 1/125,000 (figure 3) where the four grassland units are clustered into two, the dry and moist grasslands on one part and the wet and damp grassland on the other part. This clustering is necessary because the surface of each unit is too narrow to be-drawn. In some cases more complex units are drawn, with mosaïcs of small crop areas and moist and damp grassland. The drawing take into account the coloured dots, but their information has only a statistical sense. The abundancy of red, salmon or blue dots, for example, means that the concerned area is characterized by an abundancy of dry, moist or wet grassland, and not that at the geographic coordinates of a given pixel stands exactly a dry, moist or wet grassland.

Satellite remote-sensing images and coloures composites have to be taken for what they are ; i.e. data on spectral properties of earth objects taken with a small resolution. Bi-univocal relationships between radiance or spectral reflectance values and a precise object have still to be proved.

Because of great variations of luminance values from one date to the other, due to sun elevation, meteorological conditions, atmospheric influence...comparison of 1976 and 1977 classifications is done manually and not computer assisted. Another reason for this manual comparison lies in the fact that some areas are classified into different units in 1976 and 1977. Only field knowledge enables to decide to which unit affect it.

Figure 4 (a) and (b) show enlargement (1/25 000) of a small surface south of Vitteaux. Boundaries obtained from 1976 and 1977 classifications are drawn. Areas classified into the same unit each time really belong to that unit. It is true for units with a high contrast with the neighbouring

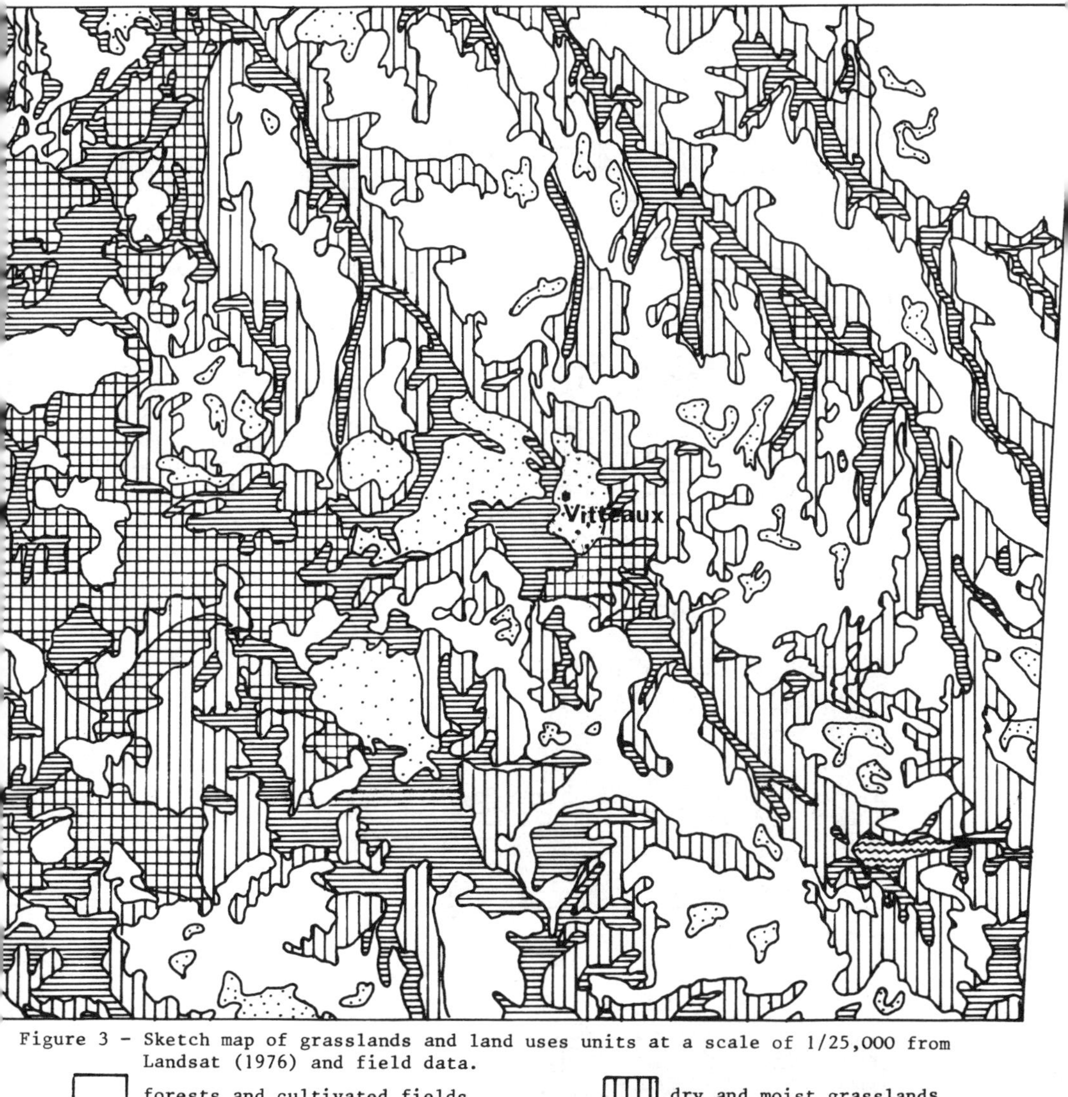

Figure 3 - Sketch map of grasslands and land uses units at a scale of 1/25,000 from Landsat (1976) and field data.

forests and cultivated fields

dry and moist grasslands

cultivated fields and dry grassland

moist and wet grasslands and some cultivated fields

wet and damp grasslands

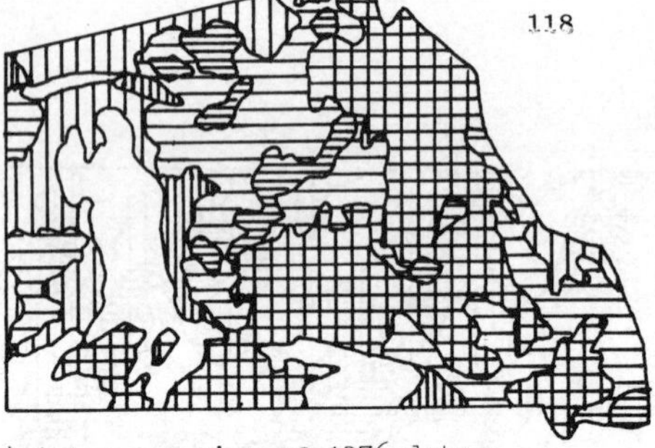

a) interpretation of 1976 data

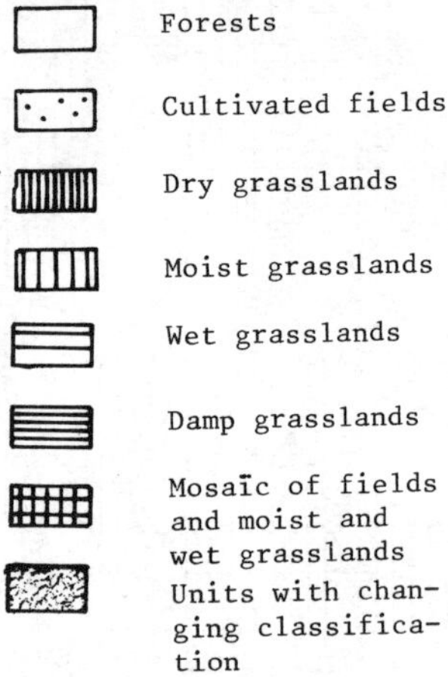

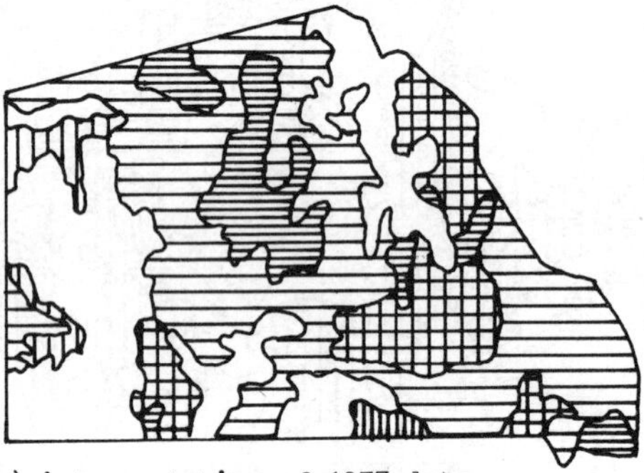

b) interpretation of 1977 data

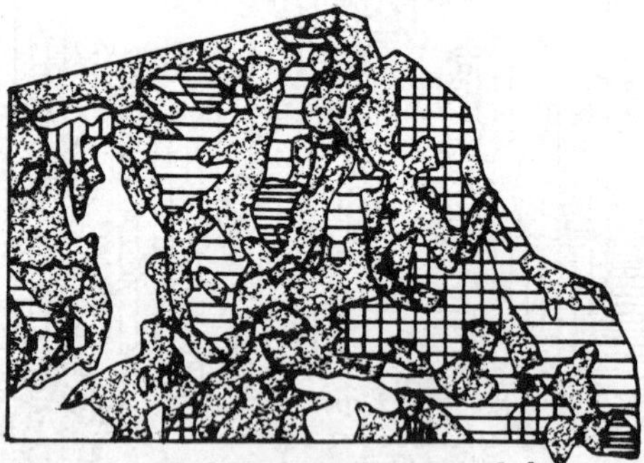

c) superposition of both sets of data.

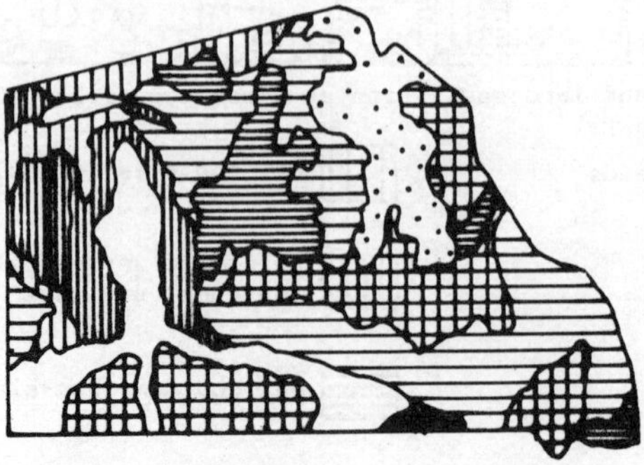

d) interpretation of both sets of data

ones : forested areas, dry or damp grasslands.

Boundaries and units varying from one classification to the other concern units with a great luminance variation between April and September. It is the case for cultivated fields with variations due to crop calendar. The luminance variation may be due also to climatic differences between 1976 and 1977. The former being very dry and the latter very rainy, a grassland may appear as a moist grassland the first year and a damp grassland the second.

Figure 4 (c) shows the mere superposition of both sets of data with non changing units referenced.

Figure 4 (d) shows the result of interpretation of both sets of data. Boundaries are more precise than with interpretation of only one scene. It is particularly the case for forests, moist and wet grasslands. Units are better defined, it is the case for cultivated fields when combining both classifications. Of course interpretation allows to go farther into units definition and mapping, because man is intelligent which the computer is not.

A quick evaluation of units classified by a mere superposition of landsat data gives 41.4 per cent rightly classified against 58.6 per cent non classified. If units deduced from comparison of both classifications had been kaken in account, the percentage of rightly classified units would have been higher and more than 50 per cent. Nevertheless, the thematician is necessary to acheive a classification with sufficient confidence in the results.

Figure 5 show a map obtained from map figure 4 (d). A rough evaluation of grassland productivity, based on field data, shows that the most productive grassland are those situated in the best ecological conditions, i.e., moist and wet soil. New boundaries are drawn on this basis, delineating areas without any grassland production (forests and fields), areas with a good productivity, and areas with a low productivity. The last unit concern areas with mixed fields and grasslands.

5. CONCLUSION

From this example some conclusions may be drawn, some of them are true whatever the test-site may be :

1) Landsat data may be used to obtain a map of grasslands at a small scale.
2) The choice of a supervized instead of a non supervized classification gives a better definition of the units.
3) The choice of a classification instead of one other is based on the luminance range on the four channel which varies with date. So the selected classification method may change with date.
4) The stronger links between botanical composition, agronomical value of grasslands and ecological parameters, the more detailed and precise the map is.
5) Comparison of two sets of satellite data gives more precise units and boundaries, and avoid misclassifications, than interpretation of only one.
6) According to the large variability of satellite data with date, and in the absence of correction of atmospheric variations, manual comparison of multitemporal data is suggested.

It is also true that the methodology developped in this experiment works particularly well because it is a much contrasted area where man's

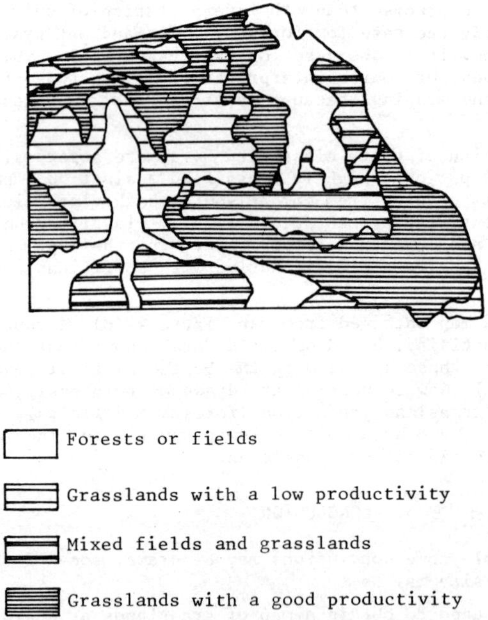

Figure 5 - Map of grassland productivity on a basis of grasslands units.

influence is rather low compared with ecological influences. It has to be tested and certainly modified if applied to other regions.

Comparison of field and satellite data stresses the poor Landsat resolution. Many boggy patches in dry grasslands on slopes, with a surface of about 500 m^2, cannot be detected and mapped, and yet they are very important for the farmer and the agronomist. Many details affecting grassland monitoring are thus observed in the field and the actual Landsat data are of no use for their detection and mapping. Thematicians and planners need new satellite data with a better resolution which SPOT and Landsat D are expected to give.

These two last satellites may also help to solve the difficulty of obtaining satellite data in these grasslands and rainy European countries. The satellites' repetitivity over the same point give more chance for obtaining some data.

At last, in the proposed example, a rough evaluation of productivity is tempted. It is mapped at a rather large scale : (1/25 000) in accordance with spatial distribution of the mapped parameter. Much is expected from satellites with higher resolution in order to match effectively and with success field and remote sensed data. But for this matching of different data man is necessary, and more particularly the thematician who knows exactly what to look at, what to map, and for what purpose.

7. REFERENCES

CROMALIN : Process and equipment manual.

DELPECH, R. (1975) : Contribution à l'étude expérimentale de la dynamique de la végétation prairiale. Thèse Doctorat. Orsay.

DENEGRE, J., LUMMAUX, J.C., POULAIN J. (1980) : Traitement d'images spatiales : méthodes et applications à l'I.G.N. France. Bull. Inf. IGN 80/3 (42), 18-24.

GIRARD, C-M. (1981) : Estimation of phenological stages and physiological states of grasslands from remote-sensing data. Oecol. Planta.(submitted for publication).

GIRARD, CM, GIRARD MC. (1975) : Applications de la télédétection à l'étude de la biosphère. Masson, Paris, 186 p.

LONCHAMP, J.P. (1977) : Etude phytoécologique et essai de typologie des prairies permanentes de l'Auxois. Thèse de Doct. en Sc. Biologiques. Fac. Besançon. 123 p. plus annexes.

PEARSON, R.L., and L.D. MILLER (1972) : Remote mapping of standing crop biomass for estimation of the productivity of the shortgrass prairie, Pawnee National Grasslands, Colorado. pp. 1357-1381. in : Proc. 8th Int. Symp. on Remote Sensing of the Environment. Univ. of Michigan, Ann Arbor, Michigan.

PEARSON, R.L., C.J. TUCKER and L.D. MILLER (1976) : Spectral mapping of shortgrass prairie biomass. Photog. Eng. and Rem. Sensing. 42 : 317-323.

THALEN, D.C.P., J.A. LEEMBURG, and R. BECK (1980) : Estimating aerial biomass in semi-natural vegetation from spectral reflectance measurements. 1) Preliminary experiences. Acta Bot. Neerl. 29 (5/6) : 565-578.

TUCKER, C.J. (1980) : A spectral method for determining the percentage of green herbage material in clipped samples. Remote. Sens. Environ. 9 : 175-181.

S.C.E.E.S. (1980) : Annuaire de statistique agricole 1980 (résultats 1979).

THE NORFOLK BROADS - MONITORING THE CHANGES IN VEGETATION COVER USING LANDSAT DATA

J R Baker, Experimental Cartography Unit, NERC, Swindon

In a collaborative programme between NERC's Institute of Terrestrial Ecology (ITE) and Experimental Cartography Unit (ECU) (Drummond and Fuller, 1981), aerial photographs taken in January 1980 and ground truth data are being combined to produce a set of large scale computer drawn maps of land use types in the Norfolk Broads. From a small scale product formed by combining and generalising these maps, a set of polygons bordering significant vegetation classes has been derived and these have been used as training areas in a supervised maximum likelihood classification of two Landsat multispectral scanner scenes combined together as a multitemporal data set. The two scenes were acquired at quite different times within the annual growth cycle of vegetation (January and May 1980), enabling relatively similar types to be reliably separated in the classification procedure. The resultant small scale base image has been further compared with Landsat scenes acquired in July 1975, June 1976 and October 1978 to study the changes of vegetation type from year to year, and in particular the reduction of marshland area at the expense of pastureland brought about by drainage over the five years covered by available data. The addition of future Landsat imagery will enable the small scale base map of land use within the Norfolk Broads to be updated on a yearly basis. Thus one of the major strengths of Landsat data, its regular and synoptic cover, will have been exploited to the full. The numerical analysis for this work was performed on the Model 70F image analysis system at ECU using algorithms within the System 101 software package of International Imaging Systems.

THE USE OF SATELLITES IN WATER MANAGEMENT

R W Herschy
Department of the Environment (Water Data Unit)
Reading

Abstract: The objectives of water management are to provide a supply of water for domestic and industrial use, irrigation, energy, navigation and recreation, to control floods and droughts and to prevent pollution. To achieve these objectives, information is required on all aspects of the water cycle on an operational basis. At the present time most of this information is provided by low technology with high labour costs. In some cases the present methods appear satisfactory using existing technology but in others the methods are clearly unsatisfactory. The potential use of satellites in providing information for these water management functions is attractive. However considerable confusion exists over the possible benefits which accrue from using a satellite system. Since water management is fundamentally an operational undertaking, a satellite system would require to be operational also rather than experimental. There is still a two-way communication gap between those engaged on research and the potential user. It might be argued however that those engaged on operational functions are not aware of the potential benefits of satellite technology.
The paper highlights some of the problems in this area and attempts to identify the requirements of satellite technology for application to water management functions.

1. INTRODUCTION

Satellites are capable of providing information on the water cycle by means of imagery. In addition, water data can be transmitted by satellite from ground sensors using data collection platforms. The prospective user of satellites however requires to know now which satellites are available operationally and what may be available to him in the future. Moreover, before the prospective user even considers a satellite system, he requires a guarantee of continuity of satellite availability. No user would wish to transfer from what may well be considered a cumbersome, but workable, manual system, to a satellite system which may prove to be only temporary.

It is clear that considerable research has already been carried out in different parts of the world into elements of the water cycle although, as far as is known, none of these research activities has yet led to the formulation of fully operational systems. Research is therefore still necessary into elements of the water cycle but it should not continue to be predominant indefinitely. Every effort should be made to transfer research technology into operational systems. A problem concerning possible satellite use in the UK and Europe is cloud cover and experience shows that on average there is only one chance in seven of being able to see any part of the United Kingdom from a satellite under such conditions. Apparently after two years of operation less than 2% of Landsat 1 and 2 imagery of the

Paper presented at a conference on Matching remote sensing technologies and their applications convened in London by the Remote Sensing Society, 16-18 December, 1981

UK was cloud free and fewer than three in ten images had a cloud cover of less than 40%. Moreover many features in the UK such as lakes, reservoirs and rivers are small and often beyond the present resolution of the imagery. It is also highly unlikely that satellites offering one or two passes in 18 days would offer the necessary information for water management purposes. On the other hand geostationary satellites providing images every half hour would not perhaps produce sufficient detail for operational purposes in Europe although their application to larger continents might prove acceptable. It is reported for example that phenomena such as hurricanes, volcanic eruptions, desertification and forest fires, which all have an effect on the water cycle, can be detected by geostationary satellites (Robson et al, 1981).

2. REQUIREMENTS FROM SATELLITE IMAGERY FOR WATER MANAGEMENT

Information from satellites for water management functions on an operational basis is required world-wide on the following elements:-

a rainfall
b surface water
c snow
d soil moisture
e evapotranspiration
f floods and droughts
g sediment
h groundwater
i water pollution
j hydrological forecasting

Table 1 gives some guidance on the precision and frequency of measurement requirements of satellite remote sensing systems for monitoring elements of the water cycle.

Table 1. Requirements of satellite remote sensing for monitoring elements of the water cycle

	ground instrument	remote sensing	
	precision	precision	frequency
Precipitation (point)	2 mm	5 mm	daily
Precipitation (areal)	10 mm	15 mm	daily
River stage	3 mm	50 mm	hourly
Lake stage	10 mm	50 mm	daily
Snow depth	20 mm	50 mm	daily
Snow cover	10%	10%	daily
Water equivalent snow	2 mm	5 mm	daily
Suspended sediment	1 mg/l	20 mg/l	weekly
Soil moisture	10%	20%	weekly
Evapotranspiration	2 mm	5 mm	daily

Note: the values given for ground instruments refer to the precision of the instrument and do not in any way infer accuracy of measurement.

a. rainfall

Considerable research has been carried out into the use of satellite imagery for areal rainfall measurement and in many cases good results have been obtained. Even in the UK one test produced a forecast which was correct on over 70% of occasions. Most work however has been carried out abroad in locations which had few raingauges and it is in such areas where most benefit would be gained.

The mapping of rainfall distribution from satellites include statistical techniques, cloud brightness techniques, infrared techniques and active and passive microwave techniques and although no one method has been generally accepted as yet, methods for rainfall estimation will continue to be developed (Barrett and Curtis, 1976).

b. surface water

The application of remote sensing from satellites for surface water monitoring at present lies mainly in making inventories of surface water bodies such as rivers, lakes and marshes within a given region. The area covered by open water is delineated by various techniques using the particular radiation characteristics of the water. Decreased reflectivity of soils moisturized at the surface facilitates the delineation of recently flooded areas, if these are barren. Characteristics of river channel, such as width, depth, roughness, degree of tortuosity and braiding is more difficult since at present the best obtainable satellite resolution is that of a LANDSAT image of 79 x 79 m. However LANDSAT imagery in the near infrared bands can be used with success to identify water bodies and also for detecting temporal variation of large rivers and lakes by the processing of computer compatible tapes using various digital or optical techniques (Wiesnet, 1979).

The development of accurate radar altimeters however such as used on SEASAT and capable of measuring water surface elevations of large water bodies may introduce new possibilities in the surveying by remote sensing of certain surface water characteristics. The possibility of estimating runoff however, even from large rivers would seem to be outside the realms of possibility at present.

c. snow

No wholly successful method has been found to measure snow. Snow lying on a catchment can be an important resource or may be a danger and cause flooding. In catchment model studies, snow is an element which cannot be entered with confidence, if at all. The measurements required are volume and water equivalent and the work reported from the US, USSR and Norway should be followed up (Wiesnet, 1979).

In Norway it is claimed that an operational system for snow measurement from satellites has now been developed (Østrem, 1981). It has been found that using the satellites TIROS-N and NOAA-6, when the snow starts to melt in the spring the density of melting snow is almost the same from year to year and that snow cover will produce the same amount of melt-water. It has been shown that there is a definite correlation between snow cover expressed as a percent and subsequent melt-water runoff, provided the measurements of snow cover are made in the spring. The method, however, does not work if the landscape is completely covered by snow because the satellites do not, as yet, give any information about snow depth, or water content. It is estimated that an annual saving of some $36 million could be achieved if satellites could be used operationally for projecting snow packs runoff (Kish 1981).

d. soil moisture

A saturated soil will absorb no rainfall and will permit high runoff and conversely a dry soil will absorb rainfall and permit little or no runoff. At present soil moisture measurements are generally made with neutron probes by taking single point samples. This method however does not furnish a synoptic cover of an area except at prohibitive expense. Nor does point sampling give any indication of the interaction of soil moisture within a given soil layer. Infrared black and white and infrared colour films are commonly used for mapping soil moisture variations from aircraft but because the changes in soil moisture are far too dynamic, the cost of aircraft flights to monitor soil moisture on an operational basis is prohibitive. Satellites using multispectral scanners appear to offer most promise in estimating areal soil moisture content on a daily or weekly basis.

e. evapotranspiration

Evapotranspiration is estimated in the field from measurements of temperature, solar radiation and wind speed. These measurements at present are made on a single point basis and give an estimate of potential evapotranspiration. Actual evapotranspiration is difficult, if not impossible, to measure directly. The losses due to evapotranspiration can be significant and present methods of measurement are unsatisfactory. Improved methodology is required, on an areal coverage and daily basis. Satellites may have a role to play.

f. floods and droughts

Floods appear to be causing more damage to life and property than ever before and, as development continues within the flood plain, this trend will continue. Many thousands of people are affected regularly in different parts of the world and damage amounts to millions of pounds sterling annually. Any improvement in existing flood forecasting technology would save lives and reduce damage. Wiesnet (1979) reports that satellite imagery can also be used for delineating flooded areas, but complications arise when these areas are covered by vegetation. Nevertheless, although flood mapping from satellites would provide useful information, by far the greatest benefits would be improved forecasting of potential catastrophic floods on an operational basis. Repetitive imagery is required to record the maximum inundated area at all points in a drainage basin. In the headwaters imagery would have to be repeated about once every 2 hours to record the very fast rise, the peak and the recession of flood waters. Further downstream the peak is later and spread over a longer period of time. For the Mississippi, for example, imagery obtained once every 3 to 5 days would probably be adequate. Satellites like LANDSAT are therefore not ideal for flood hydrology as the repetitive cycle of 18 days is too long although they do produce data on flood extent which would be costly to obtain on the ground or from aircraft (Moore and North, 1974). An ideal sensor for imaging floods should have the following characteristics:-

a) high resolution
b) minimum distortion
c) a unique reaction to surface water
d) day or night capability
e) penetration of clouds and light rain
f) penetration of vegetation

Satellite images can also be used to infer drought situations which, on a large scale, may be classified as disasters. Areas of marginal agricultural activity can be monitored for the presence of rain bearing clouds and their absence over long periods is a clear indication of possible drought and famine. In the case of METEOSAT, for example, the appearance of the surface of the visible image actually changes in these areas according to the vegetation cover. In addition the infrared image measures surface temperature and the diurnal rate of temperature change can be related to soil moisture content (Robson et al 1981). By these means drought regions may be delineated more efficiently than present methods so that the appropriate action may be taken earlier.

g. sediment

Sediment causes reduced holding capacity in reservoirs and deposition problems in rivers. The present measurement techniques are cumbersome and costly and involve considerable time spent on site. The requirement in reservoirs is total sediment content and in rivers total sediment concentration and sediment discharge, the former on a monthly frequency and the latter on a weekly frequency. Although considerable research has been reported on sediment detection by satellite very little has been performed in quantifying sediment content in reservoirs and sediment concentration in rivers. Suitable sensors to measure these elements are therefore required.

h. groundwater

Groundwater is an important water resource in many countries but more especially in the developing countries of the Third World where thousands of people die each day because of inadequate water supplies. Groundwater can be monitored directly by heat detecting infrared film or indirectly by vegetation types or by soil or geological features. A large proportion of the world's water resources are made up of groundwater and during the present Water Decade the development of sensors to detect water table depths of from 1 to 5m would result in far-reaching world wide consequences that would revolutionize groundwater hydrology. Groundwater detection at these depths would also benefit irrigation and it is estimated that savings of up to 50% may be realized into investigations in this area (Kish, 1981).

i. water pollution

The detection of pollutants in rivers and estuaries from satellite imagery on an operational basis would have significant benefits. At present measurement techniques consist essentially of manual sampling and analysis at quite considerable expense. Dissolved oxygen can now be quantified in depths of up to 10m by using false colour infrared film with a special orange optical filter. Industrial effluent discharges can be monitored with the aid of a Fraun-Hopper line discriminator using a narrow pass band filter which analyses a certain wavelength of light. Differences in temperature between effluents and river water in which they mix can also be used to identify water pollution. Oil spills and oil seeps can be detected by multispectral scanners. Many other elements of water pollution have been measured by remote sensing from satellites including salinity in estuaries, pH of humic runoff from swamps and hazardous waste leachates (Kish, 1981), (Otterman, 1981).

j. hydrological forecasting

Hydrological forecasts are one of the most important aspects of water management. Such forecasts become increasingly necessary as the economy expands and utilization of water resources in each country increases. This is especially the case in the Third World countries. However many of the requirements relating to hydrological forecasts cannot be fully satisfied at the present stage of development of hydrology because of the inadequacy of present measuring techniques. Hydrological forecasts are necessary in connection with the rational regulation of runoff, the utilization of river energy, inland navigation, irrigation, water supplies and water pollution prevention. Hydrology is concerned with extremely complex processes taking place over vast areas of the earth's surface, in some of its strata, in the atmosphere and in the channel networks. The mathematical expression of the relationship of those processes still raise insurmountable difficulties. The feasibility of observations and measurements is also limited. For this reason hydrologists in their study of hydrological processes and the development of forecasting methods, are usually obliged to use data which are insufficiently representative, and which fail to reflect the multiplicity of factors and the irregularity of the processes involved in space and in time. The development of hydrological processes is influenced by meteorological factors, but the changes they bring about in the regime do not occur immediately. However no reliable methods of quantitative forecasting of weather elements far in advance have yet been developed. For this reason, the practical possibilities of extending hydrological forecasts depend, to a considerable degree, on the extent to which future weather conditions will affect the development of the event forecast. The accuracy and timeliness of hydrological forecasts also depend on the reliability and volume of hydrological and meteorological information and the speed with which current information can be collected. The reliability of weather forecasts is therefore of decisive importance to the accuracy of hydrological forecasts (WMO 1974). It is not suprising therefore that there is some expectations in the field of hydrological forecasting that satellite technology may solve these problems. Indeed there would seem to be no acceptable alternative.

3. REAL TIME DATA COLLECTION FOR WATER MANAGEMENT

Satellites provide a link to data collection platforms within their field of view and geostationary spacecraft provide this link for data related to the earth and its natural phenomena. This is a field which offers significant potential for the transmission of water cycle data as shown by studies in the United States (Shope and Paulson, 1980), Canada (Halliday 1978) and the United Kingdom (Herschy 1980). Data collection platforms for the collection of water data such as river levels, river flows, rainfall and evaporation have been used on an experimental basis for several years and have shown this form of data transmission to be reliable and cost effective as compared with existing forms of terrestial data transmission. There seems little doubt that data collection and transmission by geostationary satellites will be operational within a few years.

4. CONCLUSIONS

The elements of the water cycle which it would seem likely to be advantageous to secure remotely include snow depth and snow coverage, water equivalent of snow, areal rainfall, evapotranspiration and sediment concentration. Remote sensing has the great advantage of both frequency of observation and areal coverage and therefore a reduced measurement accuracy compared to single point measurement accuracy is acceptable. The water cycle is a dynamic phenomenon and its monitoring requires both spatial and temporal coverage. It has been all too clear that this cannot be accomplished by single point measurements. Hydrological forecasting, particularly of floods and droughts is also an important area where satellites have potential applications. In the present decade however satellite technology will be expected to play an important role in providing information to those engaged in the implementation of the 'Water Decade'. It is therefore in the Third World that the greatest effort should be made where some 1,500 million people are without clean water and over 5 million people die each year from water borne diseases (Brandt 1980). This should be the challenge to those engaged in satellite technology and who are able to apply this technology to water management.

5. ACKNOWLEDGEMENT

Acknowledgement is made to the Director of the Environment's Water Data Unit for permission to publish this paper. The views expressed in the paper, however, are those of the author and not necessarily those of the Department.

6. REFERENCES

Barrett, E C and Curtis L E, (1976) : Introduction to environmental remote sensing, Chapman and Hall, London 204-225.

Brandt, W (1981) : North South, Pan Books London.

Halliday, R A (1978) : The use of satellites in hydrometry, in Hydrometry, Principles and Practices (Editor R W Herschy), John Wiley and Sons, Chichester, 424-447.

Herschy, R W (1980) : Hydrological data collection by satellite, Proc. Instn. Civ. Engrs. Part 1, 68, Nov. 759-771.

Kish T, (1981) : The many uses of remote sensing, J Wat. Pollut. Control Fed. Vol. 53, No 4, 420-425.

Østrem, G (1981) : The use of remote sensing in hydrology in Norway, in Remote Sensing in Meteorology, Oceanography and Hydrology (Editor A P Cracknell), Ellis Horwood Limited, Chichester, 258-269.

Otterman, J (1981) : Remote sensing of pollution, in Remote Sensing in Meteorology, Oceanography and Hydrology (Editor A P Cracknell), Ellis Horwood Limited, Chichester, 243-248.

Moore G K and North G W (1974) : Flood inundation in the southeastern United States from aircraft and satellite imagery, Water Resources Bulletin, Vol. 10, No 5 October, 1082-1096.

Robson A, Morgan J, Herschy R W and Zschau, J (1981) : Potential for detection of natural disasters via Meteosat, International Astronautical Federation XXX11 Congress, Rome, September, 18pp.

Wiesnet, D R (1979) : Applications of remote sensing to hydrology. Operational Hydrology Report No. 12, World Meteorological Organization, Geneva, 55pp.

Shope, N G and Paulson, R W (1980) : Real-time data collection via satellite for water management, Am. Soc. Civ. Engrs. 80-561, 14pp.

Guide to hydrological practices. Geneva, 6.1-6.59., World Meteorological Organisation, (1974).

REFERENCES

[illegible]

MONITORING THE CHANING AREAL EXTENT OF IRRIGATED LANDS OF THE GEFARA PLAIN, LIBYA

J S Latham
Consultant to the Food and Agricultural Organisation of the United Nations

Abstract: This paper summarises the investigation methodology and results for the inventory of the irrigated lands of the Gefara Plain, Libya. The project had as its goal the demonstration of the utility of Landsat processed data for making improved acreage estimates when integrated with ground measurements. The results achieved would indicate that given an appropriate base year eg 1981, Landsat data could be used in a programme to monitor and continually update detail on the changing areal extent of irrigated lands of the Gefara Plain, with a much higher level of associated precision to the estimate than could be achieved through normal sampling procedures. The strategy adopted includes a stratified double sample approach with automatic processing of digital Landsat data and ground sampling.

1. BACKGROUND TO THE STUDY

The Secretariat of Agriculture together with the United Nations Food and Agricultural Organisation have recently undertaken a project aimed at modelling water use and quality for an extensive area in north west Libya. This area know as the Gefara Plain, covers approximately 18,500km^2 and extends from 11^0- 14' east to 14^0-15' east and from the Mediterranean coast to 32^0 - 15' south. The project is developing a model designed to assess and monitor ground water through the Gefara Plain aquifer. This model will then serve as an input to an overall plan for implementing water management policies. The need for such an effort is mandated by the agricultural importance of the Gefara Plain aquifer and the fact that it is already starting to suffer from problems of sea water intrusion. The aquifer is the source of water for a high proportion of irrigated lands within Libya. However, the estimated rate of withdrawal extends natural replenishment. As dependence on the Gefara Plain aquifer system continues and as new additional uses are defined, it becomes increasingly important to determine demand, to project the economic life span of the aquifer, and possibly to regulate withdrawal.

Complete data on the volume of water withdrawan and its areal distribution are available but they have a number of inaccuracies inherent in their calculation. Moverover, elements of these data are calculated from statistics which are static, and to collect data repeatedly through intensive field work programmes would be both expensive and timeconsuming. A well organised, standardised approach to the collection and presentation of land cover information, which can be implemented twice yearly, could provide an essential element in the established framework for land resource

Paper presented at a conference on Matching Remote Sensng Technologies and their Applications, convened in London by the Remote Sensing Society, 16-18 December, 1981 for land resource planning and management. Furthermore such an approach must have a balance between timeliness, cost-effectiveness and accuracy (bias and precision).

Given the background to the study it is appropriate to state the objectives of the work around which the methodology has been developed to

1. Define vegetative cover to aid in groundwater recharge evaluation
2. Define land cover classification for two seasons of winter and summer to aid in determining water use
3. Evaluate the effectiveness of satellite data as a means of providing ongoing land use mapping on a twice yearly basis

THE RATIONALE FOR THE UTILISATION OF LANDSAT SATELLITE DATA

One of the greatest potential advantages of using Landsat satellite data in this environment lies in the capability to repetitively analyse data collected over large areas and to make regional acreage estimates in a controlled fashion. It is even more valuable to do this in countries such as Libya where the existing state of knowledge about the region's irrigated land is deficient, and frequently inaccurate. In addition, the use of these remote sensing techniques, is generally found to be most valuable when it is not required to furnish the complete answer, but rather is used in conjunction with other sources of information. The proposal is to usitise satellite derived and field derived data on the areal extent of irrigated land, together with data on crop water requirements/usage as an input on the inventory side of the agricultural model being developed. Water withdrawal in this way is calculated by indirect means.

With reference to the context of the above information, it can be seen that information for the inventory of irrigated lands in the Gefara Plain, given an appropriate methodology, can probably best be achieved with the assistance of remote sensing techniques. Given that the logistics of collecting accurate field data and aggregating it to large areas is difficult and often imprecise the methodology summarised here seems even more appropriate. However, previous experience (eg Bangladesh 1978) has shown that Landsat estimates alone are likely to result in unacceptably large (and unknown) errors due to bias. This bias is largely caused by small fields, a situation which results in few if any 'pure pixels' an abundance of boundary pixels, and an inability to monitor accurately all of the small (sub-Landsat resolution) fields. This factor is undoubtedly of major concern in the Libyan agricultural scene where there are many individual fields which are will beneath the resolution of a system such as Landsat. It should be noted that the boundary pixel problem is not just a problem for small/or irregular fields. In any field, there will be boundary pixels that do not fall entirely on the field. However, the percentage of such pixels is smaller on large regular fields so acreage estimates can be exptected to be better on those types of fields in an inventory using satellite data alone.

However, there are a variety of procedures that could be used to perform an inventory of irrigated lands in this environment. These procedures can be characterised according to their cost and timeliness, an well as their accuracy (bias and precision). Field measurement procedures for example are frequently not very timely, and since they are relatively costly, generally only a small sample of the universe of interest can be measured, which results in sampling error (imprecision). Field measurements can, with appropriate safeguards, be made in such a way that there is negligible measurement error (bias).

Landsat inventory procedures have the potential to be less costly on a per unit area basis, and can be implemented in a more timely fashion. Because of the lower per unit cost of measurement, a larger sample, and frequently a complete census, can be made. This capability can potentially result in an estimate with no sampling error (imprecision). On the other hand, Landsat estimates (measurements) can have substantial(and generally unknown) measurement error (bias) associated with the boundary pixel problem previously discussed.

However, either choice of inventory procedures need not, and indeed should not, be restricted to a procedure using all of one kind of measurement. In many situations, the optimal inventory procedure may involve a mix of different types of measurement. Based on the characteristics of the Libyan irrigated lands, it was felt that a mix of measurement techniques would provide an improved estimation procedure. Essentially, it is contended that the large area Landsat statistics of irrigated land adjusted through ground measurements of irrigated acreage, would provide more precise information than could be attained by using Landsat data and field information separately.

THE SAMPLE DESIGN

The volume and method of field data collection provided a data base which could be manipulated statistically in many different ways. At the base of these date was a rigorously selected random sample and, as such, it could be utilised in conjunction with other types of data in more complicated sample methods.

Essentially a stratified random sample was undetakane on a 0.7% sample of the Gefara Plain in order to obtain ground estimates of the irrigated land on a sample unit basis. In the random sample, after subdivision of the material into sample units, the requisite number of units were selected at random from the whole population of units, within which the variable of interest was measured (y_i). The associated sampling uncertainty was decreased for a constant level of field work by stratification using the available remote sensing data in conjunction with ancillary data. In this approach the population was subdivided into groups or 'strata' before selection of the sample. These strata could have contained the same number of units or differing number of units according to

i) proportional allocation to standardise
ii) allocation to strata where the perceived density of the phenomean of interest is greatest

The latter was chosen. It was also deemed that maximum overall 'accuracy' would be attained if the strata were chosen so that the units within each stratum were as similar as possible with regard to particular criteria. Since in a stratified random sample only variation within the stratum gives rise to sampling error, it is this component of variation that requires estimation, and this can in general only be derived from differences between units in the same stratum. It was therefore, necessary that if an estimate of sampling error was required that the strata be of such a size that the sample contained two or more units from at least the majority of the strata. In this case the total size of the sample was to be sufficient to be statistically significant for results on a sample unit basis to be extrapolated/aggregated to the whole stratum (through Ny). The sample size was 0.7%.

The variance associated with a fixed number of samples would be reduced if those samples were directed toward strata homogeneous with respect to known variables that are correlated to the unknown variable being measured. The criteria for stratification for this project were selected to be:-

1. Field size stratification
2. % tree cover
3. Colour stratification (related to perceived density of irrigation)

A composite stratification of the above was constructed such that the resultant stratification was the interaction of colour strata six categories of field size strata and six categories of percentage tree cover. The composite stratification then facilitated the optimum allocation of sample units that might result in a reduction in the sampling error. The size of the sampling units equalled 156.25 hectares or 25 pixels x 25 pixels after restoration and resampling to 50 metre pixels.

With the data available from the ground survey and the corresponding satellite processed data for the same units, there are several approaches that could be taken to the preparation of estimates for irrigated land acreage (ie y or specifically yh in the hth stratum).On the ith unit in the hth stratum the ground survey has obtained yhi.Correspondingly, the satellite data has produced xhi and also xh for each stratum.

As a beginning it was desirable to obtain results based upon simple random sampling of the yhi. Such results could be produced by aggregation of all the sample results into one number the average of the yhi = $\bar{y}$. This approach would be likely to have low precision because of ignored strata differences, hence it is better to obtain ys the overall stratified average

$$\bar{y}s = \frac{1}{N} \sum_h Nh\bar{y}h \text{ in which } \bar{y}h = \sum_i yhi/nh,$$

the sample average over the nh units in each stratum where N is the potential sample size per stratum.

This estimator is unbiased under simple random sampling in each stratum, but not of minimum dispersion (variance) in this case. The departure from the optimum was due to the sample allocation and being made in accordance with stratum size, variability within the stratum and cost within the stratum. In this estimator Nh/N gives the proper weight to the sample acreage of each stratum to obtain the overall total estimate. Thus it is only ncessary to consider the lack of bias of $\bar{y}h$.

Landsat only estimates of irrigated land derived from the classification/greeness count only were also made and found to be substantially lower than those calculated from the simple random sample and outside the variance of the latter. It was not possible to calculate any associated variance for a Landsat estimate but this could be done based on sample unit (Landsat) measurements and extrapolated to the whole stratum.

THE REGRESSION ESTIMATION

The overall objective of this invesigation was to determine a procedure which might yield a more cost effective estimate of irrigated area in the Gefara Plain of Libya. Cost effectiveness measures the cost of information relative to its value (extent, accuracy and timeliness). It implies the need for a particular approach employing the operation of sampling, measurement and aggregation ie a classical sample survey structure. Particulary, since the resources of Libya as regards labour are in short supply and trained personnel are already stretched beyond their capacity, it would not be realistic to consider a complete survey twice yearly. If sampling was deemed to be an important issue, equally the success of the inventory system is based on the sampling aggregation plan utilised. Substantial differences in final estimated precision, bias and cost would occur depending on which sample design was selected. Moreover the number of parameter (eg different crop acreages) that could be estimated and the reporting level at which they would be available would be similarily affected by design. The advent of timely and relatively inexpensive remote sensing data has fostered new inventory sample design options and improved estimate performance possibilities. A stratified double sampling (two phase) design has been implemented. Specifically the survey product an estimate of irrigated agriculture based on a combination of Landsat and field data. The resulting combination estimate is called a regression estimate, and is the result of a special form of double sampling. The form is special because one 'sample' areas a complete Landsat inventory (100% sample). The most distinctive aspect of the combined regression estimate is that the estimate is based in part on the relationship between the two types of measurement (Landsat and field) on identical 'sample' units. The Landsat complete inventory (100% sample) is used to characterise the entire universe of interest with no sampling error (imprecision). This characterisation is modified (bias corrected) using the relationship between Landsat measurement and field measurements as determined on selected sample units, specifically, the estimate will be in the form

$$\dot{y}hr = \bar{y}h + bh\ (\acute{X}h - \bar{x}h)$$

where lower case letters designate quanities obtained from the sample and the upper case $\dot{X}h = Xh/Nn$ are obtained from satellite data processing over the whole stratum and $\ddot{y}h$ upper case, is the unknown quantity to be estimated. The r and the h indicate the regression estimates.

Similarly, the stratum total for y is estimated by multiplying by the total number of possible sampling units in the stratum, Nh, thus

$$Yhr = Nh\ \{\hat{y}h + b\ (\bar{X}h - \bar{x}h)\}$$

where Yhr 1 estimate of area of field type (eg irrigated land) for the area of interest.

y = field sample mean of area of field type
$\dot{x}$ = Landsat sample mean
$\bar{X}$ = Landsat population mean
N = total area/area of individual sample unit
b = shape of regression between field and Landsat

The respective variance and standard deviations for these estimates are

$$U\ (\acute{y}hr) = (S^2h/nl)\ (1-p^2xy)$$
$$S\ (\dot{y}hr) = \{S^2h/nh)\ (1-p^2xy)\}^{\frac{1}{2}}$$
$$U\ (\hat{y}hr) = N^2h\ (S^2h/nh)\ (1-p^2\ xy)$$
$$S\ (\hat{y}hr) - Nh\ 5\ (\bar{y}h)$$

when pxy denotes the stratum correlation between x + y (note the index h has been omitted from p to simply the reading).

It should be noted that the regression estimate produces two types of adjustments:-

i) the first is based on the unrepresentativeness of the sample units and is a function of the difference between $\bar{X}$ and $\bar{x}$.

ii) the other adjustment calibrates Landsat measurement space to field data space so that a Landsat consistent over/under estimate is adjusted to truth ie

$$\bar{y} \text{ to } (\bar{X} - \bar{x})$$

Essentially one is evaluating how representative the sample units are by $\bar{X} - \bar{x}$ and making adjustments to calibrate the Landsat data to known field conditions through the slope of the regression line (b). However, if $\bar{X} - \bar{x}$ is equal to 0 then there is no adjustment to the simple random sample no matter what the value of b. The adjustment being made to the simple random sample is b $(\bar{X} - \bar{x})$ and can be expressed as a percentage by

$$\frac{b\ (\bar{X} - \bar{x}) \times 100}{y}$$

Where there is no significant difference between the simple random sample and the regression estimate(as in the figures in Table 1) the latter acts as an insurance policy that the estimate produced is more precise and has an associated reduction in the level of sampling error.

Table 1. Estimate of irrigated land obtained from the alternative methodologies

Variances	Simple Random Sample	Landsat Only	Regression Estimate
	Total irrigated land - hectares		
	139006	96400	138236
95%	26637	-	18432
65%	13318	-	9216
	Irrigated field cultivation (field crops with no tree overstorey)		
	61488	39981	44636
95%	13318	-	9683
65%	6973	-	4841

The analysis clearly depicts that there are significant differences in area estimates dependent on the approach used. In particular, neither ground survey nor satellite data when used separately can provide 'accurate' answers. The most precise estimate must be one which combines the advantages of the two approaches through the use of a regression estimate providing more precise and representative estimates of conditions within each stratum.

The relative efficiency (RE) of the regression estimator based on Landsat and field data as compared with the simple random sample estimate based solely on field data is given by

$$\text{RE (regression/simple} = \frac{\text{var } (\bar{y})}{\text{Var } (\bar{y} \text{ reg})}$$

which according to Jessen (1978) is approximately equal to

$$\frac{1}{1-R^2}$$

Thus the regression estimator is always more efficient than the simple estimator for any R>0. For example, given a correlation of 0.25 the regression estimator is 0.7% more efficient, for a correlation of 0.50 the regression estimate is 33% more efficient, and for a correlation of 0.80 the regression estimator is 277% more efficient. The correlation for the individual strata are show in Table 2.

Table 2. Correlation of individual strata

Strata	Total irrigated Agriculture Landsat estimate	Total field cultivation/ Landsat greeness estimate
S1	.59192	.56132
S2	.34794	.34803
S3	.16746	.63754
S4	.76706	.866
S5	.60606	.41772
S6)	No regression estimates due to 0 hectare of ground cover	
S7)		
S8	.2796	.84852
S9	No regression estimate due to 0 hectare of ground cover	
S10	.85938	.87652

Clearly, regression estimation would reduce imprecision of all strata, but just as clearly regression estimation is more effective in some strata than in others.

A pooled correlation coeffieicent for all strata was calculated for total irrigated land this was 0.464 and for field cultivation 0.7469 resulting in pooled relative efficiency increases of 27% and 226% respectively.

COST EFFECTIVENESS

An analysis of whether the increased efficiency of the regression estimates demonstrated in the previous section is cost-effective is considerably more difficult, because of uncertainties in associating true 'operational costs' with various components of the estimation procedure.

Our best estimate of true operational costs (not the costs associated with this experiment) for a large area are approximately $75-100 per sampling unit (ha) for Landsat estimates and £200 per sampling unit for field estimates. The Landsat estimates have large fixed costs independent of area, so apparent Landsat costs per unit area would be much larger for smaller areas.

The variance of the simple random sample is given by:-

$$\text{var}(\bar{y}\ \text{sample}) = S^2 y/n$$

Thus the variance would be reduced by adding additional samples by the factor

$$\frac{n}{n + \# n}$$

A typical stratum for this experiment had approximately 10 sampling units. The increase of precision and the cost associated with achieving this increases by adding field sampling units for a typical stratum

Table 3. Potential percent increase in precision and associated costs for field sampling (for base stimate n = 10)

Percent increase in precision	Associated Additional costs
10%	200
100%	2000
1000%	20000

A typical stratum might have 1000 potential sampling units, so that Landsat estimates per stratum might cost approximately $1000. Thus any Landsat/field condition that would increase precision by more than the equivalent cost devoted to field sampling ie

$\#n = \frac{1000}{200}$ = 5 more sampling units = 50% increase in precision because of a 50% increase in sampling units in this case might be considered cost-effective. This would generally occur for Landsat/field correlations $\geq$ .58 which is less than the pooled correlation for Landsat green measure estimates of field cultivation. This is also the level of correlation for which the original precision could be achieved at less cost using regression estimation with fewer field samples. Thus, in theory for the (R = 866, N = 14) green measure/field cultivation for stratum 4 the original precision based on the simple random sampling estimate could be achieved by regression estimation with approximately 11 (14.3) fewer field samples, at a cost saving of $1,200 ($1000 for Landsat regression estimates of $2,200 for 11 fewer sample units at $200).
$(14/3 = 4.52 = \text{rel eff} = 1/1- R^2)$

Of course, much more experience with the particulars of a situation would be required for one to have much faith in the above analysis. Until then, the material presented here probably should be considered only as a guide to analysing cost-effectiveness rather than a determination of actual cost-effectiveness.

DETECTABILITY OF WETLAND FEATURES IN LANDSAT MSS DATA

S J Kristof; R P Mroczynski, Purdue University, West Lafayette

The complex ecosystems collectively known as wetlands have, until recently, been poorly understood by man. Over time man has attempted to modify these areas to other uses by extensive wetland filling and/or draining. As man's understanding of the benefits of wetlands increases through scientific investigation his desire to preserve this ecosystem also increases. Good wetlands management requires the availability of meaningful data on their areal extent, composition, and spatial distribution. Because wetlands comprise a significant portion of the earth, collecting the data can be a time consuming and expensive proposition. Fortunately, research over the past decade indicates that much of the basic data required for good wetland management can be acquired from aerospace remote sensing.

This paper reports results from various Landsat analysis experiments which dealt with the determination of environmental units within wetlands. Results will be presented for the MacKenzie River Delta of Northern Canada, the Canadian prairies, and the Texas Coastal Zone. Landsat multispectral scanner data combined with computer-assisted analysis techniques were used to identify a range of wetland conditions which included: fresh and saline marshes, emergent wetland vegetation, and multiple classes of water. The consistency of results both spatially and temporally indicates the suitability of Landsat data for wetland management purposes.

THE USE OF LANDSAT MSS DATA FOR ECOLOGICAL MAPPING

J.G. Campbell
Computer Applied Techniques Ltd.,
3 St. James Terrace
Malahide,
Co. Dublin.

Abstract: The European Economic Community has identified a need to classify and map the territory of the Community on the basis of environmental and ecological characteristics. LANDSAT multispectral scanner data have been evaluated for this purpose using a study area in North West Mayo, Ireland. Work has concentrated on land-use/land-cover mapping using computer based multispectral classification. Experimental results indicate that some land-use classes have shown unsatisfactory accuracies. There are clear indications that some of these problems can be resolved by improved processing methods; other difficulties will be absent from the data of future remote sensing missions. LANDSAT data can make a valuable contribution to ecological mapping; the level of the contribution is still, however, uncertain.

1. INTRODUCTION

1.1. Background

Since 1975 the European Economic Community has been promoting the study of methods of classification of the territory of the community on the basis of its environmental and ecological characteristics. An Foras Forbartha, Dublin has participated in this research in proposing and testing classification and mapping methodologies (Cabot, 1980); the Irish team have identified preferred methodologies, which use data from Ordnance Survey maps, aerial photographs, field survey and other documentary and specialist sources. These methods have been applied to a selected study area in North West County Mayo.

1.2. Ecological mapping - summary (Cabot, 1980):

The ecological mapping study was based upon a grid of 1 Kilometer squares (corresponding to the Irish National Grid). For each square (cell) an 'indicator sheet' was used to note measurements, land-use/land-cover types, presence of ecological indicators, subjective judgements, etc.

Paper presented at a conference on Matching remote sensing technologies and their applications convened in London by the Remote Sensing Society, 16-18 December 1981.

There follows a summary of the information gathered on the sheets.

(i) Physical Features
Measurements of altitude parameters, soil quality and assessment of landscape visual quality.

(ii) Climate
Measurements of wind, precipitation, temperature, etc.

(iii) Water
Presence, length and width of rivers , coverage of sea and lake water, ground water usage; pollution measures.

(iv) Air
Type and quantity of mineral fuels used, concentration of Sulphur dioxide.

(v) Land use
Types of crops, population density, mining activity, etc.

(vi) Infrastructure
Existence of roads, industrial development, areas of scientific interest.

(vII) Natural Environment
Measurement of the areas, within the call, of the following land-use/land-cover:-
Deciduous wood, coniferous wood, heath, blanket-peat, basin peat, marsh/fern, reseeded grassland (reclaimed peat bog), upland grassland, limestone grassland, freshwater, exposed rock, sea, sand.

A data base was compiled from the indicator sheets. The parameters (or derivative/compound measures) were plotted on the one Kilometer grid.

1.3. Study area
The study area covers about 3000 square Km. in North West Mayo, bounded on the North and West, by Atlantic Ocean, on the East by a North-South line running through the town of Ballina and on the South by an East-West line running just North of the Clew Bay coast.

The area is one of great diversity of land-forms and land-cover including
(i) rich agricultural farmland in the East;
(ii) a central upland region covered by a vast area of blanket bog (1100 square Km);
(iii) a coastal area which includes sand dunes, narrow strips of cultivation, rough grazing land, peat, etc.

1.4. Remote sensing study objectives (Campbell, 1981).
The objective was to evaluate the contribution that remotely sensed data could make to ecological mapping. Due to the lack of any alternatives, it was assumed that LANDSAT MSS data would be used in the experimental parts of the study.

1.5. Outline of this paper
This paper describes the activities undertaken in the evaluation of LANDSAT MSS data for ecological mapping, presents experimental results and draws conclusions based on experimental and other considerations.

Section 2 describes a preliminary assessment of the objectives and how several constraints necessitated review and subsequent trimming of these. Section 3 describes the choice and design of analysis tools and techniques. Section 4 outlines the experimental work program, leading to presentation of the results in Section 5 and study conclusions in Section 6.

2. REVIEW OF OBJECTIVES.

2.1. Preliminary Considerations
Based upon elementary considerations, and on the specification of the LANDSAT multispectral scanner (MSS) (Bernstein, 1976) it was possible to conclude that many of the ecological indicators could not be detected/measured using LANDSAT data. Also, certain difficulties could be anticipated. For these considerations, particularly relevant features of LANDSAT, its operation and the MSS data are :-
(a) The MSS instantaneous field of view (IFOV) is approx. 80 metres by 80 metres. Objects smaller than this are difficult or impossible to isolate from their surroundings.

(b) Feature (a) above, leads to the problem of pixel mixing (i.e. one pixel containing contributions from two or many different land cover types). This problem is particularly serious in the West of Ireland where many land-cover units are of comparable size to the IFOV.

(c) LANDSAT cannot 'see' through obstacles. Thus, image radiance is largely dependent on the 'top layer', which may be water, cloud, vegetation etc; these may obscure the features of interest.

(d) In Irish climatic conditions, LANDSAT coverage must be considered intermittent due to the intervention of cloud. From past records it appears that usable images, with partial or no cloud cover, can be expected about three times per year. There is no guarantee of even distribution throughout the seasons.

(e) LANDSAT measures the radiance of the scene within four spectral bands. Objects and phenomena must exhibit a distinct radiance spectrum in these bands, if they are to be reliably detected/measured using LANDSAT data.

2.2. Revised objectives.

2.2.1. Classes
Based upon the previous section, the objectives of the study were revised; attention was focussed on automated land-use/land-cover mapping. This reduction of scope is not altogether drastic, since a large proportion of the effort, in the conventional mapping, is devoted to determination of land-use /land-cover (field survey and air photo interpretation).

The following land-cover types were identified as suitable for mapping; together, they cover most of the study area; individually, they are well enough represented for training (archetype) pixel samples to be isolated; further, there was a reasonable expectation of spectral and spatial separability.

a) Sea and freshwater (lumped together),
b) Tidal mudflats,
c) Sand - dunes,
d) Limestone grassland,
e) Cut bog (blanket peat, hand-, and machine - cut),
f) Uncut bog (blanket peat),
g) Upland grassland,
h) Coniferous forestry, (no extensive deciduous forestry in the area),
i) Permanent pasture and reseeded grassland (reclaimed peat bog),
j) Farmland, (necessarily a mixture of pasture, cereal and root crops, and other agricultural land-uses that prevail in Mayo),
k) Heathland.

2.2.2. General mapping strategy.
In order to put the study within a firm framework a desired general mapping methodology was defined as follows :-

(i) Obtain current LANDSAT image of region to be mapped.
(ii) Do field survey over small portions of region to isolate respresentative areas of all land-use/land-cover classes. (ground truth)
(iii) Train classifier procedure on ground truth segments of the image
(iv) Classify the whole region based on (iii)
(v) Produce map from classifier output.

3. ANALYSIS TOOLS.

3.1. General methodology
In the beginning, it was decided to aim for automated land-use mapping, using digital data and computer techniques. This required the provision of interactive image processing and pattern recognition procedures.

3.2. Hardware
The computing was carried out on a DECSYSTEM-20 mainframe computer, located at University College Dublin. Unfortunately, no colour or monochrome image display device was available, except for pseudo-grey level character displays on terminals and line-printers.

3.3. Software
The authors previous experience (Campbell, 1979) eased the design and programming of an extensive suite of multivariate data analysis programs, including statistical pattern recognition procedures. Later a suite of interactive image processing procedures was found necessary (using memory based image storage and communication - for speed). The features of the combined software system - named LAMID (Laboratory for the Analysis of Multivariate and Image Data) are outlined in Appendix A.

4. WORK PROGRAMME

4.1. Choice of imagery
The criteria used in the selection of imagery were :-
(i) cloud free study area,
(ii) high sun elevation (closeness to mid-summer),
(iii) closeness to the dates when the ground truth was gathered.

LANDSAT 2 image 224/022, 9th May 1978 was selected for initial work. Time and data storage constraints caused virtually all the experimental work to be concentrated on this image.

4.2. Preprocessing of imagery.

4.2.1. Radiometric corrections.

No radiometric corrections were applied, except for those already carried out at the ground station. Careful examination indicated that sensor-banding (Sabins, 1978) was detectable but that it did not noticably affect classification results.

Corrections for topographical variations (suns incident angle) were considered but dismissed as impractible; short-cut methods (e.g. ratioing, 'equal-brightness normalisation) received limited study but were found to degrade classifier accuracy.

Equal probability quantisation (Haralick, 1973) was used for display enhancement but not in the classification process.

4.2.2. Geometric rectification.

Attempts to work with unrectified imagery progressed from being a minor source of annoyance to being a major source of errors (particularly in the combination of imagery and ground truth maps).

After attempts at short-cuts, (Hardy,1978a; Fisher et al, 1979), proper control point rectification to the Irish National Grid was carried out (Hardy, 1978b). Using 30 control points (the study area covered about 1/12 of complete LANDSAT image) a root mean square error of just over 1 pixel, in each axis, was achieved. In order to speed up the process, and because image spatial quality was not a primary concern, nearest neighbour resampling (Bernstein, 1976), was used.

The image was resampled to pixels of 50 metres (East-West) by 83.3 metres (North-South). This pixel size is reasonably close to that of the raw pixels; more importantly, 50 by 83.3 corresponds to the shape of line printer character grid (1/10 inch by 1/6 inch) for convenient display of maps and images.

4.3. Ground truth.

Ground truth information was available on the 6 inch (1:10560) maps which had been used during field survey and air-photo interpretation. Although the ground truth was extensive, covering all of the study area, it was gathered before any remote sensing study was anticipated. From a remote sensing point of view, the ground truth was deficient in some respects:-

a) The maps were compiled over a lengthy period and were not synchronised with any LANDSAT image,
b) The major classes were recorded, mostly, without any indication of incidental details or sub-class information.
   In some cases, the ground truth maps ignored details of vital importance to a remote sensing study, for example ,
   (i) density of vegetation canopy,
   (ii) age of forestry (newly cleared = forestry = mature forestry),
   (iii) type of vegetation cover in cases such as 'uncut bog', which do not explicitly refer to vegetation.

4.4. Preliminary analysis

Preliminary analysis of the image data structure was carried out using LAMID (see Section 3.3.). Procedures used included display of grey level data, clustering, examination of cluster spectra, comparison of clustered and ground truth maps, etc.

4.5. Choice of features

From the traditional pattern recognition viewpoint (Devijver and Kittler, 1978), the choice of features is not difficult. There are only four features, and to some extent LANDSAT designers have done the feature selection.

However, band ratios and other non-linear combinations of bands (see for example Sabins, 1978) have been used with some justification in special cases. Further, the spatial nature of the data has led to the use of texture measures (Mason, 1980; Haralick, 1979). Also the method of identifying 'fields' (radiometrically homogeneous and spatially connected regions) (Robertson, Fu, Swain, 1973; Bryant 1979) prior to 'per-field' classification using averages has been used; per-field classification, however, requires the existence of large fields.

In spite of the preceeding discussion, this study has concentrated on the simple 'per pixel' spectral signature approach to features. Brief exploration of ratios and simple texture measures were inconclusive.

In the time available, exhaustive research of novel methods was not possible. Also, there was a deliberate attempt to keep the procedures as simple as possible.

4.6. Classifiers

Based upon examination of the data structure and on the broad consensus in much of the literature it was decided that supervised statistical classifier procedures (Duda and Hart, 1973; Fukunaga, 1972; Devijver and Kittler, 1978) would yield nearly optimal performance. There follows a summary of the considerations for selection of a classifier from those available.

(i) Nearest Neighbour classifier
Excellent theoretical credentials (Cover and Hart, 1967); No probability distribution assumptions required; However, too slow, even when using the Condensed Nearest Neighbour Algorithm (Hart, 1968); Also, susceptible to 'rogue' (wrongly labelled) training pixels.

(ii) Pairwise Fisher classifier
Pairwise extensions of the two-class Fisher discriminant classifier (Fisher, 1936) to multiclass; Too slow for 11 classes (55 dichotomies to be resolved for each pixel); Similar theoretical basis and practical performance to (iii) below, though much slower.

(iii) Gaussian Maximum Likelihood classifier
Good theoretical basis (data sufficiently close to Gaussian distribution); Tolerable speed (1 minute computing time for 120x120 image, 4 bands, 11 classes); best performance.

(iv) Nearest Mean classifier
Chooses class based on closest mean (i.e. related to (i) and (iii) above). Very fast, but error rate unacceptable.

(v) Box classifier
Multidimensional 'density-slicing'; Very fast but poor accuracy.

Thus the Maximum Likelihood classifier was the best classifier available to the project. This agrees with must of the published literature on 'per-pixel' spectral classification of LANDSAT MSS data.

In all cases the training data were examined for major departure from Gaussian behaviour (for example, multimodal distributions). Also care was taken to use sufficiently large, uncorrelated, training data samples (Foley, 1972).

Duda and Hart, (1973), show that the Gaussian Maximum Likelihood classifier can be efficiently implemented using discriminant functions; the class mean and covariance matrix parameters were estimated from the training data sample.

4.7. Classifier training and evaluation

4.7.1. Training data extraction.

Classifier training data samples were obtained by defining 'ground-truth' areas for each class, within the image processing system; the pixels within these areas were labelled appropriately and extracted to disc-file for subsequent analysis.

4.7.2. Classifier training

The only training required by the Maximum Likelihood Classifier is the estimation of class mean and covariance statistics (Duda and Hart, 1973), followed by computation of the determinants and inverses of the covariance matrices. A priori probabilities were not included in the discriminants, that is, the a priori probability of occurrence of the classes were assumed to be equal.

4.7.3. Classifier evaluation

In normal statistical classifier design, a commonly used testing method is to classify a sample of labelled data that is distinct from the training sample. The classifier derived labels and true labels may then be compared to compile a class confusion matrix. A variation of the approach was used to evaluate the final results (classification averaged into one Km. squares). However, the main testing during development was carried out qualitatively, by visual comparison of classifier and ground truth maps.

4.8. Maps and Displays

4.8.1. Full resolution line printer maps.

The imagery was rectified, and resampled to pixels of 83.3 m. by 50m. (see Section 4.2.2.). This made it possible to produce line printer image displays and maps to a scale of 2 inches per kilometer. For grey level image display, quadruple overprint combinations of characters were used. Single print characters were more useful for classifier maps, Figure 1 shows an example.

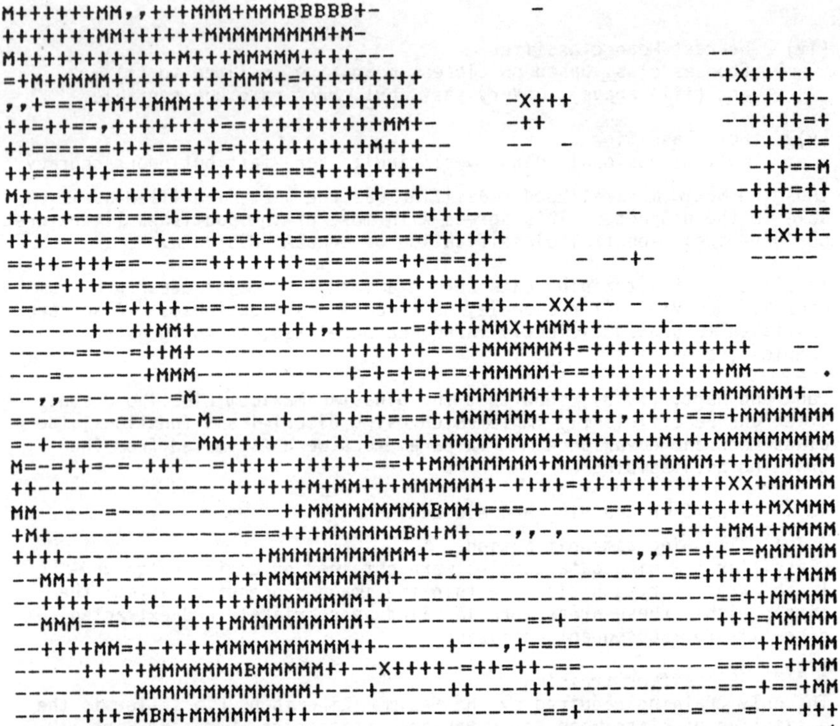

FIGURE 1: Classified subimage: Water = 'blank', mudflats = '.'
sand dunes = ',', uncut bog = '=', cut bog = '-',
upland grass = '+', forestry = 'x', pasture = 'B',
farmland = 'M'.

4.8.2. Kilometer square maps
The original Ecological Mapping project used 1 Km cells as the basic resolution element. Ecological indicators were aggregated within these cells; for example, a cell was classified as cut-bog if cut-bog occupied the major part of that cell.

One kilometer resolution LANDSAT classification maps were obtained in an analagous manner. First, individual pixels were subjected to maximum likelihood classification; then, the majority class was found by voting over the 240 pixels (20, E-W by 12, N-S) in each cell. Figure 2 shows a sample segment of the study area.

In order to avoid obvious anomalies the following ad hoc rules were augmented to the straight voting procedures: -

(a) Pasture and general farmland votes were pooled into the general farmland class (but only in cases where neither, individually, could form a majority),

(b) in the analagous situation, cut-bog and uncut-bog were pooled into the uncut bog class.

5. RESULTS

5.1. Introduction.
The primary result of the study was a map, based on 1 km. cells, showing how the study area was classified using the LANDSAT data. This map and a similar ground truth derived map, were compared to yield a class confusion matrix. Further qualitative assessment of classifier output was obtained by visual comparison of the 1:10560 ground truth maps and the full resolution 2 inch to 1 kilometer classification displays; practical difficulties made it impossible to obtain a confusion matrix at the level of individual pixels.

Additionally, the evaluation of the data was widened by plotting the LANDSAT spectra for individual classes and by the use of these and other graphics, the spectral relationships of the classes were examined. Figure 2 shows a segment of the LANDSAT map; this may be compared with Figure 3 which shows the corresponding ground truth derived map.

5.2. Class confusion matrix.
Section 4.8.2. describes how the average (or majority) class of each 1 Km square was obtained. A similar map was obtained from the ground truth. Figure 4 shows the result of compiling a class confusion matrix from these two maps. Each row of Figure 4 corresponds to a 'true' class; within the rows each column shows the number of cells classified, wrongly, as other classes or, correctly, in agreement with the true class.

5.3. Spectra, Scatter diagrams, histograms
Unfortunately, the most useful mode of presentation of MSS data, - false colour image display, was not available during the study. However, it was possible to examine the data structure using the following displays,

(a) Spectral signatures
Figure 5 shows the mean spectra, and standard deviations, of the classes, as estimated from the training sample.

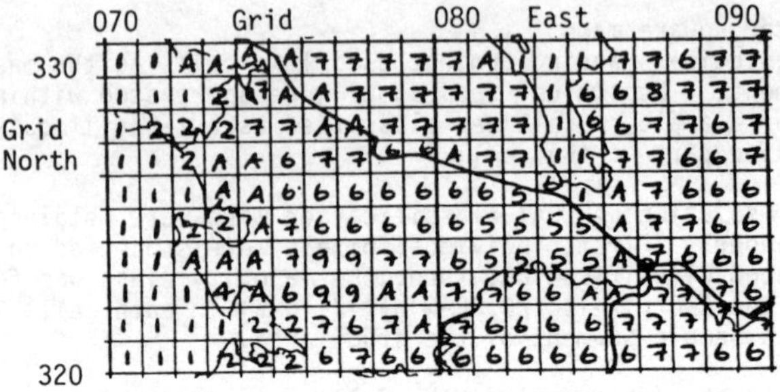

FIGURE 2: Portion of automated classifier map (see Table 1 for legend).

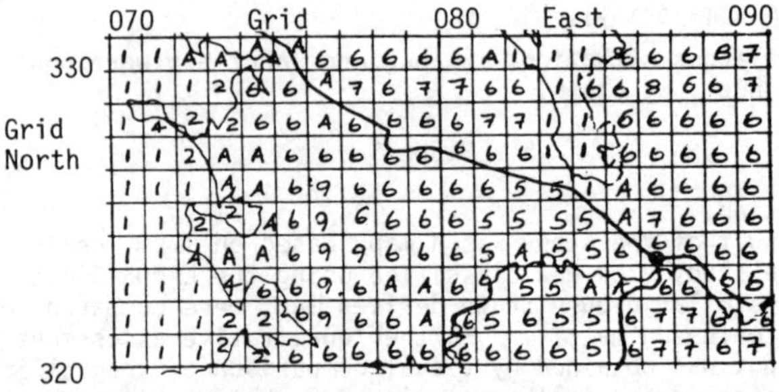

FIGURE 3: Portion of ground truth map (see Table 1 for legend).

Class Number	Name	Map Code
1	Water	1
2	Mudflats	2
3	Sand-dunes	3
4	Limestone Grassland	4
5	Cut Bog	5
6	Uncut bog	6
7	Upland Grassland	7
8	Forestry	8
9	Permanent Pasture	9
10	Farmland	A*0
11	Heathland	B+A

TABLE 1: Class Label Allocations
*0 on Scatter Plot
+A on Scatter Plot.

True Category	Total	Automated Classification Category										
		1	2	3	4	5	6	7	8	9	10	11
1	383	364	10	0	0	1	0	1	0	1	6	0
2	57	6	43	0	1	0	0	6	0	0	1	0
3	6	1	1	4	0	0	0	0	0	0	0	0
4	7	2	1	0	4	0	0	0	0	0	0	0
5	53	0	0	0	0	44	7	0	0	0	2	0
6	1153	7	5	0	6	14	569	458	1	10	78	5
7	69	0	0	0	1	2	9	56	0	0	1	5
8	69	0	0	0	2	0	17	20	27	0	2	1
9	7	0	0	0	0	0	2	1	0	4	0	0
10	530	4	4	1	9	0	2	17	1	12	480	0
11	16	0	0	0	0	0	0	14	0	0	0	2
12 undecided	16											

FIGURE 4: Class Confusion Matrix. (based on 1Km. cells.)

(b) Scatter diagrams
Figure 6 shows a scatter plot of the distribution of the classes in the plane formed by Band 5 and Band 7 values. This diagram, along with those for other band pairs, gives much insight into the class relationships; it also confirms the general form of the class confusion matrix.

(c) Histograms
Histograms of each class, in each band, were plotted and examined to ensure that the classes were suitably distributed for the use of a Gaussian classifier.

5.4. Discussion of Automated Mapping Accuracy
This section discusses the accuracy of LANDSAT derived land-use maps using the evidence contained in

(a) the 1 km grid maps,
(b) the class confusion matrix,
(c) full resolution digitally classified maps (50 m x 83.3 pixels).
(d) the 1:10560 source maps
(e) the spectra and scatter diagrams.

Some of the comments in (1) are relevant to all classes.

1. Water
Figure 5 shows that water has an entirely distinct spectrum. The 1 km maps confirm this with a very accurate classification (95 per cent). The 5 per cent disagreements are unlikely to be due to any classifier fault but to

(a) intrinsic errors in the voting over 1 km square (particularly in marginal cases)
(b) similar errors in the manual aggregation of the ground truth,
(c) changes in water level, silting and tidal effects.

Visual assessment of the full resolution maps shows water to be mapped very accurately.

2. Mudflats
This class again has a fairly distinct spectrum although the relatively large standard deviations in the bands may indicate that mud, wet sand and some dry sand are represented in the training data. The class is not sufficiently represented in the 1 km maps, for any comment, but the full resolution maps indicate good accuracy insofar as this can be measured without full tide details.

3. Sand Dunes
From the examination of the spectrum in Figure 5 and of the imagery it appears that sand dunes are characterised almost entirely by their content of dry sand, that is, they show high reflectivity in all bands. This fact leads to some suspicion of the admissability of otherwise accurate results.

4. Limestone Grassland
This category exhibits a fairly typical grassland spectrum. Comparison with the spectrum of reseeded grassland indicates that limestone grassland has higher reflectivity in the bands 4 and 5 and a lower reflectivity in the infra-red bands. This fact and the greater spread of the limestone grassland spectrum may indicate some

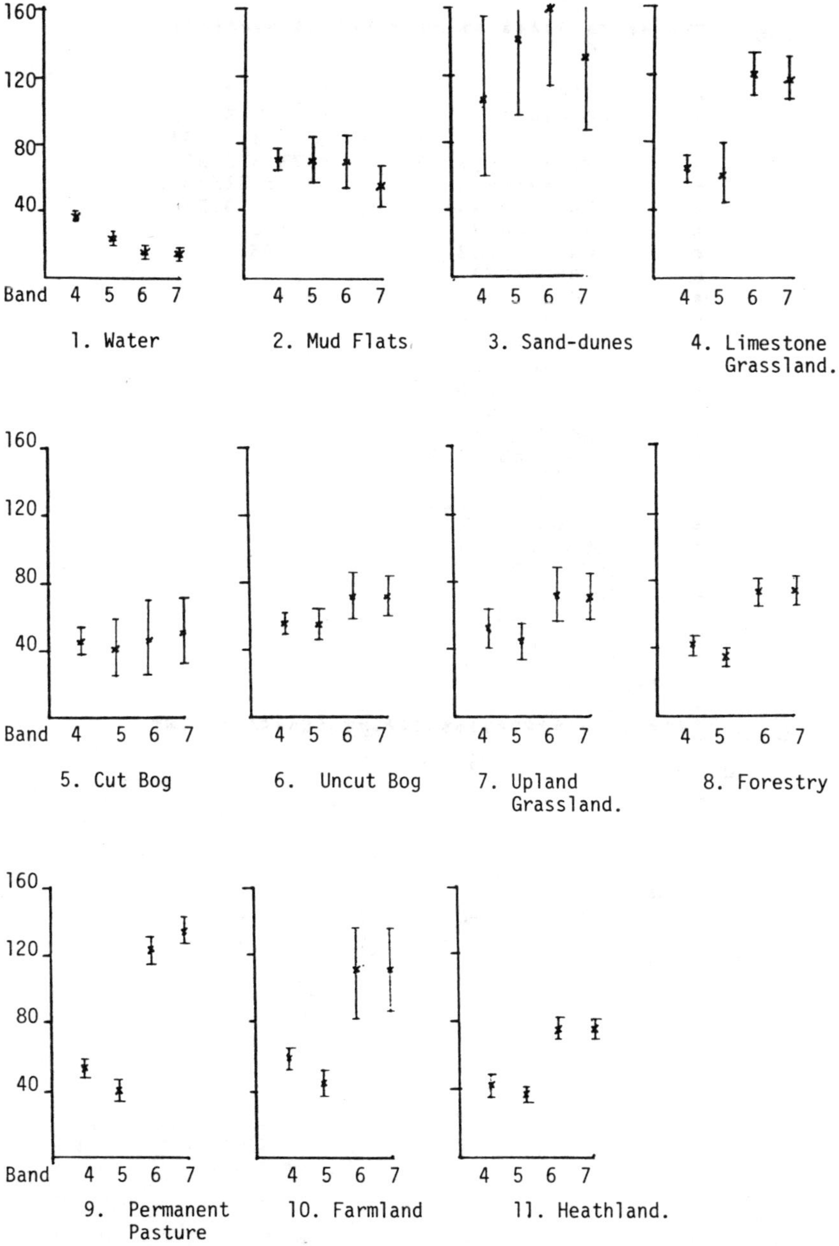

FIGURE 5. Class Spectra (Mean $\pm\sigma$).

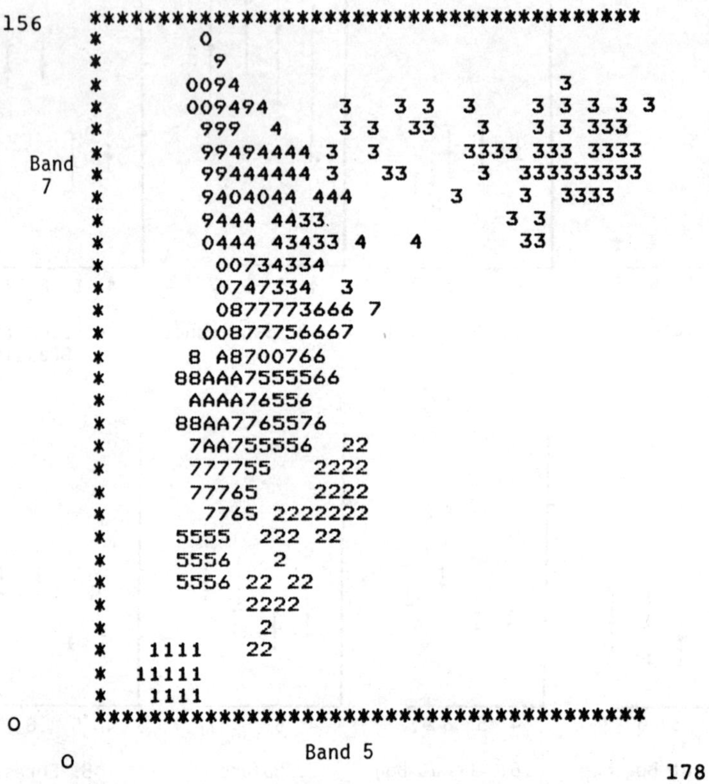

Figure 6. Scatter plot showing class distribution in Bands 5 and 7. (see Table I for labels).

influence from dry sandy areas.

This category is not well enough represented in Figure 4 for any meaningful comment to be made. However, the full resolution maps shows that

(a) some pasture/farmland is classified as limestone grassland,
(b) to a lesser extend, some limestone grass is classified as pasture (class 9) or farmland (class 10).

Both these facts could indicate that limestone grassland is not actually a valid class for detection, using just spectral information.

5. Cut bog
Extensively cut bog exhibits a characteristic overall low reflectivity spectrum. In particular, mechanised cutting is clearly visible on colour or monochrome band 6 or 7 imagery. Patchy cutting expecially hand cutting, is difficult to discriminate from uncut bog. In some areas there is an understandable confusion with uncut bog, but good accuracy can be concluded from both full resolution, and 1 km, maps.

6. Uncut Bog
There is a chaotic confusion between uncut bog and upland grass (approximately half of the (true) uncut bog is classified as upland grass). The similarity of the two spectra Figure 5 and the overlap of their distributions confirms the problem. This does not, however, explain the nature of the problem.

The uncut bog spectrum probably contains a mixture of contributions from patchy occurrences of grass, heather, weathered peat (exposed like cut peat), bare rock, rushes, water pools, moss etc. It can only be assumed that the greater ratio of grassy vegetation in the upland grassland category do not sufficiently alter the spectral signature.

This apparently unexplained failure indicates an area for further study possibly using
(a) high spatial resolution aerial imagery, or future satellite imagery,
(b) higher spectral resolution imagery,
(c) employ spectrometer equipment to examine the incidence of sub-spectra within the classes,
(d) it has been pointed out (Moore, 1981), that the choice of image (early May) has contributed to this failure, and that the use of an image sensed in later Autumn or Winter would almost certainly facilitate better separation of the uncut bog and upland grassland categories.

7. Upland Grassland
The results for upland grassland are apparently good when viewed in isolation from the results discussed in the previous paragraph.

Additionally the full resolution maps show that this class appears to be confused with some farmland. This is probably due to the similarity of upland grass to vegetation occurring in the poorer marginal farmland (for example, rushes and bracken). In the

6. CONCLUSIONS

Elementary considerations have led to the conclusion that certain of the ecological mapping indicators are definitely impossible to measure using LANDSAT data; for example, topographical details, subjective measures and features that require very high spatial resolution. Eventually the study objective was trimmed to that of land-use/land-cover mapping. This reduction in scope was not drastic, since most of the effort, in the conventional mapping procedures, is devoted to field survey of land-use.

Eleven representative land-use classes were identified and these were mapped using an automated classifier algorithm which had been trained on small regions of known land-use. The automated mapping was subjected to a very exacting test, in the sense that the whole study area was mapped, using the land-use specifications of the conventional ecological mapping. Thus, the results were not derived from hand picked data and they exactly represent what could be achieved using the current, far from perfect, procedures.

When field survey and automated mapping results were compared, it was found that certain classes could be mapped accurately; however, other classes gave unsatisfactory performance. Based on these results it would be impossible to recommend the replacement of the present sources by LANDSAT.

Nevertheless, the good results and other indications show that LANDSAT data can contribute to ecological mapping; further work is needed to increase the level of that contribution. The short study described in this paper must be considered exploratory; the full potential of LANDSAT data, for mapping areas like that described, awaits demonstration by further research. Also, future satellite remote sensing missions may deliver data more suitable for the task.

7. ACKNOWLEDGEMENTS

The work described was carried out for An Foras Forbartha, Dublin and was partly funded by the National Board for Science and Technology. The author wishes to thank the following for important contributions to the work :

Dr David Cabot (Project Manager) and Dr. Eon O'Mongain for setting up the project and for many useful discussions, the Project Steering Committee for helpful guidance, the Computer Centre, University College Dublin where the computing was carried out, R.A.E. Farnborough for use of the IDP 3000, finally Mrs. E.M. Whelan for her careful preparation of this typescript.

8. REFERENCES

BERNSTEIN, R., (1976): Digital Image Processing for Earth Observation Sensor Data. IBM J.Res.Develop. Vol.20, pp 40-57.

BRYANT, J., (1979): On the Clustering of Multidimensional pictorial data. Pattern Recognition Vol.11. pp 115-125.

CABOT, D., (1980): Ecological Mapping of the Territory of the European Economic Community. An Foras Forbartha, Dublin.

CAMPBELL, J.G., (1979): The Data Structure Laboratory, Description and User Manual. Internal Memo, Plessey Electronic Systems Research, Havant U.K.

CAMPBELL, J.G., (1981): Application of Remote Sensing to the EEC Ecological Mapping Project. An Foras Forbartha, Dublin.

COVER, T.M., and HART, P.E., (1967): Nearest neighbour pattern classification. I.E.E.E. Trans Vol.IT-13, pp 21-27.

DEVIJVER, P. and KITTLER, J., (1978): Selected Topics in Statistical Pattern Recognition. Lecture notes of course at Cambridge University, CAMS/78/1.

DUDA, R.O., and HART, P.E., (1973): Pattern Classification and Scene Analysis. John Wiley.

FISHER, R.A., (1936): The Use of Multiple Measurements in Taxonomic Problems. Ann. Eugenics Vol.7, pp. 178-188.

FISHER, L.T., SCARPACE, E.L., and THOMSEN, R.G., (1979): Multidate LANDSAT Lake Quality Monitoring Program. Photogrammetric Eng. and Remote Sensing Vol. 45, pp. 519-527.

FOLEY, D.H., (1972): Considerations of Sample and Feature Size. I.E.E.E. Trans. Vol.IT-18, pp 618-626.

FUKUNAGA, K., (1972): Introduction to Statistical Pattern Recognition. Academic Press.

HARALICK, R.M., (1979): Statistical and Structural Approaches to Texture. Proc. I.E.E.E. Vol. 67, pp 789-804.

HARDY, J.R., (1978) a : Land Cover Studies Using LANDSAT MSS data. Proc. 12th Int. Symp. on Remote Sensing of the Environment. Manilla, Phillipines, April 1978, Vol 3 , pp. 1717-1727.

HARDY, J.R., (1978) b : Methods and Accuracy of Location of LANDSAT MSS Points on Maps. J. British Interplanetary Society. Vol.31, pp 303-311

HART, P.E., (1968): The Condensed Nearest Neighbour Rule. I.E.E.E. Trans Vol. IT-14, pp 515-516.

MASON, D.C., (1979): Segmentation of Terrain images Using Textural and Spectral Characteristics. Computers and Digital Techniques (I.E.E.) Vol.2, No.6.

MOORE, J.J. (1981): Private Communication.

ROBERTSON, T.V., FU, K.S., and SWAIN, P.H. (1973): Multispectral Image Partitioning, LARS Information Note 071373, Purdue Univ., West Lafayette, Indiana.

SABINS,F.J. (1978): Remote Sensing, Principles and Interpretarion W.H. Freeman.

APPENDIX A

Laboratory for the Analysis of Multivariate and Image Data (LAMID), list of processing functions available.

A.1. Statistical Pattern Recognition (SPR):

A.1.1. Utilities :

Generate random test data.
Read ascii file into universal data storage format (UDSF).
Display UDSF file (numerical output).

A.1.2. Displays:

Scatter plot, 2-dimensions.
Histogram 1- and 2-dimensional.
Display image data on line printer or terminal.

A.1.3. Statistics:

Compute ranges and add to UDSF file.
Compute means and covariance matrices.

A.1.4. Discriminant Computation:

Fisher discriminant, 2-class.
Multiclass (pairwise) Fisher discriminant.
Least square error linear discriminant.
Principal components projection (Karhunen-Lòeve transform).

A.1.5. Transformation:

Matrix mapping using discriminants.

A.1.6. Classifiers:

Bayes-Gauss classifier.
Pairwise Fisher classifier.
Least square error classifier.
Nearest Neighbour classifier.
Condensing Nearest Neighbour rule.

A.1.7. Clustering :

Tree structured Parzen kernel clustering algorithm. ISODATA clustering algorithm.

A.2. Image Processing System (IMPS):

A.2.1. Utilities:

Generate test images.
Read from UDSF file
Write to UDSF file
Read from LANDSAT unrectified file
Read from LANDSAT rectified file
Display on terminal or line printer
Ground-truth training data specification
Compute and store ranges.
Zero Image
Copy images

A.2.2. Spatial Processing:

Laplacian edge enhancement
Low pass smoothing
Texture estimator

A.2.3. Spectral and Greytone Operations:

Principal components projection
Density slice
Normalise bands to equal variance
Threshold
Take logarithm
Ratio images
Multiply images
Linear combination of images
Compute line averages

A.2.4. Classifiers:

ISODATA clustering
Bayes-Gauss classifier
Nearest neighbour classifier
Nearest mean classifier
Least square error classifier
Fisher discriminant pairwise classifier
Mahalanobis distance classifier
Box classifier (parallepiped multi-image)
Bayes-Gauss spatial probability combination classifier
Class voting over neighbourhoods
Merge class labels.

A.2.5. Statistics

Estimate image histograms
Mean vectors and co-variance matrices of multiband multiclass images.

A.2.6. Rectification:

Least square error using control points.

ABERDEEN THERMAL INFRARED SURVEY

SOME IMPLICATIONS FOR AGRICULTURAL RESEARCH

R V Birnie
Department of Peat and Forest Soils
The Macaulay Institute for Soil Research
Craigiebuckler
Aberdeen
AB9 2QJ

Abstract: High resolution night-time thermal infrared (10-14u) imagery of the agricultural area around Aberdeen is examined. The appearance and distribution of wind-induced thermal shadowing is used to define shelter effects. Variations in tree-canopy radiant temperatures are apparently related to near-surface air temperature stratification and are used to identify frost hollows. Differences in radiant temperatures of field crops are linked to management practices.

1. INTRODUCTION

In December 1980, the Macaulay Institute for Soil Research took part in a collaborative exercise to obtain thermal infrared data of the Aberdeen area. The survey was flown by Clyde Surveys Ltd. using a Daedalus DS-1230 linescanner, and high resolution night-time thermal imagery was obtained for an area in excess of 100km2. About 25 percent of this area was agricultural and afforested land. Much of the subsequent evaluation of this rural cover at the Macaulay Institute was concerned with the value of thermal infrared surveys of this type for agricultural research. This paper presents some possible agricultural applications arising from the results of the Aberdeen survey.

2. SURVEY PROCEDURE

The survey was undertaken between 2000 and 2430 GMT on the 11th of December 1980. A total of 24 runs was made over the city area at a height of 600m giving an instantaneous

Paper presented at a conference on Matching remote sensing technologies and their applications convened in London by the Remote Sensing Society, 16-18 December, 1981

field of view of 0.85m and a cross-track swath width of 800m. Data were recorded in analogue form on magnetic tape.

During the survey period, grass surface temperatures were monitored at 14 sites. Continuous automatic recorders were installed at four of these. The remaining sites were manned and readings were taken from calibrated mercury thermometers every 15 minutes (Birnie et al., in prep.).

3. DATA PROCESSING

Analogue and digital thermal imagery were used in interpretation. The microprocessor system employed for the data processing has been described in detail elsewhere (Adams and Ritchie, in press; Adams, this volume). Types of available output from this system include colour display, printer or plotter hard copy.

4. IMAGE INTERPRETATION

Patterns in apparent surface temperatures are determined by variations in surface temperature, surface emissivity, view angle and instrument sensitivity.

Emissivity effects are not normally considered important in the remote sensing of agricultural crops. Fuchs and Tanner (1966) have shown a typical emissivity range of 0.96 to 0.98, resulting in corrections of less than 1^oC. View angle effects may be an order of magnitude greater. Kimes et al. (1980) have observed a response difference as great as 13^oC when going from a viewing angle of 0° to one of 80° over a wheat canopy. Such viewing angle effects result from different proportions of vegetation and soil within the same field of view. The main reason, however, for patterns in apparent surface temperatures on thermal images is the real variation in soil or vegetation surface temperatures. These are determined through energy exchanges by convection, radiation and evaporation, which are partly controlled by local meteorological conditions. Surface temperatures derived from thermal images should, therefore, reflect surface meteorological patterns. This is now examined with respect to the Aberdeen data.

5. DEFINING MICROCLIMATES - A. THERMAL SHADOWING

The most prominent feature on the imagery covering the rural fringe of Aberdeen is the appearance of pronounced thermal shadows in association with hedgerows and woodland. Nubler (1979) explains the development of such features through obstacles, e.g. hedges, reducing the windflow and preventing the mixing of the cold near-ground air with the warmer air from upper layers. The ground, therefore, remains cooler in the lee of the obstacles.

On the Aberdeen imagery, thermal shadows also appear to result from of this mechanism. Ground data indicate that a calm situation in the early evening, where radiative cooling predominated, was disrupted later by a warm SSW wind which generally increased ground temperatures except where obstacles prevented mixing.

Thermal shadowing provides a means of assessing the effectiveness of shelter belts. For example, Grace (1977) reports that the sheltered area in the lee of wind-breaks extends to a maximum of thirty times the wind-break height. From the Aberdeen imagery, it is observed that mature but leafless deciduous trees up to 20m high induce pronounced shade for 100 to 200m downwind i.e. less than ten times their height. However, incoherent shading referred to elsewhere as wind smears (Reeves et al., 1975, p.962) does continue several hundreds of metres downwind, although it can be difficult to trace its origin to individual obstacles.

Modification of local ground temperature regimes by wind-breaks influences crop growth rates (Grace,1977). By repeated thermal surveys it may be possible to assess the significance of this local modification. Under the Aberdeen survey conditions the maximum temperature difference across a shadow was 3°C.

6. DEFINING MICROCLIMATES - B. VERTICAL STRATIFICATION

High apparent tree canopy temperatures occurred over the entire image area. One 10m high coniferous wind-break had a canopy temperature of between 5.5 and 8.5°C, whilst an adjacent hedge 3m high had a canopy temperature of 4.7 to 5.5°C. Air temperature at the time was 4.6°C at 1.2m. It would appear, therefore, that the canopy temperatures approximate a vertical air temperature profile as suggested by Balick and Wilson (1980). If this is the case then it should be possible to use the radiative data for tree canopies to define local variations in air temperature and, in particular, to identify frost hollows.

The Seaton Park in Aberdeen, a known frost hollow, was used to test this hypothesis. Apparent ground surface temperatures within the Park appeared 0.75°C to 3.75°C colder than on adjacent higher ground suggesting that a localised inversion had developed within the Park area. Tree-canopy temperatures within the Park were from 4.75°C to 6.25°C compared with 8.5°C in higher areas. Of particular importance was the 'non-appearance' of a 5m high sycamore plantation, whose crowns lie 3m below the lowest point on the rim of the Park i.e. embedded within the inversion.

7. EFFECTS OF MANAGEMENT PRACTICE

The imagery covering the MacRobert Experimental Farm revealed temperature variations between grass and ploughed fields. No differences were observed between grass field trials which had been subjected to the same management. However, temperature differences were observed between red clover plots which had been subjected to early and late autumn grazing, the late grazed plots being cooler. This pattern can be explained by the increased area of bare soil within the late grazed plots. Alternatively, if the pattern reflects true temperature differences in the vegetation canopies, it indicates that the late grazed crops are more likely to suffer frost damage.

8. SUMMARY

The results of the Aberdeen study have indicated at least three agricultural applications of high-resolution night-time thermal imagery:

a) Defining shelter belt effectiveness and effects.

b) Identification of frost prone areas.

c) Monitoring the effects of differing agricultural management practices.

9. REFERENCES

ADAMS, M.J., (1981) A microcomputer system for processing and analysis of remotely sensed data. This volume.

ADAMS, M.J., and RITCHIE P.F.S., (in press) Processing and analysis of thermographic data with the aid of a micro-computer. Computer Applications, University of Nottingham.

BALICK, L.K. and WILSON S.K., (1980) Appearance of irregular tree canopies in nighttime high-resolution thermal infrared imagery. Remote Sensing of Environment 10 299-305.

BIRNIE, R.V. et al. (in prep.) Thermal infrared survey of Aberdeen City - data processing, analysis and interpretation To be submitted to the International Journal of Remote Sensing.

FUCHS, M. and TANNER C.B., (1966) Infrared thermometry of vegetation. Agronomy Journal 58 597-601.

GRACE, J. (1977) Plant response to wind. Academic Press, London, 204pp.

KIMES, D.S. et al. (1980) View angle effects in the radiometric measurement of plant canopy temperatures. Remote Sensing of Environment 10 273-284.

NUBLER, W., (1979) Wind-induced microclimatic differences detected from thermal scanner data. Tellus Newsletter 5 Joint Research Centre, Ispra, Italy.

REEVES, R.G., ANSON, A. and LANDEN, D. (1975) Manual of Remote Sensing. American Society of Photogrammetry, Falls Church, Virginia 2144pp (2 vols.).

TEMPERATURE CALIBRATION AND THERMAL INERTIA EVALUATION OF HCMM DATA OVER BELGIUM

R. Gombeer

Laboratorium voor Bodemgenese en Bodemgeografie
Katholieke Universiteit te Leuven
de Croylaan 42, B-3030 Leuven(Heverlee), Belgium

Abstract: The scene calibration method uses kinetic ground measured temperatures which are compared with radiant temperatures registered by the Heat Capacity Mapping Mission satellite. Linear regression equations were established from which real ground temperatures can be calculated with the satellite infrared digital counts. Apparent Thermal Inertia (ATI), pertaining to major land-use classes in Belgium, were calculated for a day/night recording with the inertia algorithm proposed by NASA for HCMM. Preliminary values indicate that forests, open water, urban land, grassland and bare soil can be separated by the proposed ATI-product and that even certain forest species can be discriminated.

I. A SCENE CALIBRATION METHOD FOR THERMAL DATA

1. INTRODUCTION

Comparisons between satellite registered temperatures and soil temperatures were performed on a day-infrared registration (11.46 GMT) of 16th September 1979 (scene AA-0508-12000-2). This Heat Capacity Mapping Mission scene is of good quality and shows the Belgian territory to be completely cloud-free. Other results on HCMM thermal data calibration are presented elsewhere in the framework of the Tellus-Project (JRC-CEC) (Gombeer, 1981). Ground temperatures were provided by the meteorological stations (Regie der Luchtwegen) located at 14 airfields distributed over the Belgian territory (fig.1). These soil temperatures are recorded continuously (day and night) and are measured by electrical resistance thermometers placed at different soil depths (2,5,10,20,50 and 100 cm). In this calibration study the temperatures at a depth of 2 cm (grass-covered soil) were used for comparing purposes. Satellite measured temperatures were calculated from digital infrared counts and no atmospheric correction was applied.

Paper presented at a conference on Matching remote sensing technologies and their applications convened in London by the Remote Sensing Society, 16-18 December, 1981.

Fig.1. The Meteo-stations on airfields distributed over Belgium

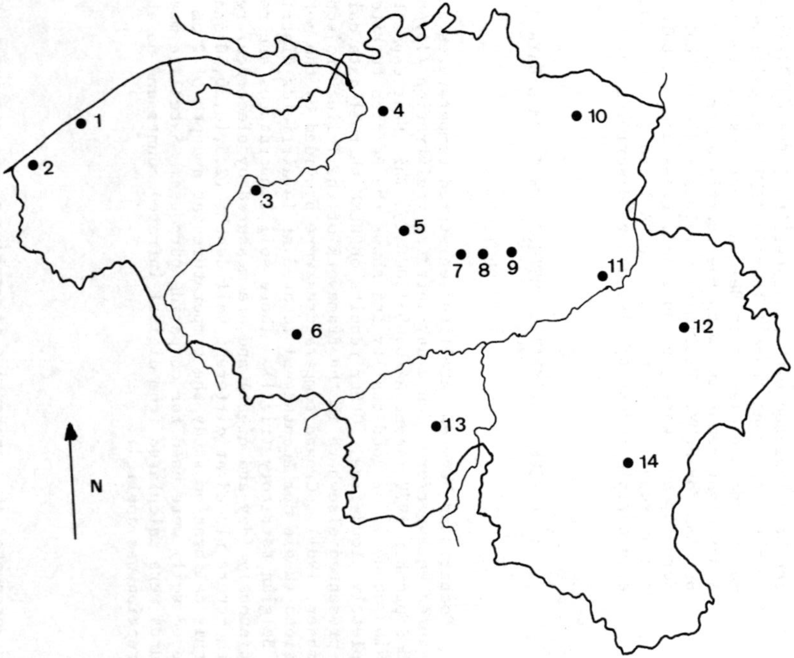

Legend

1. Oostende
2. Koksijde
3. Munte
4. Deurne
5. Melsbroek
6. Chièvres
7. Beauvechain
8. Goetsenhoven
9. St.Truiden
10. Kl.Brogel
11. Bierset
12. Spa
13. Florennes
14. St.-Hubert

Source: R.Gombeer

2. PROCEDURE

Two ORSER-programmes (Pennsylvania State University, U.S.A.) were used for computations (Turner et al., 1978) after transcription of the HCMM-data into the ORSER-format.

NMAPW-programme: The programme produces a maplike printout of classified radiation levels on which pixels can be identified by scanline and element number.

STATS-programme: The programme performs the statistical analysis for selected polygon areas which are delimited on the NMAPW-print.

The digital scene data (500 x 500 m pixels) were visualized (scan lines and line elements) by means of the NMAPW-programme as a brightness classification printout at an approximate scale of 1/200.000. Airfield areas were located and delimited through optical superposition of this printout with road and/or topographic maps using a Bausch & Lomb Zoom Transfer Scope (ZTS) (Gombeer, 1980). The scan direction of the Heat Capacity Mapping Radiometer (HCMR) was indicated on each map to be superimposed with the printout. This direction could be determined by a number of reference points which could be clearly identified both on the maps and computer lineprint. (small lakes, forest edges, small urban areas). As printouts were somewhat (scale) deformed, maps were optically stretched perpendicular to the HCMR scan direction and scale adjustments were made in order to achieve optimum congruency. For the HCMM day-scene, the delimitation of airfields was greatly facilitated using visualisation printouts of both the visible and IR-channel. In the visible channel (which is congruent with the day-IR band) the localization of airfields was achieved using the position of urban areas as reference points. Such areas stand out clearly by their low reflectance. It was necessary to use this technique for geographic areas (coastal area, Leemstreek) without any forested parts (e.g. for the airfields near Oostende, St.-Truiden, Bierset).
In forested areas (Ardennes) or especially in case when forests were located in the vicinity of the airfield, the positioning of the latter was now greatly facilitated on the day-IR printout, on which forests stand out by their low temperature range (e.g. airfield Spa, Kl.Brogel,St.Hubert).

The line and element numbers delimitating polygonal airfields on the NMAPW-printout were entered into the STATS-programme. Surface temperatures were calculated with the algorithm of Bohse et al.,(1979) and with the infrared count (mean value) of each airfield total pixel area by the STATS-programme :

$$T(I) = K_2 / \ln (K_1/I - K_3 + 1)$$

with I = infrared pixel value (0 - 255)

T = temperature in degrees Kelvin

K_1= 14421.587

K_2= 1251.1591

K_3=-118.21378

3. RESULTS

Table 1 compares the calculated HCMR temperatures (T_{sat}) with kinetic soil temperatures (T_{ground}) (grass covered soil) measured at 2 cm depth. All temperatures measured by the satellite are lower than the real ground temperature and atmospheric effects and/or sensor offset values may account for the observed differences. In the 10-12 micrometer spectral interval (weak atmospheric absorption) the relation between T_{sat} and T_{ground} may be described as follows (NASA, 1980) :

$$T_{sat} = T_{ground} (1 - E) + E\, T_{air} \tag{1}$$

where E is the surface to space atmospheric transmittance and T_{air} is a weighted (water vapour) mean temperature of the atmosphere. When both the atmospheric transmittance (E) and T_{air} do not change significantly over different areas, satellite temperatures and ground temperatures can be expressed by the linear regression equation (1). Such equation was established for the temperature data set in table 1 which was obtained from precisely positioned and delimited airfields over different parts of the Belgian territory. For the HCMM-satellite pass under consideration, real ground temperatures can now be calculated from the infrared digital counts through equation (2). This procedure is referred to as a scene calibration method (see also fig.2).

The HCMR calibration procedure was further tested using temperature measurements of river (Schelde) and coastal waters (Oostende). Results are presented in table 2 from which it can be seen that good correlations are found between ground and remotely measured temperatures. A particular good correspondence is obtained when temperatures are corrected for emissivity (0.989 for grass and 0.993 for water) although the use of any other emissivity value from literature leads to a substantial temperature error (table 2). Since the emissivities to be used for natural land surfaces may be uncertain, one may conclude that real surface temperature can be properly evaluated with the mean correction factor $\Delta\overline{T} = T_{ground} - T_{sat}$ (table 2). This factor can be easily derived from the data calibration set summarized in table 1. Obviously it has to be determined for each satellite pass.

Table 1. The regression equation of temperature data established for a HCMM-pass on 16 Sept.1979 (Day-IR) (Belgium).

Airfield Meteostation	T_{ground} (°K) kinetic	T_{sat} (°K) radiant	$T_g - T_s$(°K)
Melsbroek	297.600	289.276	+ 8.324
Deurne	296.500	287.499	+ 9.001
Chièvres	294.200	284.338	+ 9.862
Kl.Brogel	295.300	285.813	+ 9.487
Oostende	297.100	287.027	+10.073
Koksijde	295.600	287.499	+ 8.101
St.Hubert	291.500	282.149	+ 9.351
Spa	292.700	282.633	+10.067
			ΔT mean value + 9.28°K

Regression equation: $T_s = f(T_g)$

$$T_s = 1.1377\, T_g - 49.9228 \quad (r^2 = 0.92) \tag{2}$$

Tabel 2: Comparison of HCMM registered with ground measured water temperatures

Area	Date Time	Water temperature (°C) (kinetic)		
		ground measured	HCMM measured 16 Sept.1979: 11.46GMT	
			(1)	(2)
Schelde-river (Liefkenshoek)	16 Sept. 1979 11.46 GMT	19.2 ± 0.2 (at -1m level)	a) 19.54 b) 19.24 (ε=0.993) c) 20.09 (ε=0.981)	a) 19.85 b) 19.57 (ε=0.993) c) 20.46 (ε=0.981)
Coastal water (Oostende)	14 Sept. 1979 10.00 GMT	17.5	a) 17.36 b) 17.05 (ε=0.993)	a) 17.94 b) 17.66 (ε=0.993)
	17 Sept. 1979 10.00 GMT	17.0	c) 17.93 (ε=0.981)	c) 18.54 (ε=0.981)

(1) a) corrected with the mean temperature factor derived from table 1
$\overline{\Delta T} = T_{ground} - T_{sat} = +9.28°K$ without emissivity corrections

b) & c) corrected with a $\overline{\Delta T} = +8.46°K$ using an emissivity for grass of 0.989 (Becker,1980) and the two indicated water emissivities (ε) (Lorenz, 1973; Becker, 1980)

(2) a) calculated with the regression equation of table 1 without emissivity corrections

b) & c) calculated with the regression equation: $T_{sat}=1.1408\ T_{ground} - 49.9262$ ($r^2=0.92$) and applying emissivity corrections for grass and water.

Note: Noise Equivalent Temperature Difference (NETD) of the HCMM thermal channel sensor: 0.3°K at 280°K

Source: R.Gombeer

Fig.2. Regression equation of temperature data for a HCMM-pass on 16 Sept.1979(day-IR) over Belgium.

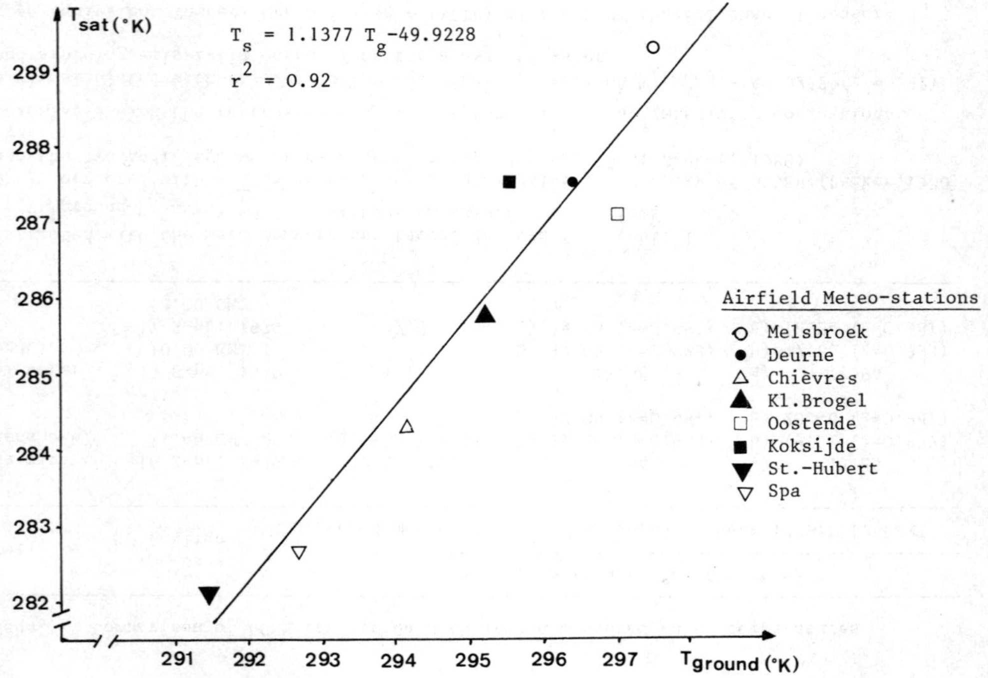

Source: R.Gombeer

II. APPARENT THERMAL INERTIA (ATI) OF LAND-USE CLASSES CALCULATED FROM DAY / NIGHT REGISTERED DATA

1. INTRODUCTION

Studies on thermal behaviour of soils can provide useful information on subsoil characteristics and on soil moisture. In such studies, the use of the thermal inertia product is particularly interesting as it is more a volume property rather than a surface property which can be sensed by remote sensing devices. It permits discrimination between different materials with similar surfaces which can be difficultly separated by their reflectance properties only. The thermal inertia effect is related to the diurnal surface temperature variation, having maximum and minimum values, which can be remotely measured by the Heat Capacity Mapping Radiometer (HCMR) through its appropriate orbit configuration (day/night passes). The second part of this paper presents some preliminary results on thermal inertia calculated for various land-use classes distributed over the Belgian territory.

2. METHOD AND PROCEDURE

The HCMM-data (VIS, day-IR, night-IR) were visualized and further analysed applicating the computer programmes NMAPW, STATS and according pixel (test area) positioning procedures (optical ZTS superpositions with maps) described in the first part (I) of this paper. Thermal inertia were calculated from a day/night recording over Belgium on September 16th, 1979, using the algorithm proposed by NASA (1980):

$$ATI = \frac{NC\ (1 - A)}{\Delta T}$$

with ATI : apparent thermal inertia
- N : integer scaling factor (N = 1000)
- C : scaling factor related to the solar declination
  For Belgium: on Sept.16th at 51°N latitude, the solar declination was calculated to be equal to 2°34'. From NASA tables a value of C=1.0422 was extracted.
- A : apparent albedo (A=I/255), with I:pixel count from the visible channel
- ΔT : temperature difference ($T_{day}-T_{night}$) of surface materials as calculated from day/night registrated data

In the ATI algorithm it is assumed that the factor (1-A) corrects for the variable energy reflection of different surface materials (given solar incidence). The factor C attempts to compensate for seasonal variations of solar input.

3. RESULTS

Table 3 summarizes calculations of thermal inertia for various land-use classes of Belgium. The following general conclusions are drawn:

a) for all land-use categories studied, T_{day} temperatures are higher than T_{night} temperatures.
b) except for bare soil surfaces, urban areas always have the highest temperature during the day. The lowest day temperatures are noticed for certain forest types (Picea).

Table 3 : Apparent Thermal Inertia (ATI) of Land-Use Classes calculated from HCMM -day/night digital data recorded over Belgium on September 16, 1979.

Land Use Class	1 − a (a=apparent albedo)	T_{day} °K	T_{night} °K	ΔT(°K) ($T_D - T_N$)	ATI Apparent thermal inertia
Urban land					
Gent	0.9426	289.0241	272.1262	16.8979	58.1360
Brugge	0.9357	287.2390	270.3583	16.8807	57.7693
Oostende	0.9269	286.6472	274.2750	12.3722	78.0794
Antwerpen	0.9475	289.1540	271.7142	17.4398	56.6224
Brussel	0.9487	288.6748	272.6836	15.9912	61.8299
Verviers	0.9263	286.7274	269.3374	17.3900	55.5140
Forest					
Fagus: Zoniën 1	0.9166	280.5825	270.5167	10.0658	94.9035
Zoniën 2	0.9174	280.9796	270.8326	10.1470	94.2263
Quercus: Neuville	0.9225	280.3871	269.5383	10.8488	88.6208
Esneux	0.9190	280.4425	269.3374	11.1051	86.2470
Picea: Hertogenwald	0.9298	279.4867	266.8554	12.6313	76.7171
Ommerwald	0.9350	278.9683	265.5110	13.4573	72.4110
Pinus: Gemeentebos	0.9272	282.1007	267.8191	14.2816	67.6624
Pijnven	0.9333	281.8388	267.2798	14.5590	66.8098
Kapellenbos	0.9263	282.1697	267.4348	14.7349	65.5172
Mixed species: Meerdaalbos	0.9167	280.8474	268.9754	11.8720	80.4737
Heverleebos	0.9205	281.8643	268.9590	12.9053	74.3372
Open water					
O-Schelde: Bergen-op-Zoom	0.9798	280.4720	278.1765	2.2955	444.8475
W-Schelde: Hontenisse	0.9724	281.7550	278.9010	2.8540	355.0929
Coastal water: Oostende	0.9674	281.2326	279.9212	1.3114	768.8152
Grassland					
Herve 1	0.8705	283.8168	267.7607	16.0561	56.5040
Herve 2	0.8733	283.7273	267.5059	16.2214	56.1081
Bare soil					
Dunes: Nieuwpoort	0.9186	287.7000	270.1875	17.5125	54.6675
Oostduinkerke	0.9049	287.8037	270.3624	17.4413	54.0720
De Panne	0.9058	287.3605	270.3786	16.9819	55.5900
Sandy surface: Leopoldsburg	0.9039	287.4403	267.8899	19.5504	48.1854

Note: Temperatures are given without atmospheric correction.

Source: R.Gombeer

c) during night, except for water bodies, agglomerations are also highest in temperature
d) T_{night} values are always highest for water surfaces
e) Temperature differences $T_{day} - T_{night}$ are smallest for the open water category
f) Albedo values are highest for grassland and for bare soil (dunes) and lowest for water surfaces

Average values for ATI are indicated in table 4. From these preliminary values it is anticipated that major land-use classes can be separated by the ATI-product and that even certain forest species (Fagus, Quercus, Picea, Pinus)may be discriminated.

Table 4. Apparent Thermal Inertia (average value) of major land-use classes (Belgium), Heat Capacity Mapping Mission Sept.16,1979.

Land-use class	A.T.I.
1.Open water	522.91
2.Forest : - Fagus	94.56
- Quercus	87.43
- Picea	74.56
- Pinus	66.66
- Mixed species	77.40
3.Urban land	61.32
4.Grassland	56.30
5.Dunes and sand	53.12

III. GENERAL CONCLUSIONS

The calibration method and its pixel positioning procedures proved to be valuable to calibrate the temperature data for successive HCMM overflights (individual scenes). From the regression equations (T_{sat} versus T_{ground}) established, the real surface temperature can be calculated. It is clear that the accuracy of the method will depend on the degree of spectral homogeneity of the located pixels, the precision of airfields positioning by optical methods and the validity of using point temperature measurements at 2 cm soil depth. It is intended to apply the calibration procedure on temperature data from other satellites to be launched with a higher geometric resolution than HCMM (500 x 500 m).

The performance characteristics of the infrared radiometer may change with time. If so, it will be necessary to compare, on a (regular) time scale, the satellite temperature data with several ground measurements (in-flight calibration). Their relation is normally expressed by a calibration curve from which a sensor offset-value can be derived after the data have been corrected for atmospheric attenuation. However, the numerical value of the offset can change according to the radiometers' performance, the surface emissivities to be used may be uncertain and atmospheric corrections (based on some model assumptions) calculated from atmospheric parameters may only constitute a further approximation to the total correction for real surface temperature. Moreover, some of the corrections may be invalid when extrapolated for a particular ground temperature interval to be measured by the radiometer. Obviously, such interval will depend not only on the time of overflight (night/day), on the season (summer/winter), but also on geographic location (desert etc.) It could therefore be envisaged to compare satellite uncorrected temperatures directly with kinetic ground temperatures. The temperature difference, $T_{ground} - T_{sat}$, will than give a constituent, but precise value of the correction factor applicable to the area or scene under consideration. Such temperature comparisons can be easily performed for many areas over the Belgian territory through the meteorological network installed at 14 airfields. As kinetic ground temperatures are routinely recorded (day/night) and archived by these meteo-stations, comparative temperature data sets can always be established for the calibration of each (occasional) satellite overpass. The number and distribution of the airfields also ensures the availability of calibration surfaces for different degrees of cloud cover (fig.1).

From average values of the Apparent Thermal Inertia (A.T.I.) product (NASA,1980), major land-use classes (i.e. open water, forests, urban land, grassland, bare soil) can be separated in Belgium. Forest species also have important differences in ATI-value and it is anticipated that the product could be used in forest inventory studies. It should be remarked that the Apparent Thermal Inertia were obtained through optical/manual registration procedures using separate digital printouts of day and night radiance data.

ACKNOWLEDGEMENTS

The author is indebted to the Joint Research Center (CEC) at Ispra (Italy) for acquisition of digital HCMM-data in the framework of the TELLUS-project.

The Regie der Luchtwegen is acknowledged for providing soil temperatures.

The Bestuur der Waterwegen (Antwerpse Zeediensten en Dienst der Kust) are acknowledged for water temperature measurements.

REFERENCES

BECKER, F., NGAI, W., and STOLL, M., (1980): An active method for measuring thermal infrared effective emissivities: implications and perspectives for remote sensing. Tellus Newsletter, 21, 1-18.

BOHSE, J.R., BEWTRA, M., and BARNES, W.L., (1979): Heat Capacity Mapping Radiometer (HCMR) data processing algorithm, calibration and flight performance evaluation. NASA Technical Memorandum, 80258

GOMBEER, R.,(1980): Interpretation of nighttime infrared digital data recorded over Belgium by the Heat Capacity Mapping Mission. I.Optical comparison of thermal map-like printouts with soil association and forest maps. Pedologie, 30,(1), 137-151

GOMBEER, R.,(1981): Summary of interpretation results from digital radiance data recorded by the Heat Capacity Mapping Mission (HCMM) over Belgium. Final Report prepared in the framework of the HCMM satellite Follow-on Investigation N° 025 (TELLUS-project) for the JRC, Ispra, Commission of the European Communities, 82 pp.

LORENZ, D.,(1973): Die radiometrische Messung der Boden- und Wasseroberflächentemperatur und ihre Anwendung imbesondern im Bereich der Meteorol gie. Ztschr. f. Geophysik, Bd.39, 627-701.

NASA (1980): Heat Capacity Mapping Mission Users' Guide. Goddard Space Flight Center, Greenbelt, Maryland, USA.

REGIE DER LUCHTWEGEN (Direktie Meteorologie, Belgium), (1979): Reference tables of ground temperatures measured at different airfields (Belgium).

TURNER, B., APPLEGATE, D., MEREMBECK, B., (1978): Satellite and Aircraft Multispectral Scanner, Digital Data User Manual. ORSER, Pennsylvania State University, U.S.A.

THE SHORT AND LONG TERM APPLICATIONS OF SATELLITE SURVEILLANCE OF THE SEA SURFACE

T D Allan, Institute of Oceanographic Sciences, Godalming, Surrey

Although marine surveillance by satellite has found an increasing number of applications in the last decade it remains true that techniques depending on sensors operating in the visible and infrared part of the spectrum are inhibited by cloud cover, making repetitive sampling difficult. One of the most significant recent advances has therefore been the development of high resolution microwave instrumentation which culminated in the launch of Seasat in 1978. Despite its short life Seasat produced enough data from its four microwave sensors to prove the concept that meaningful observations can be made from satellites that would be virtually impossible to achieve from surface platforms.

It is relevant to the European ERS-1 mission to examine in some detail the effects of the all-weather, global monitoring of sea surface parameters such as wind, waves and temperature, provided by microwave instrumentation, on day to day marine activities, and on longer term objectives such as the prediction of extreme wave heights in offshore exploration areas, and the monitoring of ocean circulation patterns.

Wave energy program

6 Months to Arctic

DALE LODGE

SURFACE EXPRESSIONS OF BATHOMETRY ON SYNTHETIC APERTURE RADAR IMAGES

D.W.S. Lodge
Space and New Concepts Department
Royal Aircraft Establishment
Farnborough
Hampshire GU14 6TD

Abstract

Images from the Seasat synthetic aperture radar show a variety of surface effects apparently linked to the sea bottom topography. This paper classifies the types of feature observed and the meteorological and oceanographic parameters pertaining at the time for some of the images available of UK coastal waters. Some comparisons are made for areas where more than one image is available and where bathometric features are not manifest on all of them.

1. INTRODUCTION

In 1978 the National Aeronautics and Space Administration of the USA launched Seasat (Born et al, 1979), the first satellite devoted to oceanographic sciences. This paper describes some of the features detected by the synthetic aperture radar carried on board, the first time such a sensor had flown in orbit around the Earth. Descriptions of synthetic aperture radars can be found elsewhere (Lodge, 1981) but a brief introduction is necessary. Side-looking radars have been flown on aircraft since the early nineteen-fifties. These have an elliptical beam shape, broad in the elevation dimension and narrow in azimuth which illuminates the ground to one side of the track of the aircraft by being directed at right-angles to the flight path and below the horizontal. Figure 1 shows the geometry. The echo from each pulse is sub-divided by sampling in discrete time intervals. Successive pulses illuminate contiguous strips on the ground. The echoes are stored and displayed as an image, that is an array of back-scatter coefficients or reflectances corresponding to areas on the ground defined by the radar range resolution, which is in turn set by the width of the range-gate time intervals, and the width of the transmitted beam in the azimuth or along-track dimension. Achieving fine resolution in range, or the across-track dimension, is in principle no more difficult from an orbiting satellite than from a low flying aircraft. However, the azimuth resolution is set by the beam width. A narrower beam requires a longer aperture. For useful performance from space, an aerial several kilometres long would be needed. Very briefly, a synthetic aperture radar uses the forward motion of the vehicle to build up the equivalent of a very long aerial. The echoes from pulses transmitted as the real aerial travels along its path are stored and subsequently coherently integrated to synthesise the return

Paper presented at a conference on Matching remote sensing technologies and their applications convened in London by the Remote Sensing Society, 16-18 December 1981.

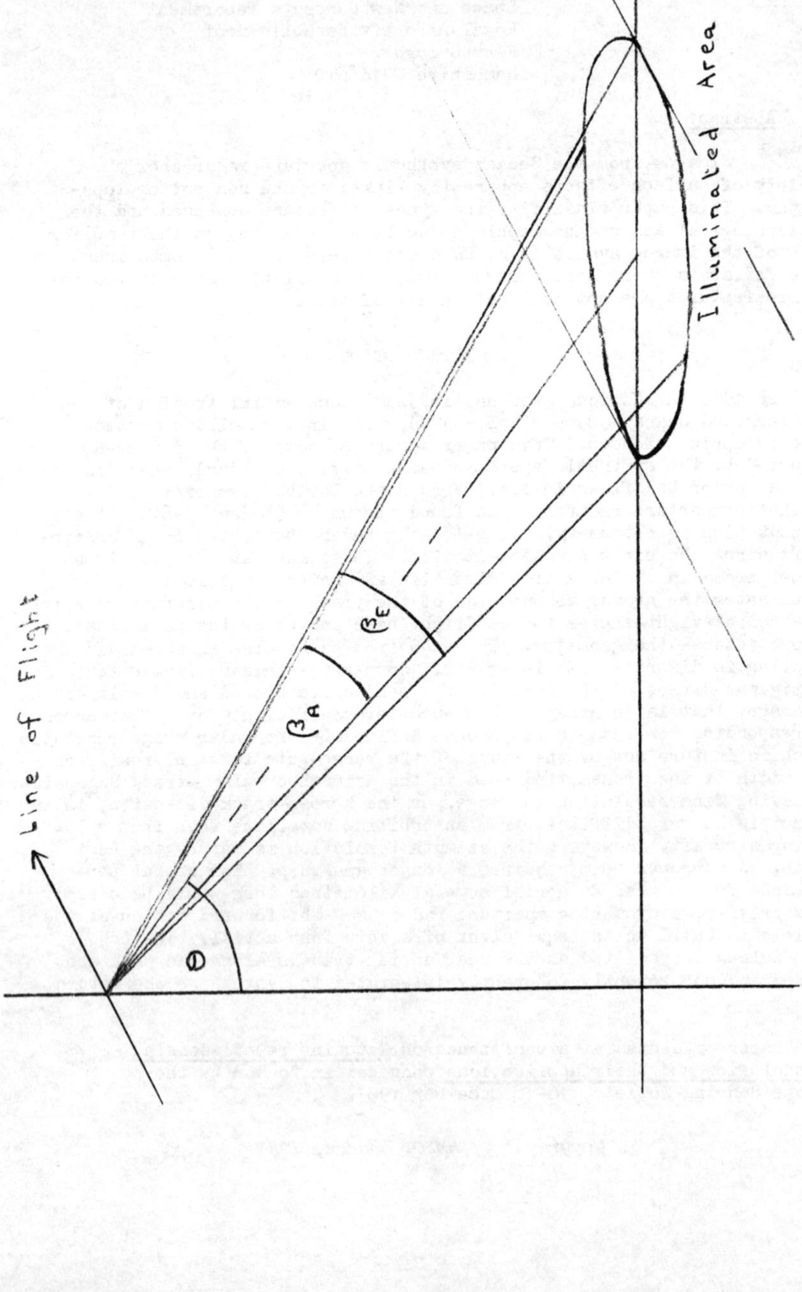

Fig 1 Geometry of side-looking radar

that would have been received from a long array of elements transmitting their pulses simultaneously. The Seasat radar used this technique to achieve a resolution on the ground of 25 x 25 m.

The complete interaction between the incident microwave radiation and the sea surface is imperfectly understood. The essential point in this context is that the necessary conditions for back-scattering of the radar pulse are those for Bragg resonance between electromagnetic and water waves of comparable length. The Seasat radar frequency was 1.275 GHz, for which the equivalent wavelength is 0.235 m. The angle of incidence was approximately 20°. Hence the radar return arose from the presence of water waves in the region of 0.34 m long. Therefore any features seen on Seasat radar images of the sea surface, whatever their dimensions, are visible because the feature in some way modulates or changes those short gravity waves in that critical region.

2. WAVE REFRACTION

The propagation of gravity waves is not affected by the topography of the sea bed unless the depth of water is less than about half the dominant water wavelength. On encountering such shallower water their period remains unchanged but their velocity, length, height, direction and shape alter (Groen, 1967). Essentially, a train of waves impinging on a coastline or area of shallow water is refracted. The relationship between the directions of the incident and refracted wave fronts is analogous to Snell's law from optics. The phenomenon can be observed on a number of Seasat images. It is particularly well defined on images of the vicinity of the islands of St Kilda and Foula made from data received on rev 1149 at about 08:20 on 15 September 1978. Waves with a length of about 200 m, corresponding to a period of about 11 seconds, are travelling in an east-west direction. But for the islands their direction would be ambiguous but in this case the refraction patterns clearly show that they were travelling from the west. The patterns themselves show from the proximity of the coastline at which the refraction effects are observed that the gradient of the sea bottom around the islands is steep. The waves would begin to be affected at a depth of about 100 m. The regular appearance of the waves elsewhere in the image shows that except for St Kilda and Foula, the water depth exceeds about 100 m.

The usefulness of wave refraction observations of this type arises because it is in principle possible to make quantitative bathometric measurements. The other observations in this paper permit only the conclusion that there are indications of relief on the sea bed. The vertical scale of that relief cannot be measured yet. However, the relationship between waves and depth is well enough understood to permit effects to be predictable and for, at least in principle, bathometric contours to be produced. Furthermore, it is not necessary for there to be an area of dry land associated with the refraction. The effect should be observable wherever there is a region of depth less than about half the incident dominant wavelength.

3. SHALLOW WATER FEATURES

The heading for this section may be misleading. It implies that limits can be set according to some physical model defining when and where the relevant features may be observed. That is not yet the case. However, the phenemona of interest have only been observed so far in water depths of less than about 50 m. Figure 2 shows a part of the image produced from data received on rev 762 at 06:46 on 19 August 1978. It shows the English Channel between Dover and Dunkerque and the approaches to the Thames estuary. The most striking features on the sea surface are the numerous striations, particularly prominent between 51° 30' north and the European coast. A cursory examination of the hydrographic chart for the area immediately showed that those surface features were strongly correlated with the position and extent of changes in depth. Over most of the area the depth rarely exceeds 40 m. Above the sand-banks the depth may be less than 10 m. These depths are relative to the chart datum, which corresponds roughly with the lowest astronomical tide, and therefore are lower limits. The 20 m depth contour has been included in Figure 2. It clearly demonstrates the correspondence between the surface roughness and the sand-banks. The discrepancies between the latitude and longitude markings show the errors introduced because the chart and the image use different cartographic projections. A conceptual model to explain these observations could be that the tidal flow over the banks causes the surface modulation. High water at Dover on 19 August 1978 was at 11:27. The image data were recorded at 06:46. The tidal streams corresponding to five hours before high water at Dover are also included on Figure 2. They show 1 to 2 knot currents flowing along the long axis of the submerged banks. One remaining observation is that the south facing edges of the outer Ruytingen and Sandettie banks are in each case much steeper than the other edge. The delineation of these banks on the image suggests that the steep edge is associated with a bright return, that is a prevalence of short gravity waves. The other edge appears as a darker return indicating smoother water. On a high definition image there is some indication of structure on the surface corresponding to the crests of the banks, which superficially appears to correspond with the form and extent of sand waves on the banks. Bearing in mind the importance of short, wind driven waves to the formation of the image, the wind velocity may also be significant. The reports by coastal stations in the area to the Meteorological Office at 06:00 were of winds in the range 6 to 12 knots from bearings of 130° to 170°. There were no reports of precipitation, which could also modulate the surface

The short life and restricted amount of data from Seasat do not leave many opportunities for comparative studies of the same area under different conditions at different times. However, three days earlier than rev 762 on 16 August 1978, data were received from rev 719. The imaged area overlapped with that described above at the eastern end of the English Channel from Dover to Cap Gris-Nez. If the conditions were similar, the Varne and the Ridge sand-banks shown outside the imaged area in Figure 2 would be expected to produce similar effects. In fact, the image of that area from rev 719 shows no features of any kind. The time of the image was 06:40. High water at Dover was at 08:56. Figure 3 shows the tidal stream two hours before high water at Dover. Over the crucial area is slack water, further evidence, albeit negative, for currents being the main cause of the surface manifestation. On this occasion most of the winds reported were from 10 to 12 knots from bearings of 210° to 230°, with no precipitation. There was one exception. The 06:00 report from Dungeness described a wind of 19 knots from 270° and showers.

On the image from rev 762, features similar to those observed in the

Fig 2 Rev 762, 20 m depth contour, tidal flow 5 hours before high water at Dover

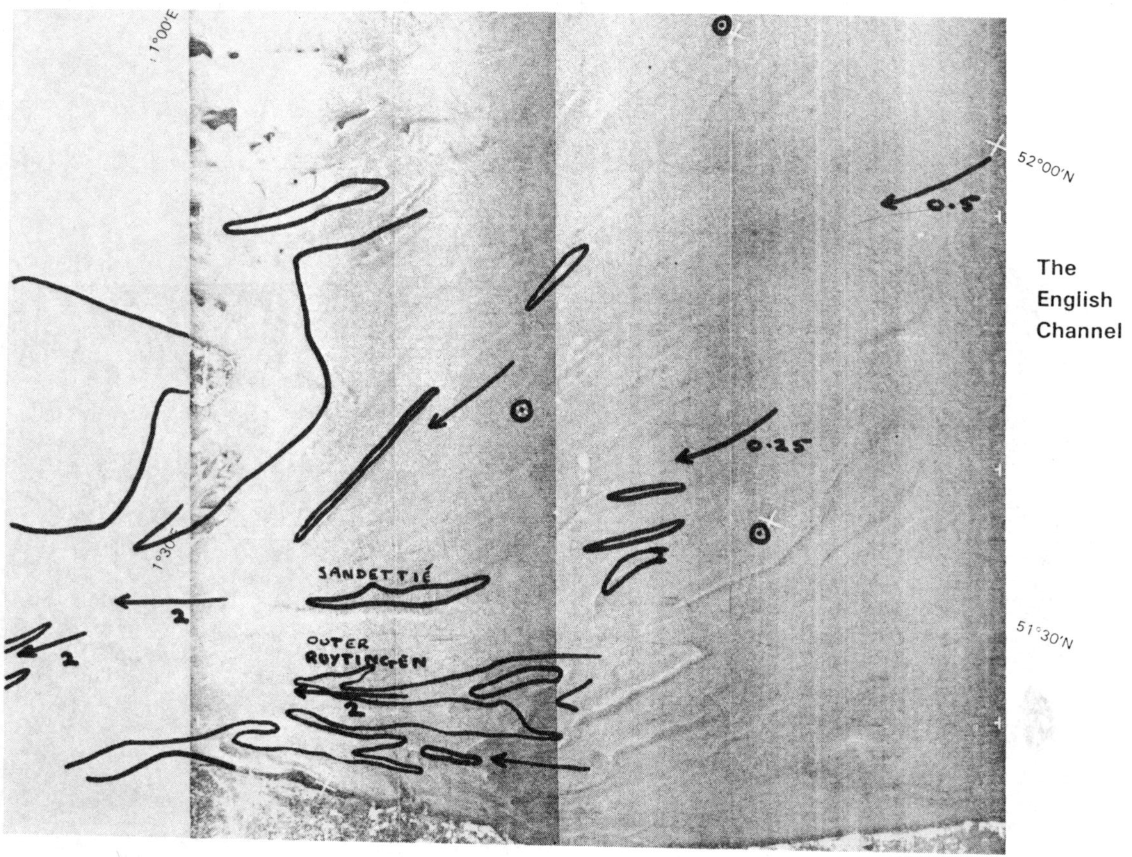

Fig 3 Rev 719, 20 m depth contour, tidal flow 2 hours before high water at Dover

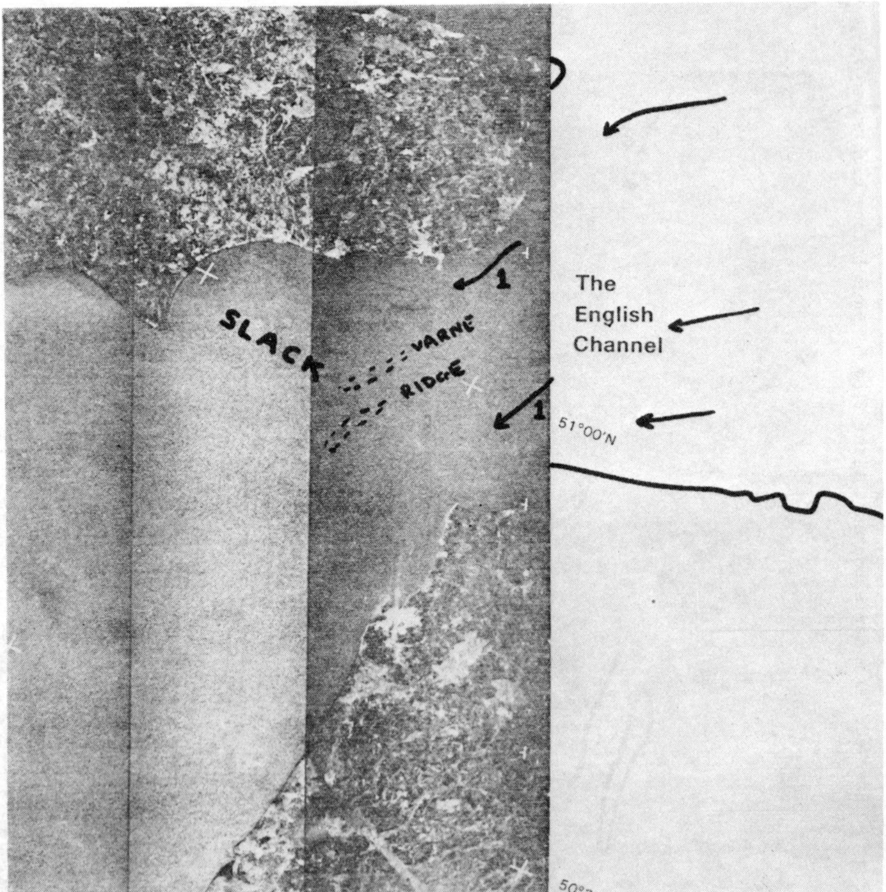

Channel were visible around the Wash. Without reference to a chart, the principle linear features were extracted from the image and are shown in Figure 4. Subsequent correlation with the appropriate chart lead to their identification as shown. The tidal stream at the time was again quite strong and parallel to the coast. The only features that did not correlate with the chart were the apparent circulatory patterns on either side of the entrance to the Wash. The tidal flow would have been directly into the Wash. It is possible that the incoming current could set up circulating streams that themselves modulated the surface. The reported winds were the same as those for the Channel.

4. DEEP WATER FEATURES

A great many of the images of the open ocean received from Seasat show prominent wave like features, propagating dispersively with wavelengths of several kilometres. Superficially they resemble internal waves. Why they should appear on synthetic aperture radar images is not clear. However, they appear also to be linked to changes in depth but in water 100 m to 1000 m or more deep. A good example can be taken from rev 719. The data were received at 06:43 on 16 August 1978. The area, shown in Figures 5 and 6 is north-west of the Isle of Lewis. The most prominent features have been extracted. In most cases they correspond with major changes in depth. Typically they occur where there is a steep gradient from 500 m deep to 1000 m to 1500 m deep. Furthermore, there do not appear to be such changes that do not have a surface manifestation. The exception is the V shaped feature nearest to Lewis that seems to correspond approximately with the 100 m depth contour. The meteorological reports indicate winds mainly of 6 knots or less although 17 knots was recorded at Cape Wrath. The direction varied from 210° to 260°. There were some reports of drizzle and rain.

The same area was covered by three other passes from which usable images were produced. Rev 1006 at 08:15 on 5 September 1978 showed no comparable features. It did however show unexplained periodic striations aligned approximately east-west and about 2 km apart. The winds varied from 0 to 13 knots and from 60° to 130°. There was some fog and drizzle. Rev 1049 at 08:27 on 8 September 1978 showed some structure in the water in the vicinity of Lousy Bank. There are waves of about 300 m length arriving from the west. The reported winds were from 9 to 18 knots except Cape Wrath which reported 12 knots at 06:00 and 3 knots at 12:00. Directions were between 160° and 230°. Most stations reported fog or precipitation in the area. Rev 1307 at 09:43 on 26 September 1978 showed waves travelling from west to east and a faint hint of a feature corresponding to the edge of the Hebridean Shelf. The winds were from 10 to 25 knots and from 220° to 290° with squalls, fog, rain and drizzle prominent once more. Finally rev 762 at 06:43 on 19 August 1978 showed clear features corresponding to the Wyville-Thomson ridge and the 200 m contour of the West Shetland Shelf, as depicted in Figure 7.

5. CONCLUSIONS

This paper has described some features observed on synthetic aperture radar images whose correlation with changes in depth is difficult to ignore. Except in the case of wave refraction there is no well-defined model that explains the observed effects. Intuitively the shallow water

Fig 4 Rev 762, tracings of major features and associated shoals

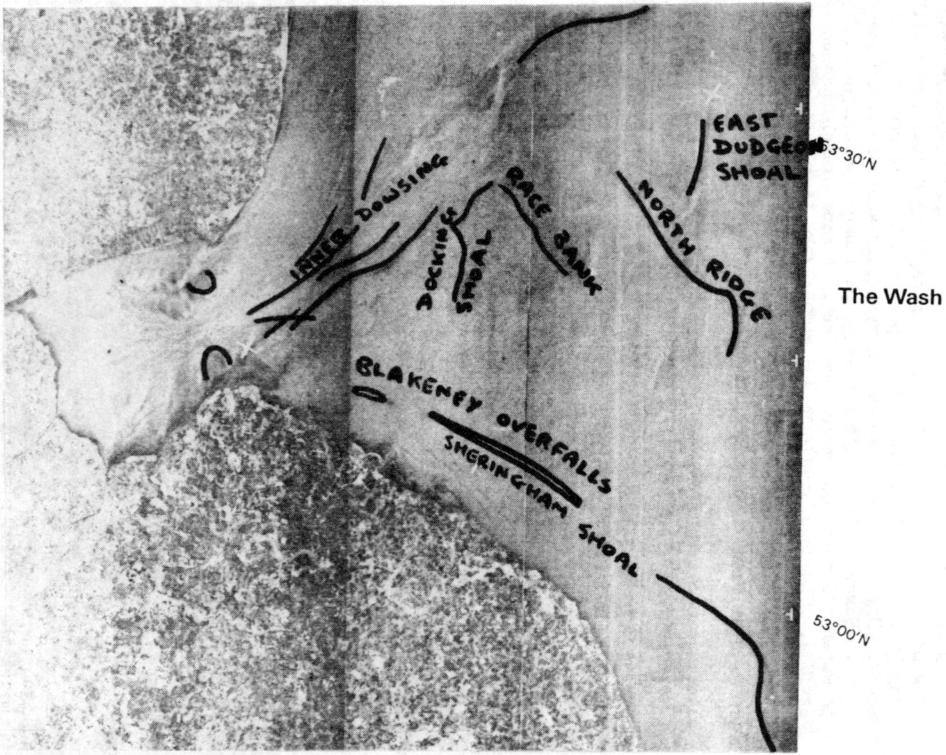

Fig 5   Rev 719, tracings of major features
and associated bathometric changes

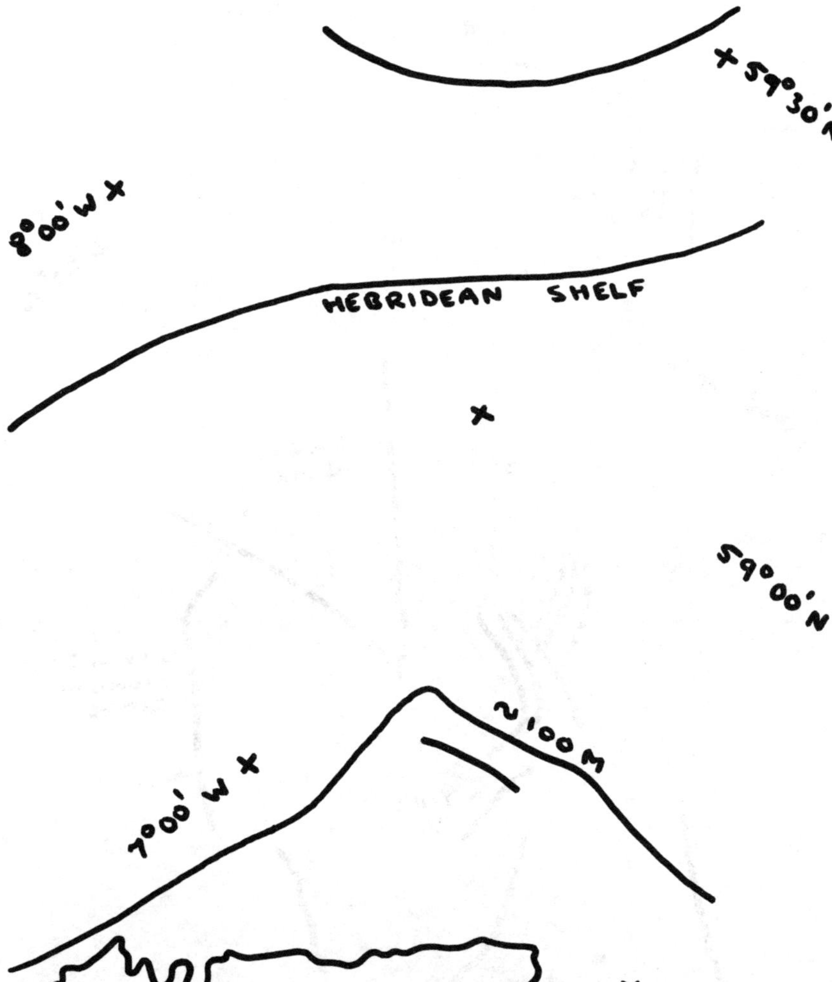

Fig 6 Rev 719, tracings of major features
and associated bathometric changes

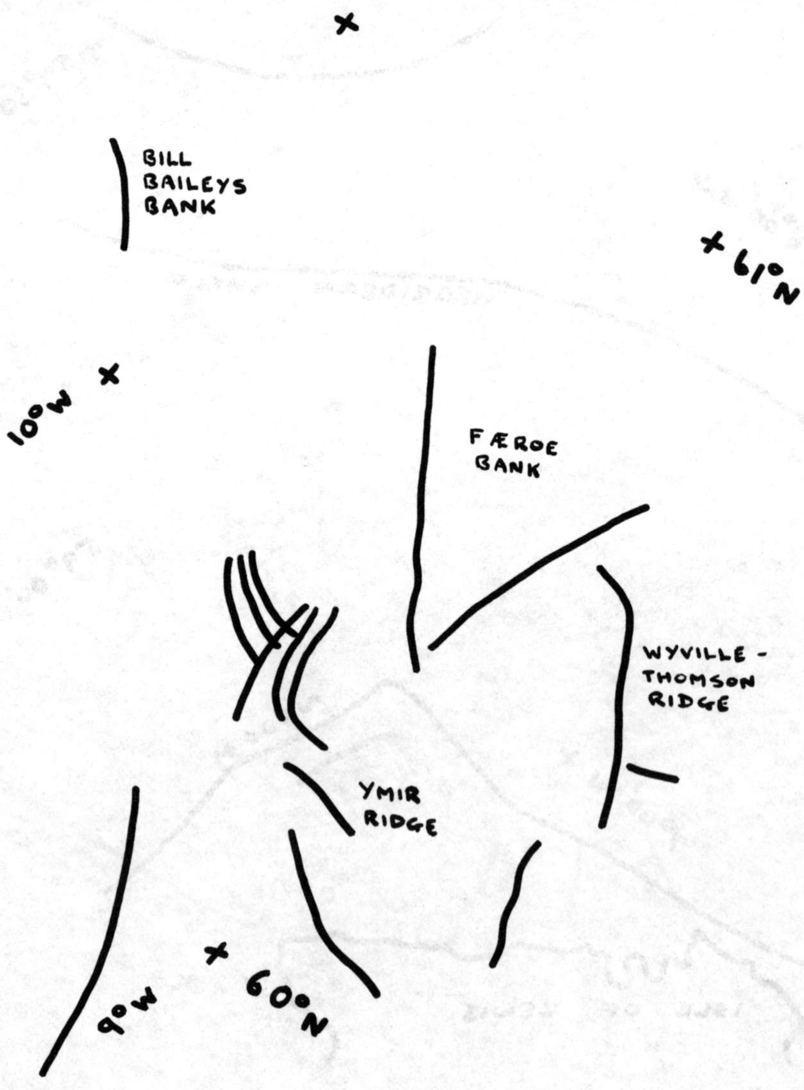

Fig 7 Rev 762, tracings of major features, associated bathometric changes and 200 m depth contour

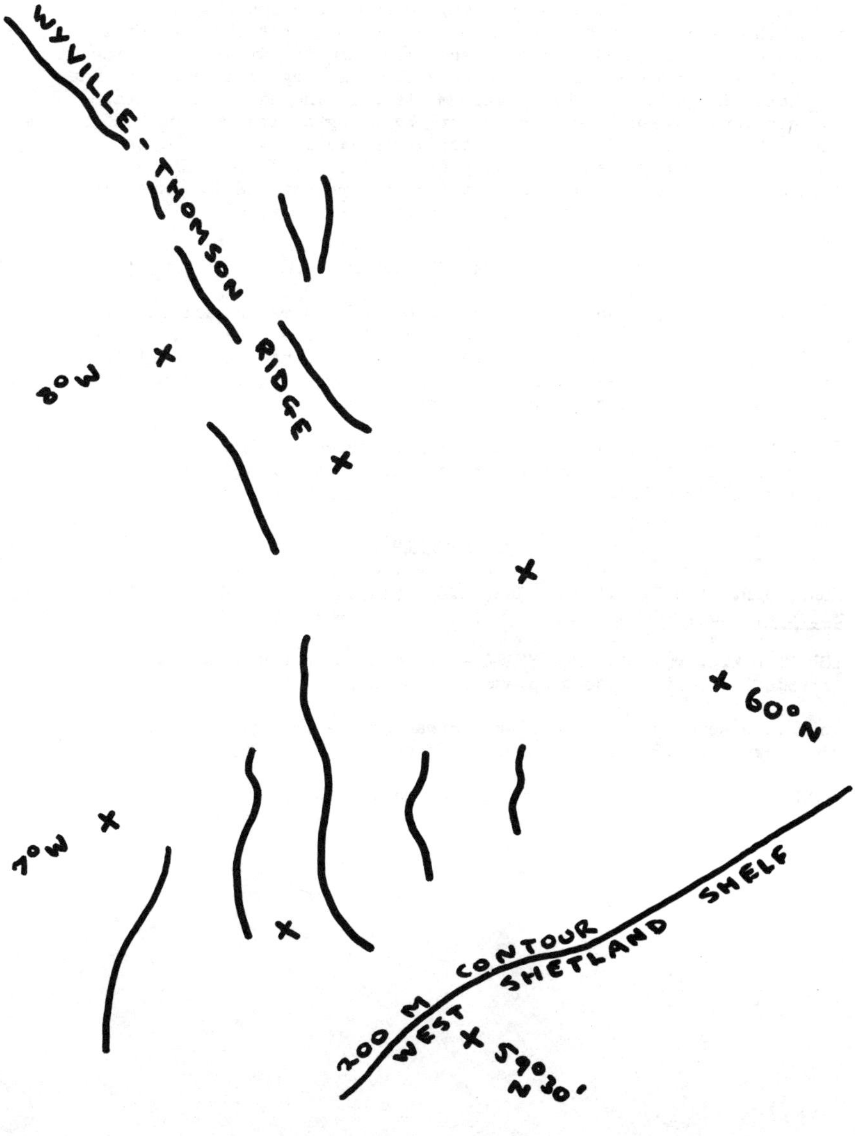

features can be accepted without invoking any significant mechanism not already understood. In deep water though, the generation of internal waves and their modulation of the short surface gravity waves requires urgent investigation. At present it seems unlikely that real bathometric measurements could be made, except again by using refraction. However, at least in shallow water it seems there is a potential tool for monitoring the position of shifting sand-banks. Although not covered in this paper, similar effects are manifest in rivers and estuaries and there the up-dating of the position of shoals may be even more important. In deep water the major significance may be by the contribution to a better understanding of hydrodyanmic interactions and the structure of the water mass. The work reported represents a first survey of a sample of the available imagery. Further developments will include more accurate correlation between images and charts by using digitally processed images where the projection can be a better approximation to that of the chart, and surveys of more Seasat imagery for further evidence. There may be usable aircraft data, such as that from the recent SAR 580 experiments, although the narrow swath width may be too restrictive.

6. ACKNOWLEDGEMENTS

The data from the Seasat synthetic aperture radar were received on behalf of the European Space Agency (ESA) by Space Department of the Royal Aircraft Establishment (RAE) at their Oakhanger ground station. Optical processing was by the Environmental Research Institude of Michigan and digital processing by Space Department of RAE.

The locating grids for the optically processed images were generated under contract to Space Department of RAE on behalf of ESA by Huntings Surveys Ltd.

7. REFERENCES

BORN, G.H., DUNNE, J.A and LANE, D.B. (1979)
Science, 204, 4400

LODGE, D.W.S. (1981) The SEASAT-1 synthetic aperture radar: introduction, data reception and processing.

In Remote sensing in meteorology, oceanography and hydrology
ed Cracknell, A.P., Ellis Horwood, Chichester

GROEN, P. (1967) The Waters of the Sea van Nostrand, London.

SEA SURFACE TEMPERATURE MEASUREMENTS AROUND THE UK DERIVED FROM METEOSAT AND TIROS-N DATA

R.W. Saunders, N.R. Ward, C.F. England and G.E. Hunt
Laboratory for Planetary Atmospheres
Department of Physics and Astronomy
University College London
Gower Street, London WC1E 6BT

Abstract: The sea surface temperature around the British Isles has been inferred from meteorological satellite infrared (11μm) data. We have combined the high spatial and radiometric resolution (1km, 0.1(K)) polar orbiter data (AVHRR on TIROS N) with high temporal resolution (half hourly) geostationary orbiter data (METEOSAT). Both types of satellite data were absolutely calibrated for this time using coincident ship data. The changes in the sea surface temperature around the British Isles for 12th July 1979 show much detail when derived from the 11μm channel on the AVHRR which is not revealed in the conventional ship data. To demonstrate the importance of combining different satellite datasets, we have made a study of an anomalously warm patch of water in the North Sea, which was evident in the images. It is suggested that the cause of this warm patch was due to low surface winds which reduced mixing in the surface layer.

1. INTRODUCTION

Until recently, the only way to obtain sea surface temperatures over large areas was to rely on ship reports. There are only a few ships which make these measurements, so that even near busy shipping lanes the number of observations in a one degree lat/long square will be small (∿10/day maximum for the North Sea).

With the advent of meteorological satellites, sea surface temperatures can be derived from the radiative temperature measured in the atmospheric windows at infrared wavelengths (i.e. 3.7 and 11μm) by a radiometer. The advantage of this method is that temperatures can be obtained globally twice a day (for the 11μm channel) over cloud free areas with a spatial resolution down to a kilometre. However, absolute values of sea surface temperature over the whole globe at any time are still not easy to obtain accurately (Williamson and Wilkinson:1980) since atmospheric absorption is difficult to estimate accurately (Maul and Sidran:1973) as it continually varies due to changing temperature and water vapour profiles.

A more accurate calibration could be obtained for TIROS N/NOAA 6 data using the coincident sounding data (TOVS) (Smith et al:1979) which provides measurements of the vertical temperature and humidity profiles below the satellite, and hence a better estimate of the actual atmospheric absorption. Alternatively, two successive observations of the same target area could be made through significantly different air masses. The atmospheric absorption can then be removed. Finally, another method now available with the AVHRR-2 instrument on NOAA 7 is to use a "split window" at 11 and 12μm. The different atmospheric absorption at the two wavelengths allows these effects to be taken into account when deducing sea surface temperatures.

The results presented here were derived by calibrating both satellites using coincident ship reports. This calibration is only valid for the particular atmospheric conditions present below the satellite for this time.

2. SATELLITE DATA

The principal reason for looking at both TIROS N and METEOSAT infrared images was that although the former has a greater spatial (∿1km) and radiometric resolution (1024 grey levels- ∿0.1°K) it makes only two passes per day, approximately twelve hours apart. METEOSAT, however, takes images every half hour from geostationary orbit allowing the development of short-lived phenomena to be observed. It has a spatial resolution of 8km for European latitudes and radiometric resolution of 256 grey levels (∿0.5°K).

The radiation emitted from the sea surface at 11μm will be from only the top few millimetres because of the strong infrared absorption characteristics of liquid water. The top metre is usually well mixed so that the observed surface layer temperature is close to the bulk water temperature. Grassl (1976) shows that the surface radiative temperature is normally only 0.2 degrees (K) lower than the bulk water temperature. Figure 1 shows an infrared 11μm image of the British Isles from TIROS N that has been enhanced to show up relative changes in sea surface temperature. This enhancement was obtained by setting the warmest sea areas to black and the coolest areas to white so that variations in sea surface temperature showed up on a scale between white and black.

This was carried out on the Interactive Planetary Image Processing System (IPIPS) (Hunt et al:1981) at University College London. A sequence of corresponding 11 micron infrared METEOSAT images (0315-1815 GMT) of the European area, which were converted into a Mercator projection, were used to study temporal variations of sea surface temperature.

A calibration for METEOSAT's infrared channel has been carried out by comparing ship sea surface temperatures and cloud top temperatures with coincident METEOSAT count values (Koepke:1980) which gives the 11μm brightness temperature in space, of the source. This was initially used to calibrate the AVHRR data by comparing radiances over high, cold cloud and over warm land from the two satellites. The ship measurements were then compared with the satellite measurements. A difference of 6 degrees (K) between them was noted due to the atmospheric emission and absorption between the satellite and sea surface. This difference was consistent with the value predicted from a full radiative transfer model using near coincident radiosonde data. Figure 2 shows the relationship between METEOSAT and AVHRR counts with ship measured temperatures for 12 July 1979. The response can only be assumed to be linear over the small range of temperatures plotted.

Figure 1: TIROS-N AVHRR 11μm infrared image of the British Isles at 1424 GMT on 12th July 1979. The image has been enhanced to show up small changes in sea surface temperature; (black-warm, white-cold). Two prominent fronts are indicated.

Figure 2: METEOSAT and AVHRR 11μm counts, plotted against ship measured sea surface temperature. This calibration will only hold for this time and intervening atmospheric absorption.

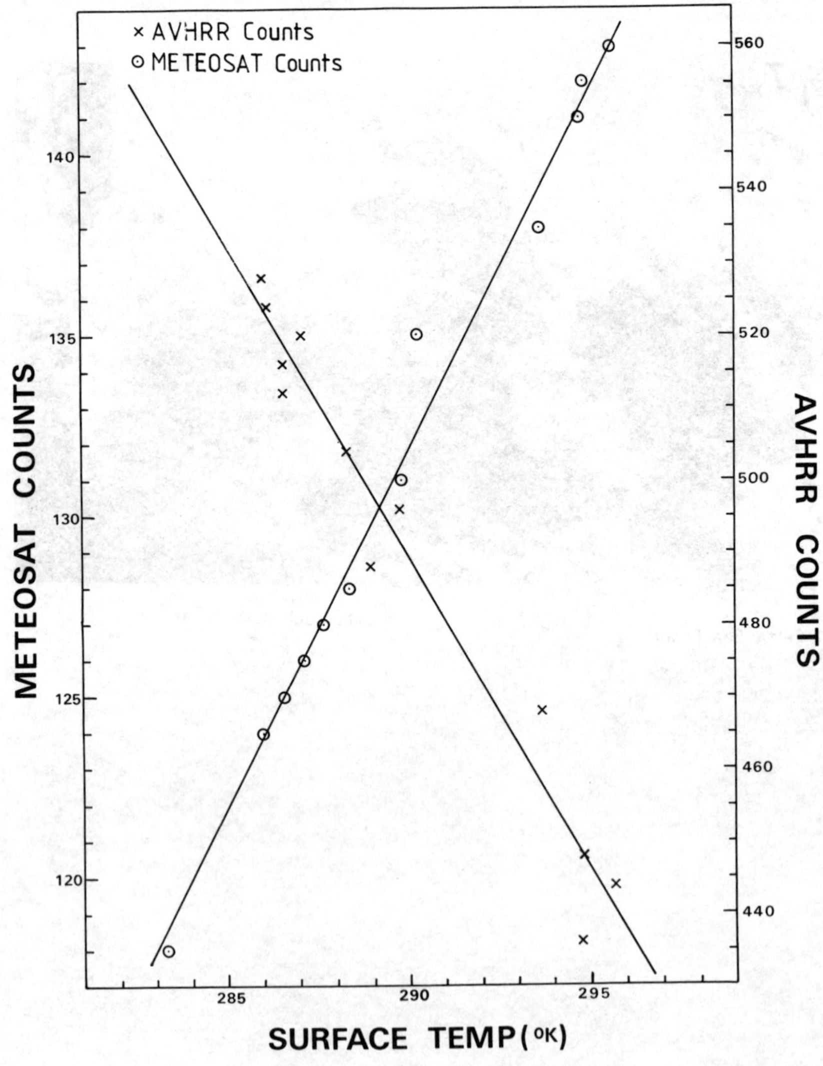

3. OBSERVED VARIATIONS IN SEA SURFACE TEMPERATURE AROUND THE BRITISH ISLES

Taking the 5 day mean of sea surface temperatures reported by ships around the UK, figure 3(a) was produced. Figure 3(b) shows the sea surface temperatures measured by the TIROS N AVHRR channel at 1424 GMT, but absolutely calibrated over uniform areas by the ship data shown in figure 3(a). The satellite data shows up more detail in the sea surface temperature structure. The orbit of TIROS N is such that it gives sea surface temperatures at approximately the time of maximum diurnal heating. The structure is best seen by referring to figure 1 (darker shades represent warmer water). Well defined discontinuities in the temperature can be seen; for instance along the line from Humberside to the Netherlands (Flamborough Head front) and across the St. Georges Channel from S.W.Wales to S.E. Ireland (Celtic Sea front). These "tidal fronts" have been predicted and observed (Pingree and Griffiths:1978) and are the result of sudden changes between stratified and well mixed conditions in the sea. The Dogger Bank is evident as a crescent shaped patch of cooler water north of the Flamborough Head front in the North Sea. Baroclinic eddies (Pingree:1978) are also evident off the west coast of Cornwall. These fronts are only present during the summer months when the net heat fluxes tend to form stratified conditions.

Another feature immediately obvious in figures 1 and 3(b) is a warm patch of water to the north of the Netherlands. The centre of this area was as much as 3.5 degrees warmer than the surrounding sea at 1424 GMT. To investigate how the phenomenon varied with time, the METEOSAT images were studied. A plot of the sea surface temperature, measured at the centre of the warm patch by METEOSAT as a function of time of day, is shown plotted in figure 4. Note the peak after noon which suggests that direct solar heating is responsible. The anomaly was also visible in the same position on the TIROS N image of the previous day (1436 GMT).

The surface winds were plotted for noon and they showed that the observed warm area appeared to be at the centre of a ridge of high pressure where there will be relatively low wind stress. This suggests that surface wind is an important parameter when considering how much mixing there is in the surface layers. Also, the fact that the area is in a stratified regime during the summer months (Pingree and Griffiths:1978) will help to inhibit mixing over a deep layer.

Given that calm conditions reduce mixing at the surface, a rough calculation was attempted which assumed that only a shallow layer ($d \sim 1m$) remains well mixed. Neglecting heat losses due to conduction, the energy balance equation is given by:

$$\rho d\, Cp \frac{dT}{dt} = S(z)(1-A(z))(1-\exp(-K_\nu d)) - \varepsilon(\sigma T^4 - f(T,q)) - LF - SF \qquad \text{........ (1)}$$

Heating Term	Solar Absorption	Net IR Flux	Latent	Sensible

where ρ, Cp, A, K_ν, ε are the density, specific heat, albedo, energy absorption coefficient for solar radiation and emissivity of sea water respectively. S is the incident solar radiation at the sea surface for a solar zenith angle, z, and LF and SF are the latent and sensible heat

Figure 3: Isotherms of sea surface temperature around the British Isles for 12th July 1979 from (a) ship reports 1200 GMT and (b) AVHRR 11μm data 1424 GMT.

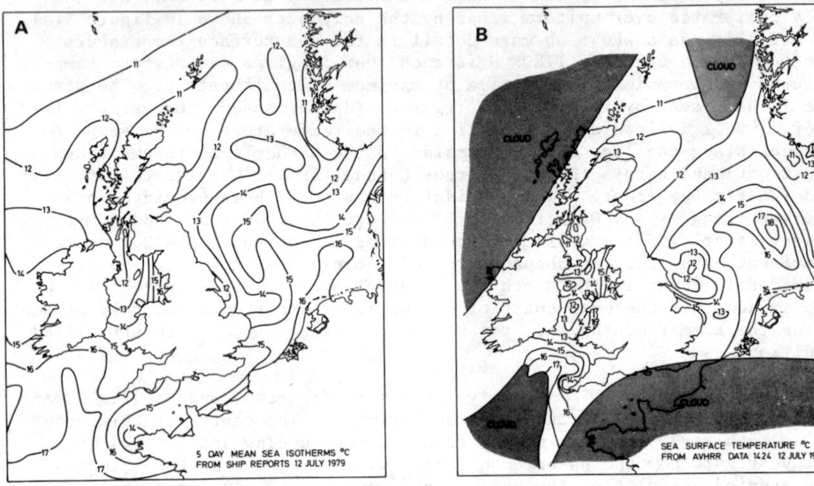

Figure 4: Calculated temperature values (for mixing depths of 0.6 and 1.0m) and observed values at the centre of the anomaly as a function of time.

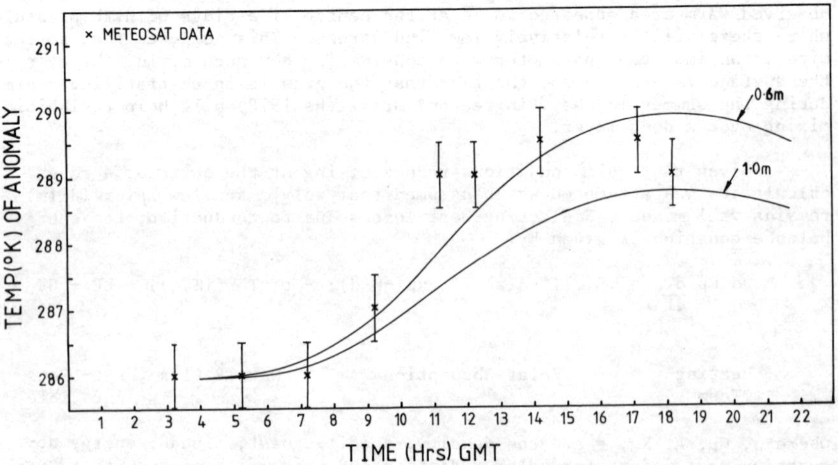

fluxes, q is the specific humidity of the air. Considering the solar absorption term in (1), Ivanoff (1977) states that the first metre of clear sea water absorbs 55% of the solar energy flux. He also gives expressions for the incident solar radiation S(z) and the albedo A(z) as functions of solar zenith angle for the North Atlantic Ocean. The infrared emission term in (1) will be the balance between emitted upward radiation from the sea surface and downward radiation from the atmosphere f(T,q). Budyko (1956) gives values for net infrared heat loss from the surface as a function of air humidity and sea temperature. The losses due to sensible and latent heat were calculated from Krauss (1972) and found to be negligible ($SF \sim 0$ $LF \sim 10\text{-}20 W/M^2$) for the observed conditions. The net absorbed flux was calculated from sunrise in half hourly intervals. The temperature rise is shown plotted in figure 4. The best agreement with the METEOSAT data is obtained when a mixing layer of depth 0.6m is assumed.

Further evidence that surface wind stress influences diurnal sea surface temperature changes can be shown by comparing TIROS visible and infrared images (Saunders et al:1981). Under certain conditions of solar illumination, the visible TIROS N image shows striking variations in reflected brightness over the sea surface. This is due to the sun-Earth-satellite angle for this area being near to that which gives direct specular reflection from the sea surface. Calm sea remains dark because there will only be a small range of angles for which direct reflection is possible but rougher seas will present a range of angles for the sea surface to directly reflect solar radiation. Therefore, over a small range of observation angles, there is a sensitive measure of sea surface roughness from the visible channels. By comparing visible and infrared images, it was shown that calm sea areas on the leeward side of land, for instance, are warmer than the surrounding rougher areas at the time of maximum diurnal heating.

4. CONCLUSIONS

Infrared images from conventional meteorological satellites can be used to obtain maps of sea surface temperature which give more detail than the conventional observations. In order to retrieve absolute values, accurate to $\pm 1^oK$, it is necessary to have either a few coincident in situ measurements or a value for the atmospheric emission obtained from coincident observations or calculated from measured temperature and water vapour profiles. The importance of combining polar orbiter data (AVHRR) with geostationary data (METEOSAT) is demonstrated by the latter showing the diurnal variation of sea surface temperature.

The existence of tidal fronts predicted and observed by Pingree and Griffiths (1978) is well shown in the enhanced AVHRR imagery. The position of the warm patch of water in the North Sea was found to be well correlated with an area of low surface wind stress which reduced surface mixing to a layer about 0.6 metres deep, deduced from energy balance considerations.

5. ACKNOWLEDGEMENTS

The authors would like to acknowledge the help of Mr. Baylis of the Department of Electrical and Electronic Engineering, University of Dundee, for providing the TIROS N data, ESA for providing the METEOSAT data and support for NRW. RWS is supported by NERC and CFE by a SERC grant.

6. REFERENCES

BUDYKO, M.I., (1956): The heat balance of the Earth's surface (translated by N.I. Stepanova), Washington US Weather Bureau.

GRASSL, H., (1976): The dependence of the measured cool skin of the ocean on wind stress and total heat flux. Boundary Layer Meteorology 10 465 - 474.

HUNT, G.E., et al (1981): The UCL interactive planetary image processing system (IPIPS) and its application to studies of planetary atmospheres. (To be submitted to Phil. Trans. Roy. Soc.)

IVANOFF, A., (1977): Oceanic absorption of solar energy. Modelling and prediction of the upper layers of the ocean, (ed. by E.B. Krauss) 47 - 71, Pergamon Press.

KOEPKE, P., (1980): Calibration of the METEOSAT IR channel by ground measurements. Contributions to Atmospheric Physics 53 442 - 445.

KRAUSS, E.B., (1972): Atmospheric Ocean Interaction, Oxford University Press, 134 - 166.

MAUL, G.A. and SIDRAN, M., (1973): Atmospheric effects on ocean surface temperature sensing from NOAA satellite scanning radiometer. Journal of Geophysical Research 78 1909 - 1916.

PINGREE, R.D. (1978): Cyclonic eddies and cross frontal mixing. Journal of the Marine Biological Association UK 58 955 - 963.

PINGREE, R.D. AND GRIFFITHS, D.K. (1978): Tidal fronts on the shelf seas around the British Isles. Journal of Geophysical Research 83 4615 - 4622.

SAUNDERS, R.W., et al (1981): Satellite observations of sea surface temperatures around the British Isles. (submitted to Bulletin of the American Meteorological Society).

SMITH, W., et al (1979): The TIROS N operational vertical sounder. Bulletin of the American Meteorological Society 60 1177 - 1187.

WILLIAMSON, D.J. AND WILKINSON, G.G., (1980): Sea surface temperature measurements from METEOSAT. Second METEOSAT Users Meeting, Report Annex J.

REMOTE SENSING OF WATER QUALITY FOR ESTUARINE ENVIRONMENTS

Alberto L. Geraci°, Robert N. Colwell°°, and Siamak Khorram°°

Remote Sensing Research Program, Space Sciences Laboratory
University of California, Berkeley, U.S.A.

Abstract: The objective of this paper is to present a review of remote sensing technology in relation to its capabilities for the inventory, monitoring, and management of water quality in estuarine environments. After briefly describing environmental problems in coastal zones, significant contributions of remote sensing in this area are identified. Then, the use of remote sensing techniques in mapping the water quality parameters of main interest is shown, including its use for the inventory and monitoring of water pollution in estuarine environments, with emphasis on oil pollution. In the following pages the use of a data bank, storing remotely sensed measurements together with information from process plants and urban agglomerations in coastal zones, is also proposed to help in the making of management decisions.

1. INTRODUCTION

As Colwell, 1978, indicated, the human demand for most kinds of natural resources is rapidly increasing while the supply of many of them is rapidly diminishing and the quality of others is rapidly deteriorating. Management is greatly facilitated if information about natural resources is periodically available. Remote sensing techniques might be used to facilitate the obtaining of such information.

Results of experiments in coastal zone areas indicate that remote sensing technology may be used to detect pollution features and water quality parameters. Large area synoptic views from aircraft and satellite

---

Published as Special Report, Series 21, Issue 23, at the Space Sciences Laboratory of the University of California, Berkeley, U.S.A., October 23, 1980. Performed under California Space Group CSI-RS80-1, University of California, San Diego, U.S.A.

°Machinery Institute, Department of Engineering, University of Catania, Italy.

°°Space Sciences Laboratory, University of California, Berkeley, U.S.A.

platforms, equipped with remote sensors, provide information that is not readily available by the use of conventional measurement techniques.

This paper deals with the current state-of-the-art of remote sensing information capabilities for water quality in estuarine environments. Beginning with a review of remote sensing literature pertaining to water quality in coastal zones, problems in estuarine environments are described and areas where remote sensing could be helpful identified.

Chronicling the significant developments in recent years and discussing the possible usefulness of various remote sensing techniques for water quality, the parameters of main interest are examined in relation with how those techniques can aid the inventory, monitoring, and management of water pollution in coastal zones, with emphasis on oil pollution. Additional studies relative to the establishment of correlations with the agglomerations bordering the coastal zone are proposed.

2. ESTUARINE ENVIRONMENTS

In recent years the increase of population and industry bordering the coastal zone has seriously degraded estuarine environments. Draeger, 1974, described the system, in which there is the confluence between the sea and the discharge of a water body; i.e. there is a dynamic combination of outflow and tidal currents, mixing and interacting between themselves.

Marine and estuarine pollution is one of the most dangerous kinds of pollution, because it affects the ecological balance governing life. The danger is greater according to the level of pollution reached, the diversity of polluted discharges, and the difficulty of enforcing regulations. Water-ways are sources of products of extreme importance in the world, and play a vital role by supplying a great quantity of the oxygen indispensable to life.

Increasing amounts and diverse types of information are needed for management of estuarine environments, and the monitoring of water quality in such environments is an area where there is a great need for remote sensing.

3. WATER QUALITY

The increase of water pollution in coastal zones has led to the development of more sophisticated means for water quality management. New techniques have been developed and refined to the point at which they are being involved in many water quality management programs.

Many investigators have described remote sensing technology as a tool for mapping water quality (Hom, 1968, Van Lopik, 1968, Welch, 1969, Wezernak, 1970, Draeger 1974, Brooks, 1975, Scherz, 1975, Johnson, 1976, and Khorram, 1979). Also it has been described as a tool for location the origin of a water pollution problem (Ward, 1974).

Remote sensing techniques are being expanded more and more in solving many problems facing management of the water quality in the coastal zone. Several projects conducted recently are helping to solve coastal zone problems, like oil and hazardous material spills. As Hill, 1979, indicated, verification of estuarine water quality models has been nearly impossible using traditional sampling procedures. Remotely sensed data should always be utilized for selecting sampling stations in large area investigations, and water quality parameters should be mapped remotely.

Remote sensing technology has the potential for studying estuarine water quality. According to the studies done by Link, 1974, environmental conditions are important in water quality problems. These conditions vary with the various kinds of pollution, and involve the hydrology of the region, determining water streams, and the layout of urban and industrial agglomerations, determining polluted discharges in the estuarine environment. Link also proposed a general method of studying river basin water quality, consisting in a rapid evaluation of the water resource characteristics of the system, and a computer-assisted study of the critical areas.

In recent years many investigators have reported the application of remote sensing technology to water quality parameters mapping and oil spills monitoring. But the objective of these investigators has also been the study of nearshore circulation dynamics. Paulson, 1968, discussed the application of thermal infrared imagery and aerial photography for estuarine circulation studies. He reported that imagery which provides the synoptic view together with temperature difference measurements between sea and outflow can provide useful information about estuarine circulation dynamics. Investigations on the use of multispectral instrumentation for the same purpose were also made by Mairs, 1973.

Several remote sensing techniques, like multispectral imagery, radar, lidar, and other reflectance and fluorescent techniques have been used for water quality mapping. Goldberg, 1972, indicated molecular spectroscopy as the most promising among these techniques. Since sensitivity in water quality detection is the main factor (because the solute concentrations involved are low, usually 1 to 100 mg/l), he reported that a lower limit of 10 mg/l can be reached with laser-Raman spectroscopy.

Landsat (formerly called ERTS) multispectral imagery has been correlated with optical in situ measurements of the harbor water by Egan, 1974. As a result he indicated that MSS computer compatible tapes can be utilized in coastal zone investigations of water quality. He concluded that MSS band 7 (0.8 to 1.1 μm) is useful in establishing the land-water interface location, and that the blue band (0.43 to 0.50 μm) is best indicative of water quality followed by the green one (0.53 to 0.60 μm) and lastly red one (0.63 to 0.70 μm).

As Drake, 1974, reported, significant information for water quality management was obtained from Side Looking Airborne Radar (SLAR), that simultaneously imaged the terrain with X-band and L-band radar wavelengths. According to experiments in Florida and southeastern Michigan, he described that radar, particularly at X-band and shorter wavelength, is an excellent indicator of the water-land boundary.

Johnson, 1978, reported that plumes resulting from ocean dumping of sewage sludge and industrial wastes have distinguishing spectral characteristics. Scherz, 1973, and Klemas, 1978, indicated that plumes in the coastal zones have higher radiances due to suspended materials. According to the studies done by Bowker, 1977, and Klemas, 1977, usually sewage sludge plumes disappear from surface waters within about 4 to 8 hours, while acid-iron waste plumes have been observed for several days after.

Johnson, 1978, and other investigators reported that qualitative analysis techniques may be used for the location, identification, and mapping of plumes resulting from ocean dumping of waste materials, and quantitative analysis techniques are used in addition to calibrate remotely sensed data, and to extend the results to the remotely sensed scene. Many researchers (Yarger, 1973, Williamson, 1973, Klemas, 1974, Rogers, 1975, and Johnson, 1976, and 1977), as reported by Johnson, 1978, used results of quantitative analysis in the form of calibrated regression equations to calibrate satellite and aircraft multispectral scanner data, and to map water quality parameters in the coastal zone. Johnson, 1978, also reported that these techniques, using a water quality parameter as the dependent variable and radiances from multispectral scanners as the independent variables, have been applied to the determining of suspended solids concentrations in sludge plumes (Johnson, 1977), and iron concentrations in acid-iron industrial waste plumes (Ohlhorst, 1978).

Other investigators in France (Morel, 1980) have indicated the need for high precision in the spectral determination of the emerging radiation, whatever the method developed to map water quality parameters.

4. MAIN PARAMETERS

Many investigators of water pollution problems repeatedly indicated that the water quality parameters selected for measurement should be such as to determine all of the relevant physical, chemical and biological properties of water. As Keyser, 1967, reported, the measurement of temperature is important because the rates of most biological and chemical reactions increase with temperature. Also temperature itself can be a pollutant, if the water is heated above its normal level by the hot discharge of a process plant.

Whipple, 1973, indicated anomalies in water body temperature as characteristic of interactions between streams. As Draeger, 1974, reported, a detector sensing in the 8 to 14 μm wavelength range is appropriate for most water applications, because water in the temperature range 0 to 25 °C emits maximum radiation at approximately 10 μm. Many investigators have reported that temperature measurements have been made in both the infrared and the microwave spectral ranges.

Johnson, 1980, indicated that thermal IR sensors have accuracies of 1 to 2 °C, and microwave remote sensors of 1 °C when they are flown on aircraft platforms. He also reported that the increased resolution provided by the recently launched satellites has increased the utility of satellite water temperature measurements, and the most widely used thermal scanners are on the satellites which have been used to map major water features.

Satellite imagery of the English Channel and North Sea coastal waters was interpreted by Kaminiski, 1972, to correlate temperature measurements with neighboring industrial and urban agglomerations. Many similar studies of thermal water pollution have been performed using infrared scanning from satellites.

Remote sensing techniques have also been used in an attempt to determine the specific parameter of water salinity. Many investigators have reported that, although saline water may show a little difference in spectral signature from that of fresh water, salt concentrations in the water may be determined by certain indicators, such as the vegetative species on the tidal flats and estuary shorelines. They have also reported that models of the estuarine salt variations may be plotted by a knowledge of the tolerance of estuary vegetation to saline water conditions, and they have accordingly made a complete vegetative map from color infrared photography. According to the studies done by Anderson, 1970, water salinity, through the identification of wetland plant species whose salinity ranges are known, may also be estimated. This reportedly is best done from high altitude color and color infrared photography to record energy reflected from subsurface targets in a spectral region centered around 0.575 μm, with a heavy absorption of light at about 0.675 μm.

As Johnson, 1980, indicated, salinity may be determined by a microwave radiometer. Both temperature and salinity values are provided by dual inversion of microwave brightness measurements in two bands.

As Keyser, 1967, reported, turbidity (the quantity of undissolved solids in the water) is important for two reasons: (1) it indicates the amount of undissolved solid material that is transported in suspension by the water, that later may settle out on the stream bottom; and (2) the degree of turbidity indicates the quantity of sunlight that can penetrate the water.

Many investigators have reported that imagery acquired in the photographic region, either from aircraft or satellites, can show the parameter of turbidity. Scherz, 1973, showed imagery as a valuable engineering tool for overall water quality investigations. He had reported that it is possible to analyze aerial photographs (and the reflectances which they show) to map turbidity.

He has also reported that the photos can be analyzed by two different methods. The best method -- one which requires good control on the absolute reflectance determinations from the photos and precisely knowing skylight conditions -- is to determine the absolute quantity of reflectance of the water indicated by the photos and to use a reflectance-turbidity curve to map turbidity. The second method is to use the ratios of the reflectances at two wavelengths in order to correlate them with turbidity, by using curves derived in the laboratory or from ground sampling.

Weisblatt, 1973, reported that data from several Landsat passes were used to measure turbidity in Galveston and Trinity Bay, and that MSS channels 5 and 6 yielded the most accurate measurements of turbidity in the 20 to 120 ppm range. Yarger, 1973, reported that works with Landsat data on two reservoirs in Kansas have provided reliable predictions of suspended loads up to 900 ppm. But Miller, 1974, indicated that turbidity measurements may not provide as much detail as required since they do not necessarily separate the organic from the inorganic particles in suspension.

Other researchers in France (Morel, 1977) have reported that in turbid waters the presence of dissolved and suspended material increases both absorption and scattering in the water. Aggording to the studies done by Morel, 1980, the detection of turbid areas should easier, since these cause an increase in the signals recorded at all wavelengths in the visible part of the spectrum and the spectral differentiation is less necessary. Morel also reported that a frontal structure between oceanic waters, with a reflectance ratio of about 1% at 0.55 µm, and coastal turbid waters, with a reflectance ratio of about 5%, is marked by a strong change in brightness. Even if the contrast is reduced by the additional atmospheric term, it still remains easily detectable. This possibility has been widely exploited by using the Landsat MSS band 4 to map coastal turbidity.

Suspended solids constitute another important environmental parameter for monitoring water quality in river water sheds. As Johnson, 1977, indicated, uncontrolled sediment runoff reduces the volume of water in which photosynthesis can take place. Johnson also reported that this parameter (i.e. suspended solids) has been recognized as a natural tracer that may be used to measure flow and distributions in a water body, thereby providing information on pollutant concentrations and dispersions.

According to the studies done by Yarger, 1975, correlations between Multispectral Scanner Computer Compatible Tape data and water samples from Kansas reservoirs coincident with Landsat passes indicate that MSS reflectance levels are useful for quantitative measurements of suspended solids up to at least 900 ppm. Yarger has reported that MSS band ratios derived from CCT's can measure suspended solids with 67% confidence level accuracy of 12 ppm over the range 0 to 80 ppm and 35 ppm over the range 0 to 900 ppm. He has also reported that suspended solids "contour maps" can be easily constructed for water bodies larger than appoximately 4000 m^2.

As Johnson, 1978, reported, chlorophyll is important for monitoring pollution effects in coastal zones. Chlorophyll is not visible to the human eye until it reaches concentrations of about 15×10^{-3} mg/l (Draeger, 1974). Clark, 1969, indicated that a sensor to measure chlorophyll as an index of biological activity should have a sensitivity to concentrations of about 1×10^{-4} mg/l.

As Draeger, 1974, reports, camera systems using false-color infrared films are helpful in detecting chlorophyll, because of the red appearance of the film when large amounts are present (Chandler, 1970, Clapp, 1972, Ham, 1967, Scherz, 1971, and Welch, 1971). Correlations between the chlorophyll content in water and the ratio of green to blue backscattered light, and between surface temperature and chlorophyll concentrations were noted (Arversen, 1973).

5. OIL POLLUTION

Pollution in marine and estuarine environments is caused by a wide range of factors, including waste disposal, oil spills, radioactivity, heat, and pesticides, affecting various living organisms.

As Norris, 1973, suggested, before the location of an outfall, surveys involving biological, chemical, and physical parameters should always be made, to avoid dangers from this source.

The ever increasing number of oil spills, due to leaks from pipelines or accidental ship discharges makes more serious the resulting pollution. Studies indicate that the effects of the oilspill of 1969 near Santa Barbara, California, continue to show (Strachan, 1972, and Foster, 1972). As Reish, 1974, indicated, the effect of an oil spill depends not only on the quantity, but also on the location and fraction of the oil. As Baker, 1974, reported, during 1973, the U.S. Environmental Protection Agency proposed effluent limitation guidelines and new source standards for the petroleum refining point source category, for the following refinery classifications:

- Topping
- Topping and Low-Cracking
- Topping and High-Cracking
- Petrochemical
- Lube
- Integrated

Waste-waters generated in drilling and production, transportation and storage, and refining were also discussed. Wastes were categorized as to source, volume, and composition.

Particularly important is chlorophyll "a". Johnson, 1978, indicated that high concentrations of it may be due to such sources as sewage treatment plants and industrial wastes, and low concentrations of it may be due to toxic substances from industrial sources.

As Duntley, 1971, indicated, spectral responses at different wavelengths can be related to changes in chlorophyll "a" concentrations. According to the studies done by Johnson, 1975, and 1977, regression equations from the analysis may be used to map quantitative distributions of chlorophyll "a". Johnson, 1978, also indicated that results of experiments with aircraft multispectral scanners over the James River, Virginia, and the New York Bight Ocean area, show the existence of significant quantitative relationships between remotely sensed data and chlorophyll "a" measurements.

Khorram, 1979, demonstrated the use of remote sensing for mapping the main water quality parameters, including salinity, turbidity, suspended solids, and chlorophyll, in the San Francisco Bay and the associated delta area. Ocean Color Scanner (OCS) data and color and color infrared photography, as acquired from a NASA U-2 aircraft, were combined with in situ data, as acquired from boats, for the assessment of the water quality parameters. Khorram also demonstrated the use of remote sensing for locating the estuarine area of highest biological activity that is characterized by high turbidity values and high concentrations of suspended solids, and chlorophyll, and that according to the studies done by many investigators from different parts of the world these attributes are a persistent feature of moderately stratified estuaries.

When an oil spill occurs, before appropriate corrective action can be taken, a knowledge of its nature, thickness, areal extent, direction, and rate of drift must be promptly established (Hollinger, 1974). There are many techniques for the remote sensing of oil spills in marine and estuarine environments, operating in different portions of the electromagnetic spectrum ranging from the ultraviolet to microwave regions (Fantasia, 1974).

Many investigators have reported that photographic techniques have a high potential for oil detection but they are also restricted to fair weather and daylight operation. Ultraviolet photography appears to have the greatest potential for oil detection. Panchromatic, infrared and color photographs do not lend themselves well because of the low contrast between oil and water. Several researchers have indicated that multispectral scanners and infrared imagers are effective in detecting and mapping oil slicks if they operate in the UV, blue, red and 8 to 14 μm regions of the spectrum. However, these scanners are limited to clear weather and daylight conditions with the exception of thermal IR scanners which can operate during periods of darkness. Radar and passive microwave techniques have been shown very promising for the detection of oil sliks, providing day/night coverage in all weather conditions except for heavy rain.

As Fantasia, 1974, reported, the techniques have been implemented by systems that are either passive or active, represented by aerial photographic cameras, infrared thermographs, low light level television systems, radar mappers, and microwave radiometers. Fantasia also reported that all these systems can detect oil spills under a particular set of conditions, in terms of water state, illumination, weather, visibility, etc., suited to each individual technique.

Ultraviolet is again reported by Estes, 1972, to be the most useful wavelength region for delineating oil in water. This is based on experiments conducted with both multiband photography and multispectral scanners, and (as reported by Thomson, 1974) is due to the fact that upwelling light from below the surface is at maximum in the wavelength range of 0.5 to 0.55 μm, depending on the particular water body. At shorter wavelengths this upwelling component is small in comparison with the light reflected from the oil surface, which has a higher reflectance than water.

Horrath, 1971, indicated that multispectral scanner observations over a controlled oil spill identified the band 0.36 to 0.38 μm as giving the best delineation of the oil slick.

Many investigators have indicated that petroleum oils contain various aromatic compounds which fluoresce strongly and thereby provide a signature useful for remote detection and identification.

Eldering, 1974, reported that Multispectral Active/Passive Scanner (MAPS) was used to obtain corrected total luminescence (excitation-emission) spectra of various petroleum products, crude oil, and natural sea water. The water surface films formed by petroleum products, from gasoline to crude oil, showed a maximum excitation sensitivity at approximately 0.275 μm. MAPS, which was developed under U.S. Environmental Protection Agency and Coast Guard sponsorship, incorporated a computer-controlled xenon flash lamp, excitation and emission monochromators, and azimuth and elevation scanning.

Other researchers in the U.S.S.R. (Kondratyev, 1974) have reported that surveys were made over the Caspian Sea in order to investigate the possibility of using a passive polarization technique in the visible spectral region. The measurements were conducted from a helicopter equipped with a spectrograph, a polarimeter and cameras, in the wavelength range 0.4 to 0.8 μm at heights up to 200 m. Kondratyev has reported, as a result, that oil spots on sea water are most pronounced in the photographs taken in normal-polarized light (photographs of the oil-polluted area were taken in the rays of the normal and parallel-polarized components, a view angle of 45°, and azimuth angle of 180°, a solar zenith angle of 45°). The oil/water contrast is considerably intensified, because in normal-polarized light the sea surface appears flatter and less detailed in comparison with photographs taken in non-polarized light.

Thermal infrared imagery of oil slicks, as many investigators have reported, has been found to be an effective tool for the identification of oil on sea water under certain conditions. Thermal sensors provide day and night surveillance capabilities because they do not depend on reflected energy for image formation. Operating in a scanning mode they also supply the important real-time data acquisition capability.

As Thomson, 1974, indicated, in general the oil tends to have a colder radiometric temperature, due to the emissivity difference between oil and water. At 10 μm the emissivity of water is 0.993 and that of oil 0.972 (Buttner, 1965). Thomson, 1972, found oil to appear colder than the surrounding water during an oil spill infrared survey at Chedabucto Bay. Similar observations were made during the Santa Barbara oil spill (Chandler, 1970).

An Airborne Thermovision System (ATS) was used in Italy to determine levels of oil pollution in eastern Sicily's coastal waters. ATS, detecting infrared radiation from 2 to 5.6 μm, was operated from an Italian Navy helicopter in conjunction with near visible infrared photography (see figures 1 to 12). Jonian Sea surveys showed that infrared photography and thermography, used simultaneously and operated from helicopters, can provide useful information on coastal water pollution (Geraci, 1978, and Terranova, 1980).

Ramblie, 1974, indicated that a two wavelength active infrared sensor works reliably in a field environment to detect the incursion of oil on a body of water. It detects the anomalous dispersion that occurs in petroleum products as a result of the carbon-hydrogen stretch bands near 3.4 μm.

As Ketchel, 1973, reported, low light level television has proved to be an useful sensor for oil detection in night surveillance or under poor lighting conditions. The television monitor provides a real time data display which can be used to observe surface slicks in the UV and visible regions.

As Thomson, 1974, reported, Side Looking Airborne Radar (SLAR) systems are based on the fact that the decrease in surface roughness, caused by an oil slick, reduces the radar cross section, and consequently a slick appears as a non scattering area. Van Kuilemburg, 1975, indicated from scatterometer observations of oil slicks that oil is always detectable, but also that oil type and thickness are of little influence on the radar echo. Experiments using SLAR operating in the vertical polarization mode showed its usefulness as a tool for the detection and mapping of oil slicks. However, in practice the horizontal mode may be sufficient for this purpose.

Studies of the fluorescence properties of oils and oil slicks on the sea surface, including remote fluorometry of oil spills, as reported by Fantasìa, 1974, were undertaken by U.S. Department of Transportation, Cambridge, Massachusetts. As a result of these studies, an airborne laser oil spill remote sensing system based on oil fluorescence has been developed. The system was operated from U.S. Marine Corps helicopter, showing that detection of oil spills is feasible in marine and estuarine environments, based on laser-excited oil fluorescence.

Gross, 1974, indicated that crude and refined oil, and other hazardous materials (e.g. phenols) in solution or dispersed in water to concentrations on the order of 5 to 500 ppm, can be excited by UV laser irradiation. This can produce luminescence signal levels that are sufficiently high to identify and estimate the concentration of such material remotely. Pollutants like heavier aromatics, oil and grease, as reported by Gross, 1975, were recognized by a field investigation of UV laser induced luminescence signatures of site water at scattered sites in Los Angeles Harbor.

Sheives, 1974, reported the applicability of a dual polarization laser backscatter system for remote identification and measurement of subsurface water turbidity and oil on water. The system incorporated both a single scatter return and a multiple scatter return from an aqueous suspension of scatterers illuminated by a vertically polarized electromagnetic plane wave. The experiments, reported by Sheives, were conducted with a

Figure 1. The helicopter (Agusta Bell 204/B ASW) supplied by an Italian Naval Air Station.

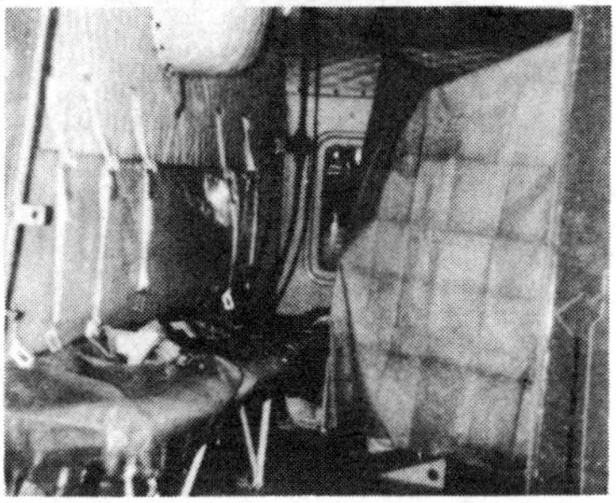

Figure 2. The aerial photographer fastened by a safety harness allowing free movements.

Figure 3. A black and white photograph taken during the surveys.

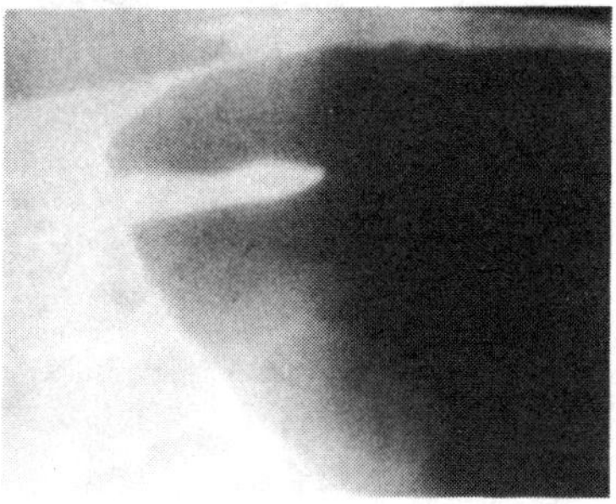

Figure 4. A color infrared photograph taken during the surveys.

Figure 5. The thermovision camera and the display units mounted together as a single unit.

Figure 6. The conventional camera installed by the side of the thermovision camera.

Figure 7. The support frame attached on the floor.

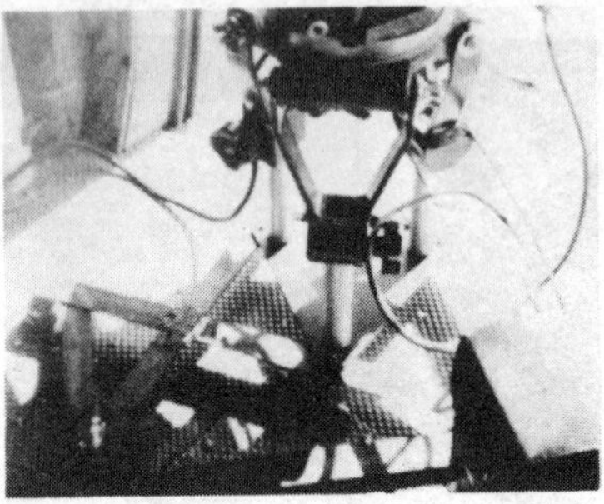

Figure 8. The color monitor fastened above the radar support.

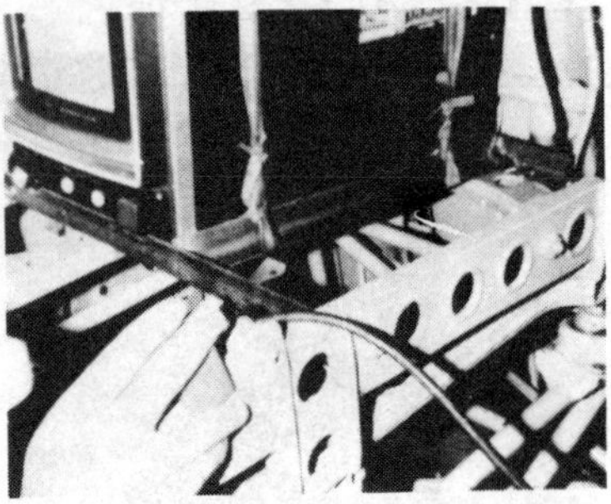

Figure 9. The camera adapter used for photographing the color monitor.

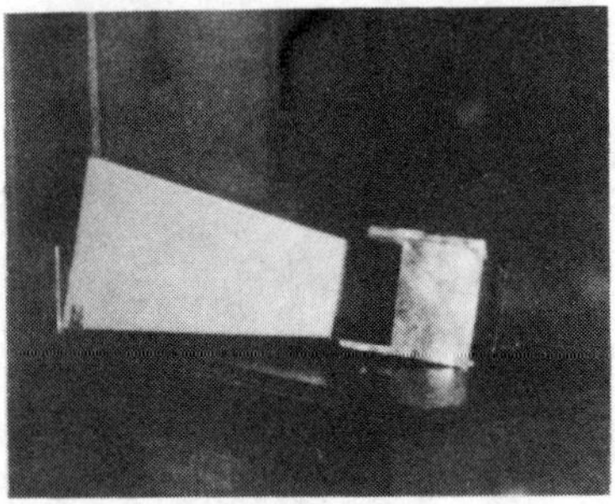

Figure 10. The ac/dc inverter used for supplying from a battery the color monitor.

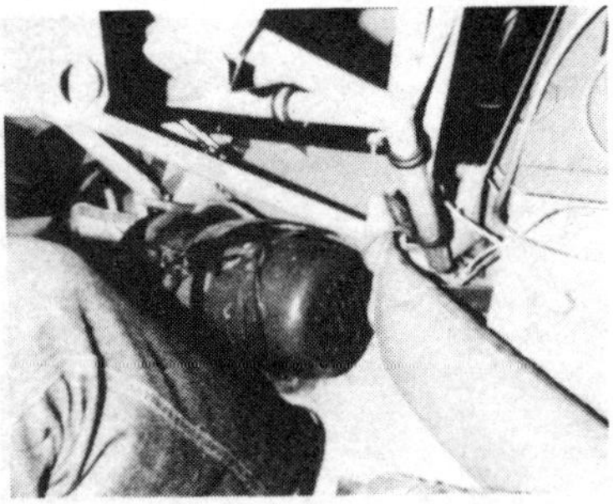

Figure 11. A thermogram photographed from the color monitor during the surveys.

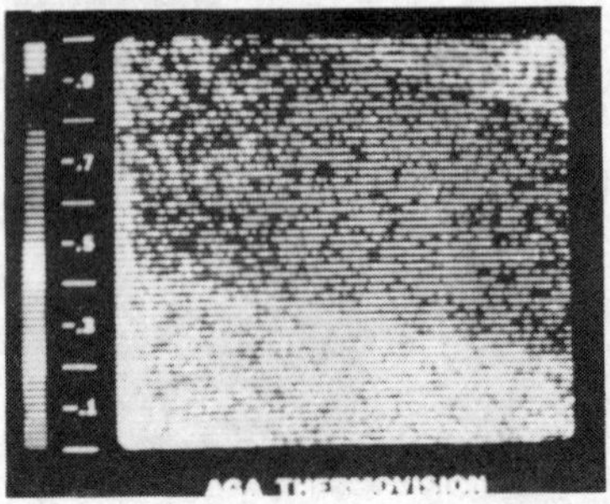

Figure 12. A thermogram photographed from the color monitor during the surveys.

lidar polarimeter transmitting a linearly polarized laser beam and receiving both the vertical and horizontal polarized backscattered energy, and show that the analytical model developed for describing the backscatter from turbid water is valid. The results were also extended to include the effect on the backscatter of a layer of oil on the turbid water. Sheives indicated that there is a reduction of the backscattered power because the difference in the index of refraction of oil relative to water alters the transmission coefficients at the air/oil boundary.

Many investigators have repeatedly reported that the determination of an oil-film thickness is important, because the film thickness along with areal extent allows the volume of the oil slicks to be estimated. The volume of oil is essential for assessing the impact of the oil spill on marine life and environment.

Hollinger, 1974, indicated that multifrequency passive microwave radiometry has the potential for determining oil-slicks having a thickness greater than about 0.05 mm. The apparent microwave brightness temperature, as reported by Hollinger, is greater in the region of an oil slick than in the adjacent unpolluted sea by an amount depending on the slick's thickness. By using two or more frequencies, thickness ambiguities introduced by oscillations in brightness temperature can be removed. In this way it is possible to delineate the thick region of oil on water from microwave observations and derive the total volume of it from microwave measurements.

The early detection and monitoring of oil in the marine and estuarine environment will significantly improve predictions of its location, nature, areal extent, and behavior over subsequent time periods. The increased use of satellite imagery in oil pollution detection will be of importance in this aspect. The treatment required and a determination of the responsibility of any damage will depend on the reliability with which the oil can be monitored under prevailing weather conditions.

6. CONCLUDING REMARKS

The concerns expressed above deal with the need for a program of action on the environment. The purpose of an environmental policy is to improve the quality of life and the living conditions of the people, by preserving and, where possible, enhancing the natural environment.

Remote sensing is of great importance in facilitating action designed to reduce and prevent pollution, providing an effective means of monitoring urban and industrial discharges. Particularly, its capability to identify critical areas in coastal zones by monitoring pollution sources can greatly increase the effectiveness of a governmental agency that is responsible for pollution control.

The results presented in this paper show how remotely sensed measurements and related observations can provide useful information on water quality for estuarine environments. Events described and matter discussed indicate (1) the need to carry out national and international programs for the inventory, monitoring, and management of water quality in coastal zones, and (2) the increasingly important role of remote sensing technology in this effort.

7. LOOKING AT THE FUTURE

In the perspective of time and viewed in a multidisciplinary approach, remote sensing has already given significant contributions to water quality in estuarine environments, and it will very likely give more in the future. However, the possibility of finding a meaningful correlation between remotely sensed measurements in coastal zones, and the urban and industrial agglomerations bordering them, should be explored more.

Pollution modifies the properties of some materials, and it is know how. Remote sensing can use this knowledge for the inventory and monitoring of pollutants as urban, agricultural, and industrial discharges in the estuarine environment. But water quality remotely sensed mesurements, and information coming from process plants and urban agglomerations in coastal zones should be stored together to establish a bank of data.

With the help of this bank of data, those measurements and information should be processed together in order to make correlations between them. The final resulting information could be used to develop appropriate regulations and procedures to control pollution sources in estuarine environments. Information derived from remote sensing data processed in this way could be very useful in the making of management decisions.

8. REFERENCES

AMERICAN SOCIETY OF PHOTOGRAMMETRY, (1960): Manual of Photographic Interpretation, American Society of Photogrammetry, Washington, D.C.

AMERICAN SOCIETY OF PHOTPGRAMMETRY, (1975): Manual of Remote Sensing, American Society of Photogrammetry, Falls Church, Virginia, I and II.

ANDERSON, R.R., (1970): Spectral Reflectance Characteristics and Automated Data Reduction Techniques which Identify Wetland and Water Quality Conditions in the Chesapeake Bay, Third Annual Earth Resources Program Review, NASA Manned Spacecraft Center, Houston, Texas, III.

ARVERSEN, J.C., et al. (1973): Remote Sensing of Chlorophyll and Temperarature in Marine and Fresh Waters, Astronautica Acta, XVIII, 229-239.

BAKER, D.A., (1974): Petroleum Processing Wastes, Journal Water Pollution Control Federation, 6, 1298-1301.

BOWKER, D.E., et al., (1977): The Use of Landsat for Monitoring Water Parameters in the Coastal Zone, Proceedings of the AIAA Joint Conference on Satellite Applications of Marine Applications, New Orleans, Louisiana, American Institute of Aeronautics and Astronautics, 193-198, 1977.

BROOKS, D.J., (1975): Landsat Measures of Water Quality, Photogrammetric Engineering and Remote Sensing, 10, 1269-1272.

BUTTNER, J.K., et al. (1965): Determination of Infrared Emissivity of Terrestrial Surfaces, J. Geophysics Res. 70-1329-1377.

CARDONE, V.J.: Satellite Observation Systems and Environmental Forecasts, Progress and Prospects, CUNY Institute of Marine and Atmospheric Sciences at The City College.

CHANDLER, P.B., (1970): Remote Sensing of Oil Polluted Sea Water, North American Rockwell Corporation Report.

CHANDLER, P.B., et al. (1970): Study of Evaluate the Utility of Aerial Surveillance Methods in Water Quality Monitoring, California State Water Resources Control Board, Sacramento, California, Publication No. 41.

CLAPP, J.L., (1972): Application of Remote Sensing to Water Resources Problems, Fourth Annual Earth Resources Program Review, NASA Manned Spacecraft Center, Houston, Texas, II, 46.

CLARK, L.G., et al. (1969): Remote Measurement of Ocean Color as an Index of Biological Productivity, Proceedings of the Sixth International Symposium on Remote Sensing of Environment, Environmental Research Institute of Michigan, University of Michigan, Ann Arbor, Michigan, II, 991-1002.

COLWELL, R.N., (1978): Remote Sensing as an Aid to the Inventory and Management of Natural Resources, The Canadian Surveyor, 2, 183-203.

DAVIS, A., et al. (1973): Remote Sensing of Water Quality Using Laser Induced Fluorescence, Inland Water Directorate, Environment Canada, Ottawa, Ontario.

DRAEGER, W.C., et al. (1974): Usefulness of Remote Sensing Techniques for the Environmental Monitoring of the Sacramento-San Joaquin Delta, Department of Water Resources, The Resources Agency, State of California, Contract B51010.

DRAKE, B., et al. (1974): Feasibility of Using Multiplexed SLAR Imagery for Water Resources Management and Mapping Vegetation Communities, Proceedings of the Ninth International Symposium on Remote Sensing of Environment, Environmental Research Institute of Michigan, University of Michigan, Ann Arbor, Michigan, I, 219-249.

DUNTLEY, S.A., (1971): Optical Methods of Water Pollution, Proceedings of the Environmental Quality Sensor Workshop at Western Environmental Laboratory, Environmental Protection Agency, Washington, D.C., II, 15-27.

EGAN, W.G., (1974): Boundaries of ERTS and Aircraft Data within which Useful Water Quality Information Can Be Obtained, Proceedings of the Ninth International Symposium on Remote Sensing of Environment, Environmental Research Institute of Michigan, University of Michigan, Ann Arbor, Michigan, II, 1319-1343.

ELDERING, H.G., et al. (1974): Multi-Spectral Remote Fluorimeter for Detection of Oil Films, Proceedings of the Ninth International Symposium on Remote Sensing of Environment, Environmental Research Institute of Michigan, University of Michigan, Ann Arbor, Michigan, III, 1709-1710.

ESTES, J.E., et al. (1972): The Multispectral Concept as Applied to Marine Oil Spills, Remote Sensing of Environment, 2, 141-163.

FANTASIA, J.F., et al. (1974): Development of an Experimental Airborne Laser Remote Sensing System for the Detection and Classification of Oil Spills, Proceedings of the Ninth International Symposium on Remote Sensing of Environment, Environmental Research Institute of Michigan, University of Michigan, Ann Arbor, Michigan, III, 1711-1745.

FOSTER, M., (1972): The Santa Barbara Oil Spill: Dosage of Crude Oil on Shore and Initial Effects on Intertidal Organisms, California Cooperative Oceanic Fisheries Investigation, 16, 130.

GERACI, A.L., (1977): Il Disegno degli Impianti Chimici, Cooperativa Universitaria Editrice, Catania, Italy.

GERACI, A.L., (1978): Jonian Sea Water Exploration and Evaluation Interim Report, Machinery Institute, Department of Engineering, University of Catania, Italy.

GERACI, A.L., (1980): Remote Sensing versus Marine Pollution, The First Week of Photography, Kodak, Terrasini, Italy.

GOLDBERG, M.C., et al. (1972): Applications of Spettroscopy to Remote Determination of Water Quality, Fourth Annual Earth Resources Program Review, NASA Manned Spacecraft Center, Houston, Texas, III, 81.

GROSS, H.G., et al. (1971): Luminescence Induced by UV and Visible Lasers for Remote Active Sensing of Materials from Ground, Air, and Space, Proceedings of the Seventh International Symposium on Remote Sensing of Environment, Environmental Research Institute of Michigan, University of Michigan, Ann Arbor, Michigan, II, 869-888.

GROSS, H.G., et al. (1974): Crude and Refined Petroleum Oil Structured Luminescence Signatures Induced by UV Laser or Lamp and their Remote Sensing Applications, Proceedings of the Ninth International Symposium on Remote Sensing of Environment, Environmental Research Institute of Michigan, University of Michigan, Ann Arbor, Michigan, III, 1747-1759.

GROSS, H.G., (1975): Los Angeles Harbor Field Investigation of Oil and Background Luminescence Signatures, Proceedings of the Tenth International Symposium on Remote Sensing of Environment, Environmental Research Institute of Michigan, University of Michigan, Ann Arbor, Michigan, I, 253.

GUINARD, N.W., et al. (1970): The Remote Sensing of Oil Slicks by RADAR, Chedabucto Bay Case Study, USGS Office of Research & Development Project.

HAM, L.W., (1967): Aerial Detection at Clear Lake and Lake Tahoe, Proceedings of Third Annual Conference of Remote Sensing of Air and Water Pollution, Cartwright Aerial Surveys, Inc., Sacramento, California.

HILL, J.M., (1979): Impacts of Land Use on Estuarine Water Quality, Summaries of the Thirteenth International Symposium on Remote Sensing of Environment, Environmental Research Institute of Michigan, Ann Arbor, Michigan, 44-45.

HOLLINGER, J.P., (1974): Passive Microwave Sensing of Oil Slicks, Proceedings of the Ninth International Symposium on Remote Sensing of Environment, Environmental Research Institute of Michigan, University of Michigan, Ann Arbor, Michigan, III, 1761.

HOM, L.W., (1968): Remote Sensing of Water Pollution, Journal Water Pollution Control Federation, 40, 1728.

HORRATH, R., et al. (1971): Optical Remote Sensing of Oil Slicks: Signature Analysis and Systems Evaluation, Final Report, Project No. 724104. 2/1, Willow Run Laboratories, University of Michigan.

HORRATH, R., et al. (1971): Remote Sensing of Southern California Oil Pollution Experiment, National Technical Information Service, PB-203 194.

HORSTEIN, B., (1973): The Visibility of Oil-Water Discharges, Environmental Protection Agency, Proceedings of Joint Conference on Prevention and Control of Oil Spills.

HOWARTH, P.J., et al.: Remote Sensing for Coastal Studies in Canada, Department of Geography, McMaster University, Hamilton, Ontario.

JOHNSON, R.W., (1975): Quantitative Suspended Sediment Mapping Using Aircraft Remotely Sensed Data, Proceedings of the NASA Earth Resources Survey Symposium, Houston, Texas, I-C, 2087-2098.

JOHNSON, R.W., (1976): Quantitative Sediment Mapping from Remotely Sensed Multispectral Data, Fourth Annual Remote Sensing of Earth Resources Conference, Tullahoma, Tennessee.

JOHNSON, R.W., (1977): Application of Aircraft Multispectral Scanners to Quantitative Analysis and Mapping of Water Quality Parameters in the James River, Virginia, COSPAR Space Research, Pergamon Press, XVII, 25-31.

JOHNSON, R.W., (1977): Multispectral Analysis of Ocean Dumped Materials, Proceedings of the Eleventh International Symposium on Remote Sensing of Environment, Environmental Research Institute of Michigan, University of Michigan, Ann Arbor, Michigan, II, 1619-1625.

JOHNSON, R.W., (1978): Mapping of Chlorophyll "a" Distributions in Coastal Zones, Journal of the American Society of Photogrammetry, 5, 617-624.

JOHNSON, R.W., et al. (1978): Application of Remote Sensing to Monitoring and Studying Dispersion in Ocean Dumping, First International Ocean Dumping Symmosium, Kingstone, Rhode Island.

JOHNSON, R.W., et al. (1980): Remote Sensing for Water Quality and Biological Measurements in Coastal Waters, Photogrammetric Engineering and Remote Sensing, 1, 77-85.

KAMINISKI, H., (1972): Detection of Waste Water Effluents and of their Surface Spread in the English Channel, North Sea, and Baltic Sea, through the Determination of the Surface Temperature of the Sea by Means of Infrared Pictures Taken by Satellites, Bochum.

KETCHEL, R.J., et al. (1973): Development of U.S. Coast Guard Prototype Airborne Oil Surveillance System, Proceedings of Joint Conference on Prevention and Control of Oil Spills, U.S. Coast Guard, Washington, D.C.

KEYSER, A.H., (1967): Water Quality Characteristics and their Measurements, Instrumentation, Honeywell, XX.

KHORRAM, S., (1979): Remote Sensing Analysis of Water Quality in the San Francisco Bay-Delta, Proceedings of the Thirteenth International Symposium on Remote Sensing of Environment, Environmental Research Institute of Michigan, University of Michigan, Ann Arbor, Michigan, 1591-1601.

KLEMAS, V., et al. (1974): Correlation of Coastal Water Turbidity and Circulation with ERTS-1 and Skylab Imagery, Proceedings of the Ninth International Symposium on Remote Sensing of Environment, Environmental Research Institute of Michigan, University of Michigan, Ann Arbor, Michigan, II, 1289-1317.

KLEMAS, V., et al. (1977): Satellite and Current Drogue Studies of Ocean-Disposed Waste-Drift, Journal Water Pollution Control Federation, 5, 757-763.

KLEMAS, V., et al. (1978): Skylab/EREP Applications to Ecological, Geological, and Oceanographic Investigations of Delaware Bay, NASA CR-144910.

KONDRATYEV, K.Y., et al. (1974): The Application of Spectrometric and Polarization Techniques for Remote Sensing of Oil on Sea Water, Proceedings of the Ninth International Symposium on Remote Sensing of Environment, Environmental Research Institute of Michigan, University of Michigan, Ann Arbor, Michigan, III, 1793-1802.

LINK, L.E., et al. (1974): Utilization of Remote Sensing in River Basin Studies, Water Resources Bullettin, 73029.

MAIRS, R.L., et al. (1973): Remote Sensing of Estuarine Circulation Dynamics, Photogrammetric Engineering and Remote Sensing, 9, 927-938.

MEEKS, D.C., et al. (1971): Microwave Radiometric Detection of Oil Slicks, Final Report No. 1335-2 Microwave Division, Aeroject-General Corporation, El Monte, California.

MILLER, W.L., (1974): The Economic Impact of Remote Sensing Data as the Source of Nonpoint Pollution Monitoring and Control, Proceedings of the Ninth International Symposium on Remote Sensing of Environment, Environmental Research Institute of Michigan, University of Michigan, Ann Arbor, Michigan, II, 1001-1011.

MOREL, A., et al. (1977): Analysis of Variations in Ocean Color, Limnology and Oceanography, 4, 709-722.

MOREL, A., (1980): In-Water and Remote Measurements of Ocean Color, Boundary-Layer Metereology, 18, 177-201.

NASA LYNDON B. JOHNSON SPACE CENTER, (1978): Skylab-EREP Investigations Summary, National Aeronautics and Space Administration, Washington, D.C.

NORRIS, D.P., (1973): Marine Waste Disposal - A Comprehensive Environmental Approach to Planning, Journal Water Pollution Control Federation, 45, 52.

OHLHORST, C.W., (1978): Quantitative Mapping by Remote Sensing of an Ocean Acid-Waste Dump, NASA TP-1275.

PAULSON, R.W., (1968): Estuarine Studies, Second Earth Resources Program Review, NASA Manned Spacecraft Center, Houston, Texas.

PAULSON, R.W., (1968): Preliminary Remote Sensing of the Delaware Estuary, U.S. Department of the Interior, Geological Survey Interagency Report NASA 128.

RAMBLIE, Jr., G.S., (1974): A Pratical Sensor, Proceedings of the Ninth International Symposium on Remote Sensing of Environment, Environmental Research Institute of Michigan, University of Michigan, Ann Arbor, Michigan, III, 1685-1694.

REISH, D.J., et al. (1974): Marine and Estuarine Pollution, Journal Water Pollution Control Federation, 6, 1437-1451.

ROGERS, R.H., et al. (1975): Application of Landsat to the Surveillance and Control Eutrofication in Saginaw Bay, Proceedings of the Tenth International Symposium on Remote Sensing of Environment, Environmental Research Institute of Michigan, University of Michigan, Ann Arbor, Michigan, I, 437-446.

SCHERZ, J.P., (1971): Monitoring Water Pollution by Remote Sensing, Journal of the Surveying and Mapping Division, ASCE, 97, 307-320.

SCHERZ, J.P., et al. (1973): Lake Superior Water Quality near Duluth from Analysis of Aerial Photos and Erts Imagery, Remote Sensing and Resource Management, American Water Resources Association, Urbana, Illinois, 147-160.

SCHERZ, J.P., et al. (1973): Aerial and Satellite Photography. A Valuable Tool for Water Quality Investigations, Proceedings of the American Society of Photogrammetry Fall Convention, Lake Buena Vista, Florida, 883-905.

SCHERZ, J.P., et al. (1975): Classifying and Monitoring Water Quality by Use of Satellite Imagery, International Conference on Environmental Sensing and Assessment, Las Vegas, Nevada.

SHEIVES, T.C., et al. (1974): Remote Measurements of Water Pollution with a Lidar Polarimeter, Proceedings of the Ninth International Symposium on Remote Sensing of Environment, Environmental Research Institute of Michigan, University of Michigan, Ann Arbor, Michigan, III, 1695-1708.

STOERZ, G.E., et al. (1969): Airborne Fluorometer Applicable to Marine and Estuarine Studies, Marine Technology Society Journal, 6.

STRACHAN, A.R., (1972): Santa Barbara Oil Spill - Intertidal and Subtidal Surveys, California Cooperative Oceanic Fisheries Investigation, 16, 122.

TELEKI, P.G.: The Use of Satellite Data in Ocean Engineering, NATO Advanced Research Institute on Earth Observation and Information Systems for Resource Management and Environmental Control.

TERRANOVA, V., et al. (1980): Relazione Preliminare su una Indagine tramite Termografia e Fotografia all'Infrarosso dell'Inquinamento del Mare Jonio tra Taormina e Siracusa, Bollettino dell'Accademia Gioenia, XIV, 37-56.

THOMSON, K.P.B., et al. (1972): A Remote Sensing Survey of the Chedabucto Bay Oil Spill, Inland Waters Directorate, Scientific Series No. 26.

THOMSON, K.P.B., et al. (1974): Remote Sensing of Oil Spills, Environmental Impact Control Directorate, Economic and Technical Review, Report EPS 3-EE-74-2.

VAN KUILDENBURG, J., (1975): Radar Observations of Controlled Oilspills, Proceedings of the Tenth International Symposium on Remote Sensing of Environment, Environmental Research Institute of Michigan, University of Michigan, Ann Arbor, Michigan, I, 243-250.

VAN LOPIK, J.R., (1968): Pollution Surveillance by Noncontact Infrared Techniques, Journal Water Pollution Control Federation, 40, 425.

WARD, R.C., (1974): Routine Surveillance Alternatives for Water Quality Management, Journal Water Pollution Control Federation, 12, 2645-2652.

WEISBLATT, E.A., et al. (1973): Classification of Turbidity Levels in the Texas Marine Coastal Zone, Machine Processing of Remotely Sensed Data Conference Proceedings, LARS, 3A-42.

WELCH, R., (1969): Remote Sensing Systems in Water Quality Control Program, Proceedings of the First Annual Symposium of the International Remote Sensing Institute, Sacramento, California, 2, 46.

WELCH, R., (1971): Remote Sensing for Water Pollution Control, Photogrammetric Engineering and Remote Sensing, 12, 1285-1286.

WEZERNACK, C.T., et al. (1970): Pollution Surveillance and Data Acquisition Using Multispectral Remote Sensing, Journal American Water Resources Association, 6, 920.

WHIPPLE, J.M., (1973): Surveillance of Water Quality, Photogrammetric Engineering and Remote Sensing, 2, 137-145.

WILLIAMSON, A.N., et al. (1973): Sediment Concentration Mapping in Tidal Estuaries, Third Earth Resources Technology Satellite-I Symposium, Washington, D.C., NASA SP-351, 651-658.

WOLF, W.L., et al. (1978): The Infrared Handbook, Environmental Research Institute of Michigan, University of Michigan, Ann Arbor, Michigan, Office of Naval Research, Department of the Navy, Washington, D.C.

YARGER, H.L., et al. (1973): Quantitative Water Quality with ERTS-1, The Third ERTS-1 Symposium, Washington, D.C., 651-658.

YARGER, H.L., et al. (1973): Water Turbidity Detection Using ERTS-1 Imagery, Symposium of Significant Results Obtained from the Earth Resources Technology Satellite-I, GSFC, New Carrollton, Maryland, NASA, SP-327, 651-658.

YARGER, H.L., _et al._ (1975): Quantitative Water Quality with Landsat and Skylab, _NASA Earth Resources Survey Symposium_, Houston, Texas, I-A, 347-370.

Part B

GENERAL PAPERS

AN ASSESSMENT OF SIMPLE PLOTTING INSTRUMENTS FOR RESOURCE MAPPING

M J C Weir
I T C, Enschede
The Netherlands

Abstract: Despite developments in orthophoto mapping and the production of other forms of geometrically correct imagery, the conventional line map remains an important end product for many resource surveys. Simple plotting instruments of the "sketchmaster" variety -operating on the camera lucida principle- have long been, and remain, a useful means of compiling such maps from aerial photographs. The use of these instruments by photo-interpreters is generally justified on the basis of low capital cost and simple operation and maintainance.
This paper traces the development, and reports on investigations into the operating cost and accuracy of the present generation of simple instruments. The influence of operator skill and terrain configuration is discussed. Their performance for resource mapping is assessed and compared with more rigorous photogrammetric methods.

1. INTRODUCTION

Although the presentation of information within a spatial reference system is an essential component of most resource surveys, field scientists involved in such surveys have generally been concerned more with the acquisition of resource data from aerial photographs and other forms of imagery than with the photogrammetric and cartographic problems of portraying it in map form. The availability of rectified satellite imagery greatly alleviates these problems when applied to small scale thematic mapping. Similarly, the techniques of differential rectification can provide the field scientist with a geometrically correct aerial photograph, the orthophoto, which may be used for more detailed surveys at larger scales. While there is a growing interest in the potential of orthophotography, the conventional line map remains an important end product for many resource surveys.

Paper presented at a conference on Matching remote sensing technologies and their applications convened in London by the Remote Sensing Society, 16-18 December, 1981

Graphical stereoplotting, unversally employed for topographic mapping, can of course also be used in resource mapping and various authors have described its application in, for example, geological, soil, forest and vegetation surveys (Pillmore, 1979; Stove and Hulme, 1980; Frei, 1976; Goodier and Grimes, 1970). However non-rigorous methods of transferring information from aerial photographs onto topographic base maps are widely employed as well, and for more than four decades instrument manufacturers have been marketing a variety of simple instruments designed for this task. Although the use of such instruments by photo-interpreters is generally justified on the basis of low capital cost and simple operation and maintainance, their accuracy and efficiency should also be considered and compared with that of more rigorous methods.

2. HISTORICAL DEVELOPMENT

The principle of the camera lucida, whereby a double prism or semi-reflecting mirror is employed to permit the simultaneous observation of two images, has long been applied in the construction of simple instruments for plotting planimetric detail from aerial photographs onto a base map. The technique finds its simplest expression in the "sketchmaster", comprising a photograph holder together with an eyepiece and prism or mirror mounted over the base map. Instruments introduced by Zeiss in 1936 and that designed by Buckmaster for the U.S. Geological Survey in 1941 are probably the best known. However a variety of other designs, particularly those developed at the French I.G.N. are worthy of note.

An important drawback of simple sketchmaters is the fact that they employ monocular viewing which is obviously tiring for the operator. This difficulty was overcome in the Duoscope, a Canadian instrument designed by the Dominion Forest Service (Seely, 1941). The Duoscope was basically a mirror stereoscope constructed so that one eye views the photograph and the other the map. A logical development of this instrument was the Multiscope (Spurr and Brown, 1945) in which semi-reflecting mirrors are incorporated into a standard mirror stereoscope, so permitting the stereoscopic image to be viewed superimposed over a base map. In 1961 Hilger and Watts developed the Stereosketch, an instrument basically similar to the Multiscope and employed in the first instance by the Directorate of Overseas Surveys for drawing hill shading directly from aerial photographs (Anon., 1963)

There was no difficulty in fitting these instruments with a simple tilt mechanism to provide a perspective correction for the errors due to inclination of the caméra axis. The effect of terrain relief, however, could only be compensated through a frequent scale correction, although a sketchmaster developed at I.G.N. did permit affine rectification. This was achieved by the use of tracing tables, each with a vertically adjustable platen, centred over known height points on the base map in a manner similar to that employed by the Ordnance Survey for rectification in undulating terrain.

The Bausch and Lomb Zoom Transfer Scope, introduced in 1971, and the O.M.I. Stereo Facet Plotter, introduced in 1976, are fitted with zoom lenses which enable the operator to continuously adjust the magnification of the photo image to cater for scale variations within the photograph. A further correction for relief is provided by differentially rotatable prism systems incorporated into the optical train of either the photograph (Bausch and Lomb) or the map (O.M.I.). These systems impart a scale change in one direction only, resulting in an affine deformation of the image. The amount and direction of affinity can be varied depending on the degree of terrain slope and its direction relative to the exposure station.

3. CRITERIA FOR ASSESSMENT

Although advances in optical design have resulted in greater flexibility of operation, all simple instruments of the camera lucida type have serious disadvantages when compared with conventional stereoplotters. In the first place they cannot be used for original mapping but require a topographic base map with adequate control detail. In developing countries, for whom these instruments can provide a useful form of intermediate technology for resource mapping, suitable base maps are not always available, while in Europe human activities such as afforestation or major construction work can render existing maps unreliable until revised editions become available.

A second disadvantage is that, even when stereoscopic viewing is possible, plotting is carried out from one photograph only rather than from the three dimensional model of the terrain as reconstructed in a stereoplotter. Consequently, contours and other forms of height information which might be useful in e.g. erosion studies or the planning of access and extraction routes for forest exploitation cannot be acquired.

Finally, the planimetric accuracy obtained with simple instruments, which do not operate on rigorous photogrammetric principles, is naturally inferior to that of stereoplotters. Furthermore, attempting to obtain a "best fit" to the topographic base is not an entirely objective operation, with obvious consequences for the final accuracy.

On the other hand, the low capital cost and the simple procedures for operation and maintainance of simple instruments can be attractive. Their use as a type of intermediate technology has already been mentioned, but similar arguments apply to small resource survey organizations in developed countries who might require some in-house plotting facilities but feel unable to justify the operation of a more complex photogrammetric unit.

With these considerations in mind, a series of experiments have been carried out to investigate the usefulness of simple instruments for resource mapping. The majority of the tests were made with a Bausch and Lomb Stereo Zoom Transfer Scope, but an O.M.I. Stereo Facet Plotter was also used. Experience had suggested that the similar design of the two instruments is reflected in similar operation and so a comparative evaluation -which would necessarily be concerned with the quality of the individual optical, electrical and mechanical components- has been avoided. Instead, the intention has been to assess the performance of these instruments under typical operating conditions. Detailed statistical analysis, which would require data from an extensive range of terrain conditions, types of photography, base map characteristics and using a number of operators was beyond the scope of this investigation.

4. ACCURACY

The potential of both the Zoom Transfer Scope and the Stereo Facet Plotter for topographic map revision has led a number of organizations to assess their accuracy for this purpose. McGivern et.al. (1972) report on a comparison between revision with a stereoplotter (Balplex) and with a Zoom Transfer Scope. 62% of the sample areas showed no significant difference between the two revisions while a further 28% had differences not exceeding 0.025 inch (0.64 mm). The maximum discrepancy was 0.067 inch (1.69 mm).

The preliminary report of the OEEPE Fribourg map revision test includes results obtained by the National Board of Survey, Helsinki, using a Stereo Facet Plotter. Standard deviations of 0.26 mm and 0.20 mm were found for house corners and road axes, results which "seem surprisingly good" (Spiess, 1980).

In order to further investigate the accuracy of both these instruments, a selection of topographic details (roads, tracks and streams) was plotted using a Wild B8S. A model from the OEEPE Reichenbach block (1:15,000 wide angle photography) was used. The model includes many sharply defined field corners, 45 of which were plotted for use as check points. A duplicate of the original pencil plot minus the check points was produced from an internegative. Since all three films were punched, the two positives could be mounted exactly in register.

The 45 check points were plotted from diapositives and the discrepancy (d) with the photogrammetric plot measured. An indication of accuracy is given by the root mean square error

$$\text{rmse} = \sqrt{\frac{\Sigma(d^2)}{45}}$$

A series of six tests was carried out. One operator repeated the test for both photographs of the model on both the Stereo Zoom Transfer Scope and the Stereo Facet Plotter. rmse values of 1.15 mm and 0.95 mm were obtained for the former and 1.21 mm and 0.84 for the latter (figure 1). Two other operators repeated the test with one photograph in the Stereo Zoom Transfer Scope, obtaining rmse values of 1.32 mm and 1.28 mm.

Since the tests were carried out with photography of hilly, but not steep, terrain, the values of about 1 mm rmse obtained are likely to be exceeded under more difficult conditions. However these results, for what are considered average conditions, are adequate for most resource survey mapping, confirming the suitability of the instruments for this purpose.

5. OPERATING TIME AND COST

Orientation

Operating times, for orientation and plotting, can be used as a measure of the efficiency of instrument and operator. Terrain condition and operator skill may be expected to have a strong influence on the orientation time. In order to investigate this, three operators made 75 orientations, 25 in each of three different terrain types - flat, hilly and mountainous (table 1). All tests were made with a Stereo Zoom Transfer Scope.

Figure 1. Discrepancies

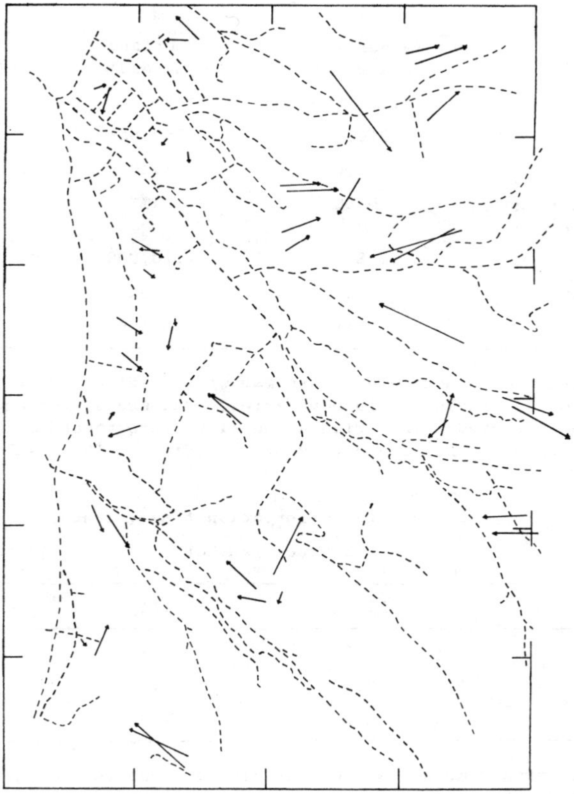

------ topographic base (4x reduction of 1:5000 original)

⟶ 10 mm represents 1 mm discrepancy

Example results from O.M.I. Stereo Facet Plotter.

Source : author

Table 1 Photography for orientation tests

Source : author

Test	1	2	3
Location	Netherlands (Almelo)	Scotland (L. Ard)	Scotland (Cairngorms)
Scale	1:20,300	1:25,700	1:41,200
p.d. (mm)	210.83	152.86	88.37
Flying ht. (Z)	4300	3950	3650
Elevation range as % of Z	0.5%	4%	25%
Base map	1:8,000	1:10,000	1:16,000
Enlargement	2.5x	2.5x	2.5x
Category	Flat	Hilly	Mountainous

In each case a monochrome base map with a selection of topographic detail - roads, tracks, field boundaries, drainage system and isolated buildings - was used. The orientation times were recorded to the nearest 0.1 minute using a photographic laboratory timer. The results are shown in table 2.

Table 2 Mean orientation time (minutes)

Source : author

Operator	Flat	Hilly	Mountainous
1	1.8	2.8	4.4
2	2.1	2.9	4.5
3	1.2	1.9	1.5

Orientation time alone does not give a complete indication of the efficiency of the different operators for the three terrain types. Each operator was free to select the terrain facets to be fitted, so that there were notable variations in the size of these facets with obvious consequences for the total orientation time required per photograph or per map sheet. Therefore, in addition to recording the time taken, the operators indicated the boundary of the area fitted at each orientation. These areas were later measured with a polar planimeter (table 3) and used to weight the orientation times according to the proportion of the field of view used (table 4). A square facet of 15,300 mm^2 on the maps was found to be the maximum area which could be observed with ease.

Table 3 Mean facet area (mm^2 on map)

Source : author

Operator	Flat	Hilly	Mountainous
1	10157	5352	5549
2	6784	2945	1515
3	6750	5538	2025

Table 4 Weighted mean orientation time (minutes per 15300 mm^2)

Source : author

Operator	Flat	Hilly	Mountainous
1	2.9	9.4	14.4
2	5.7	18.9	55.6
3	3.2	5.6	11.9

The results suggested that there are appreciable differences both among terrain types and among operators. In order to confirm this, the data were analysed using a factorial design (table 5).

Table 5 Test of significance

Source : author

Source of variation	d.f.	Sum of squares	Mean squares	F
Terrain (T)	2	21441.8	10720.9	102.37
Operators (O)	2	17864.6	8932.3	85.29
Interaction (TO)	4	14755.1	3688.8	
Residual	216	22621.5	104.7	
Total	224	76683.0	342.3	

The calculated F values indicate highly significant differences for both terrain types and operators.

The data from table 4 can be used to estimate the orientation time per model area for comparison with that required in a conventional stereoplotter. With a 23 cm x 23 cm photo format and typical forward and lateral overlaps of 60% and 20% respectively, each exposure results in a gain in coverage or a neat model area of 92 mm x 184 mm ≃ 17000 mm^2 at photo scale.

Assuming a typical enlargement from photograph to map of 2.5 x as used in the tests and taking the average times for the two most efficient operators (1 and 3), then orientation times of approximately 0.3, 0.9 and 1.5 hours per model area are required for flat, hilly and mountainous terrain respectively. Since relative and absolute orientation in a stereoplotter normally takes 1 - 1.5 hours, assuming that only an approximate levelling is made when plotting planimetry only, simple instruments have no significant advantage as far as orientation time is concerned. However, the instrumental costs, about one quarter those of a topographic plotter, are more favourable.

If plotting is carried out at a scale close to that of the original photography, then the orientation time for simple instruments, which is area dependent, is, on a per model basis, very efficient.

Plotting speed

To determine the plotting speed, data on the time required to plot annotated vegetation and forest boundaries of known length was obtained for three operators, using both the Stereo Zoom Transfer Scope and the Stereo Facet Plotter. The plotting speeds were found to vary between 1.4 and 2.7 mm/sec. on the map By comparison, Einevoll (1964) reports speeds of between 1.7 and 2.2 mm/sec. for stereoplotting. Einevoll's results were obtained for soil boundaries plotted at 1:5000 from 1:16,000 annotated diapositives using a Wild A8. The results indicate that a plotting speed of 2 mm/sec. can be used to estimate production times for both simple instruments and stereoplotters.

6. COST

Production rates (orientation and plotting), together with basic cost factors for personnel and equipment, allow the total cost, e.g. per model, to be estimated. In Western Europe operator costs (salary, insurance etc.) are currently about $15 per hour. A simple instrument, basic price $8,000 in 1980, costs $1.5 per hour while a topographic plotter, basic price $33,000 in 1980, costs $6 per hour. Their cost factors include tax, installation and interest charges and assume that the equipment is used for 2000 hours per year and written off after five years.

A model of a hilly forest area in Austria was plotted with a Stereo Zoom Transfer Scope and a Wild B8S in order to compare the two approaches. 1/30,000 wide angle infra-red photography was used. A 1:10,000 topographic map, previously plotted in the B8S from the same photography, served as a base map for the Zoom Transfer Scope and as control for absolute orientation in the B8S. Three categories of mature forest - coniferous, broad-leaved and mixed - and young stands were delineated on a transparent overlay using a mirror stereoscope. The overlay was then used as a guide for the operator. 10,100 mm of line were plotted.

On the basis of the suggested production rates the following costs are estimated (table 6).

Table 6 Cost estimation
Source : author

	Stereoplotter	Simple instrument
Orientation	1 to 1.5, say 1.25 hours	1.3 hours*
	1.25 ($15 + $6) = $26.25	1.3 ($15 + $1.5) = $21.45
Plotting	10,100 mm at 2 mm/sec.= 1.4 hrs	10,100 mm at 2 mm/sec.= 1.4 hrs
	1.4($15 + $6) = $29.40	1.4($15 + $1.5) = $23.10
Total	$55.65	$44.55

* 0.9 hours per model area at 2.5 x enlargement adjusted for 3 x enlargement.

In fact 62 minutes were required for orientation and 95 minutes for plotting on the B8S, resulting in a cost of $55.0. With the Zoom Transfer Scope 67 minutes were required for orientation and 72 minutes for plotting, resulting in a cost of $38.3.

A second test was made with the Loch Ard photography used in the orientation trials. In this test interpretation of vegetation types - coniferous and broadleaved forest, heather, bracken and rough grassland - was undertaken directly in the plotting instruments. Orientation times of 68 minutes and 59 minutes were recorded with the B8S and Zoom Transfer Scope respectively.

6500 mm of boundary were plotted in 91 minutes with the B8S. This slower plotting speed of 1.2 mm/sec. is to be expected due to the additional time necessary for interpretation. Only 5200 mm of boundary were plotted in the simple instrument in 62 minutes (1.4 mm/sec). As expected, a more complete and detailed plot was obtained from the stereoplotter, but at a cost ($56) 70% higher than that from the simple transfer instrument ($33).

7. CONCLUSION

In flat areas, or when plotting at scales close to that of the original photography, orientation with simple instruments is relatively fast. However, in hilly and mountainous terrain the operating time of these instruments does not differ appreciably from that of stereoplotter when an enlargement from photograph to map of about 2.5 times, typically employed in stereoplotting, is used. Although the equipment costs for stereoplotting are about four times those of simple instruments, this difference is decreased markedly when the cost of the operator is taken into consideration.

In the absence of well-defined planimetric accuracy tolerances, the performance of simple instruments would appear adequate for many resource mapping tasks. However, there are instances, e.g. the production of large scale (1:10,000) maps for forest management, where the application of rigorous techniques with a higher and more consistent accuracy, is advisable. Although one should not attach too much significance to the actual figures, the two trials described above suggest that the additional cost of stereoplotting, found to be about $20 per model, is not exorbitant when compared with the total costs of a resource survey project. Certainly organizations whose work load is sufficient to keep one or more instruments and their operators continuously employed should give serious consideration to the application of a more rigorous approach to resource mapping by photogrammetric means.

9. ACKNOWLEDGEMENT

The author wishes to thank Messrs Bergaoui and Lukman for their assistance with this investigation.

9. REFERENCES

ANON., (1963): Some recent developments in hill shading from air photographs in the Directorate of Overseas Surveys, Survey Review XVII No. 127 1 - 11

EINEVOLL, O., (1964): Experiments with photogrammetrical plotting of soil classification boundaries in large scale maps, Norwegian Agricultural University, Vollebekk, 9 pp

FREI, C., (1976): Methoden und Aufwand für eine Bestandeskartierung, Berichte Eidg. Anstalt für das forstliche Versuchswesen No 147, 19 - 48

GOODIER, R., and GRIMES, B.H., (1970): The interpretation and mapping of vegetation and other ground surface features from air photographs of mountainous areas in North Wales, Photogrammetric Record 6(36) 553 - 566

McGIVERN, R.F., et al, (1972): Planimetric map revision with the Bausch and Lomb Zoom Transfer Scope, Presented paper ACSM/ASP Joint Annual Meeting, Washington D.C.

PILLMORE, C.L., (1979): The history and function of the U.S. Geological Survey Photogrammetric Laboratory for Geologic Studies, Proceedings of the 45th Annual Meeting of the American Society of Photogrammetry, Washington D.C., 465 - 468

SEELY, H.E., (1941): The Duoscope, Photogrammetric Engineering VII (1) 18 - 21

SPIESS, E., (1980): Revision of topographic maps - results of the Fribourg test by Commission D of the OEEPE Int. Archives of Photogrammetry XXIII (B4), ISP Comm. IV, Hamburg

SPURR, S.H. and BROWN, C.T., (1945): The Multiscope: a simple stereoscopic plotter, Photogrammetric Engineering XI (3) 171 - 178

STOVE, G.C. and HULME, P.D., (1980): Peat resource mapping in Lewis using remote sensing techniques and automated cartography, Int. Journal of Remote Sensing Vol 1 No. 4 319 - 344

CONCERNING THE REPRODUCTION OF PHOTOMAPS OF SATELLITE IMAGERY

R.W.Dixon-Gough
Department of Land Surveying
North East London Polytechnic
Forest Road
LONDON E17 4JB

Abstract: Satellite imagery offers real opportunities to the cartographer as a base for topographic and special purpose maps, particularly of sparsely mapped areas. Two main problems exist with such bases. Firstly, the image must be converted from a continuous line to a printable form and, secondly, the image needs to be cartographically enhanced to simplify the data for the benefit of the map user. Conventional methods used to convert the image to a printable form lead to loss of resolution, whilst any attempt of enhancement invariably increases the amount of time spent preparing the photomap and thus its overall cost. This paper examines some techniques,aimed at improving the resolution and enhancing the appearance of the printed photomap,that are being developed within the Department.

1. INTRODUCTION

Photomaps have provided a challenge to cartographers ingenuity since the early 1960's. They offer means of producing maps quickly and at a negligable cost when compared with conventional line maps at a similar scale. Until the introduction of high resolution satellite imagery, photomaps were primarily produced at large to medium scales, i.e. up to 1:50 000, with 1:25 000 photomaps often being considered as the optimum scale (Southard,1975). This does not discount certain considerations that have been made concerning photomaps at small scales (1:250 000) of sparsely mapped areas but due partly to technical difficulties and partly to the cost of flying high altitude coverage, such schemes have rarely been implemented.

The ready availability of cheap, high quality satellite imagery has meant that cartographers can now produce small-scale topographic base maps from photographic images, for a variety of tasks. Many of these tasks may be in fields hitherto untapped by conventional photomaps such as atlas maps and bases for tourist and route planning maps.

Paper presented at a conference of Matching remote sensing technologies and their applications convened in London by the Remote Sensing Society, 16-18 December, 1981.

Investigations are currently taking place within the Department of Land Surveying aimed at producing low cost, coloured base maps from satellite imagery. During 1981, sufficient imagery has been obtained to cover the Principality of Wales. Following advice given by remote sensing specialists, photographic prints were obtained in Band 7 and one of those prints forms the basis for discussion within the paper.

2. METHOD OF REPRODUCTION.

If a reasonable quantity (usually in excess of 10) of copies are required of any map, it is more cost effective to print the map rather than produce it by techniques such as photography. Therefore, the image of the map must be transferred to a printing plate. With line detail, this is a simple process but when continuous tone images are involved, the tone must be broken down into a number of discrete units capable of being printed. The normal method is to produce a half-tone of the subject which reduces the negative into a regular array of dots, the size of each dot being inversely proportional to the amount of light reflected from the corresponding area within the continuous-tone subject. One advantage of the half-tone is that it is a widely accepted technique and the photographer and printer are able to exercise a high degree of control over its production and reproduction. The number of dots per cm on half-tones can vary from about 40 dots per cm to about 100 dots per cm. Obviously the greater the number of dots per cm, the finer the screen and the higher the resolution of the image. Research has shown that the optimum screen density for the reproduction of photomaps is in the region of 80 dots per cm (van Zuylen,1974). Any increase in screen density is offset partly by irregularities in the paper and partly due to difficulties in printing. Regrettably, however, the standard screen used to reproduce photomaps is 55 dots per cm. which means that a photomap originating from a photographic image having a resolution of about 450 lines/cm will drop to below 60 lines / cm.

Because of this grave disadvantage, research has been undertaken by a number of bodies concerning continuous tone printing. There are basically two approaches. The initial approach consisted of photographically modifying the continuous line image to separate it into a series of unsymmetrical random dots based upon a photographic emulsion. Research by the U.S.Geological Survey led to the introduction of phototoning (Clarke and Pumpelly 1968) in which a random dot was produced by placing a continuous tone negative and positive together and contacting them onto a high contrast film. The result is a high resolution image capable of being printed on conventional printing plates. A similar result has been obtained by photolysis in which a random dot is produced by the degeneration of silver halides caused by an over-exposure of a continuous tone negative onto high contrast film. (Eliakin 1974 and Whittington,1978). The second avenue of research lay in the development of a printing plate capable of printing continuous tone images(Keates 1978). This printing plate offered many advantages although has a basic disadvantage of being a specialised plate which will cause the problem of non-standardised working and the possibility of having to use unfamiliar chemistry.

With this in mind, attempts are being made within the Department of Land Surveying to directly reproduce a continuous tone image using conventional printing plates, with some degree of success. It has been found that providing the density range of a continuous tone negative is sufficiently compressed, it is possible to hold a continuous tone image within the grain structure of the printing plate. Great care must be exercised during printing and the best results have been obtained when the application of printing ink is kept to a minimum. Results have been quite satisfactory and an example of this printing may be found on Map.1 at the back of the paper, where it may be seen that the resolution is far superior to that of the 55 dot/cm and the 80 dot/cm on Map 3 and 2 respectively.

3. THE APPLICATION OF COLOUR.

The most economic form of photomap is one printed in black and white with the minimum amount of cartographic enhancement, that is, the minimum amount of extra detail such as names, grid, border, etc. The greater the degree of enhancement, the longer it will take to produce and the greater the cost of production.

In some instances, however, it is necessary to add colour, partly to make the map more attractive to the buyer but more importantly to make it easier to read and understand. The colour is usually added by scribing for line features and by cutting holdout masks in peelcoat materials to allow certain portions of the overall half-tone image to be exposed on the relevent printing plate. Both processes are time-consuming and require exacting checks to ensure that the peelcoat for adjacent colours do not overlap. A further problem associated with this approach is that coloured detail will have a hard, artificial edge to a feature that often has an arbitrary boundary on the ground.

Maps 1 and 2 show examples of photomaps based on Landsat imagery that have been reproduced in colour but without any mask being cut by hand. Indeed, the only manual operations involved in producing these maps were scribing the neat line of the sheet, setting the type and cutting the border mask to prevent a half-tone image from straying beyond the neat line. These were simple tasks that took about half an hour in total. The technique used to produce the colour masks was posturisation, a very simple photographic operation which enables a continuous tone negative to be separated into a number of positives, each representing an area of different density.

Initially, a continuous tone negative was produced at a scale of 1:1000 000 from the Landsat print. The possibility of loss of resolution being encountered by the process was outweighed by its relative low cost. The continuous tone negative was brought in contact with high contrast lithographic film and a short exposure was used to expose the highlight areas of the negative onto a lith positive. The length of exposure was sufficient to record only the water features and some limited areas of vegetation. By increasing the exposure time on successive pieces of lith film, the area recorded on the positive was gradually increased. By turning some of the positives into a negative form and by using combinations of positive and negative components it was possible to isolate out a series of density slices, whose outlines match , thereby covering the entire area of the image. The advantage of the density slices lies in their ease of production and their relative complexity of shape, being dependent upon natural features rather than arbitrary lines, resulting in a far more natural effect than is found on the majority of photomaps.

To produce masks of this complexity by normal methods would be a considerable task, possibly the entire process of identification, scribing and masking taking several working days. The masks used to produce Map 5 and 6 took $3\frac{1}{2}$ hours to produce, including the production of the half tone images found on Map 2 and 3.

Map 5 depicts the density sliced masks used in conjunction with a 80 lines/cm halftone screen whils Map 6 is produced using solid land colours and a slightly different arrangement of masks.

No attempt has been made to determine whether there is any correlation between the shape of mask, and thus density of the original image, with vegetation type or any other natural phenomena, although this may be done.

It is indisputable that similar results could be obtained using Agfa-Gavaert Equal-Density film but experience has shown that the chemistry has a short shelf life once mixed and unless a reasonable amount of work is undertaken at any one time, materials will be wasted. The principal advantage of posturisation lies in the materials used, which are commonplace and found in most mapmaking establishments together with its ease of use.

4. THE EDGE-LINE TECHNIQUE

A method of photographic reproduction that does not rely upon dots, either random or half-toned, is the edge-line technique which can be described as a method of photographic generalisation in which the edges of detail are isolated out as a series of lines. This is produced by initially making a lith positive of a continuous tone negative and subsequently a lith negative from the positive. If the negative and positive images are laid together in register, they should cancel one another out, but providing a controlled movement is made to one of the components, light will be transmitted around the edge of detail.

When this technique is applied to Landsat imagery (see Map 4) an excellent impression of the ground texture is produced that may be of assistance to specialist uses.

5. CONCLUSION

In conclusion I would like to draw participants attention to two quotations:-

"That if computers had been invented before photography we would now be looking upon photography as the great new way of storing and retrieving information". A.C.Marles.

and

"Photography is a magic thing ....
Nothing will ever be so much fun". Jacques-Henri Lartique.

It is realised that density tone can be separated out using a variety of pieces of specialised hardware. The paper attempts to outline certain photographic techniques that may be used by those who do not have access to such equipment and who wish to use Satellite imagery to produce attractive maps.

6. REFERENCES

CLARKE, C.A. PUMPELLY J.W.(1968). Procedures for reproducing image reproduction films for orthophotomaps. USGS Topo Division, Office of Research and Technical Standards.

ELIAKIM H (1974) Photomap Reproduction, ITC Journal, 1974(2),217-229

KEATES J.S.(1978), Screenless Lithography and Orthophotomaps. Cartograpic Journal 15(2),63-65

SOUTHARD R.B.(1975), Practical applications of photomaps, Commonwealth Survey Officers Conference, 1975, Paper E3

WHITTINGTON, RFC(1978), The production of urban orthophotomaps for the National Mapping Bureau of Papua New Guinea by the Royal Australian Survey Corps using the technique of random dot screenless photography. Proc.Int.Symp.Comm.IV. "New Technology for Mapping", Ottawa,610-629.

VAN ZUYLEN, L (1974). Some remarks about the development of the application and reproduction of photomaps during the last four years. International Yearbook of Cartography, 14, 174-182.

AN IMAGING MICROWAVE RADIOMETER

C.R. Francis
British Aerospace Dynamics Group
Bristol Division, Bristol England

ABSTRACT

The Phase A study of an Imaging Microwave Radiometer (IMR) has recently been completed, for ESA, by a team of contractors led by B.Ae. The instrument design which was established during this study will be described in this paper. The IMR is able to measure the intensity of radiation in the microwave band at a number of specific frequencies. By a complex data retrieval process, some of the physical characteristics of the atmosphere and of the underlying surface may be extracted from such measurements.

The instrument views a limited area of the Earth at any instant; this area is the Instantaneous Field of View (IFOV). By scanning the IFOV an image may be generated, and a conical scan is used so that the incidence angle at the ground remains constant.

ELEMENTS OF THE SYSTEM

A block layout of the system is shown in Figure 1.

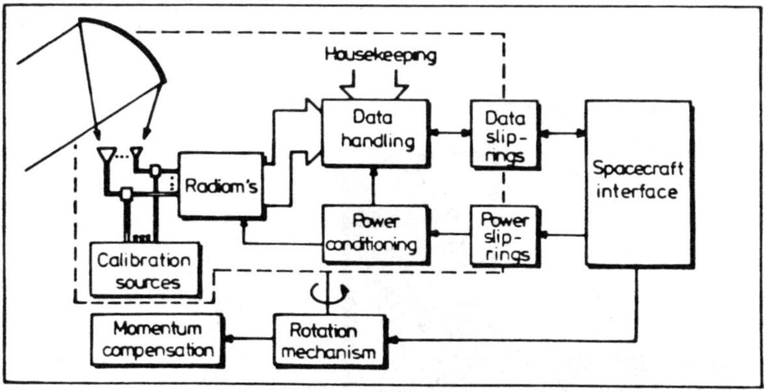

Figure 1: Block Diagram of the IMR

Paper presented at a conference on Matching remote sensing techniques and their applications convened in London by the Remote Sensing Society, 16-18 December 1981 (Also at IGARSS, Washington, June 1981)

Upwelling microwave radiation from the Earth is viewed by an offset paraboloid antenna. The depression angle of the beam (with respect to the local vertical) is 44°, so that the instrument observes microwave radiation emitted at an incidence angle of about 50°. The antenna feeds a cluster of single frequency horns. There are 10 of these covering 6 frequencies; duplication of the higher frequency horns allows multiple beams to be generated.

Periodic calibration is achieved by switching the radiometer receiver inputs from the feed horns to a calibration source. This is either a hot load, or a dedicated horn viewing cold space.

The antenna, the radiometer receivers, calibration sources, data handling and power conditioning subsystems rotate about a locally vertical axis, at a rate of 15 rpm. This enables the beam to scan a wide swath (up to about 1000 km at an altitude of 650 km, depending upon blockage of the beam produced by other payload elements) beneath the satellite, which, combined with the satellite's orbital motion, creates an image. The angular momentum of this rotating mass must be compensated by a momentum compensation device. If no such compensation is implemented, a constant torque would have to be applied to the roll axis to allow Earth viewing, by rotation of the spacecraft at the orbital rate.

The data output from each radiometer receiver is connected to the data handling unit. This formats the measurements into a data stream which, along with auxiliary and housekeeping data, is interfaced to the payload data handling system. This raw data will always be transmitted to Earth, but it is possible that some on-board processing could also be carried out to enable the broadcasting of geophysical information to small, low-cost ground stations.

The overall configuration of the IMR is shown in Figure 2, where the instrument is attached by a support structure to the spacecraft bus, the basis for which is the French Earth-observation satellite, SPOT.

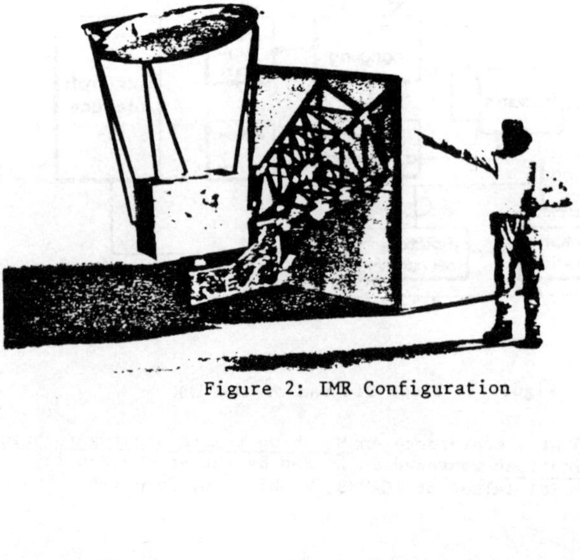

Figure 2: IMR Configuration

The IMR has the specification and performance given in Table 1. The radiometric resolution is quoted for two conditions: the integration time for all the channels is 5 ms (except at 90 GHz where it is 2.5 ms) so the radiometric resolution during this time is the value associated with each measurement. The radiometric resolution improves as the integration time increases, so it is also quoted for a value of 177 ms; this is the effective integration time associated with a radiometric measurement averaged over the footprint size at the lowest frequency.

Frequency (GHz)	6.84	10.65	15.30	23.80	36.5	90
Bandwidth (GHz)	0.25–1.0 (Variable)	1.0	1.0	1.0	1.0	4.0
Polarisation Channels	V & H	V & H	V & H	V & H	V & H	V&/orH
Radiometric Resolution (K) 5 ms	1.2	1.2	1.2	1.2	1.2	1.7[1]
Radiometric Resolution (K) 177 ms[2]	0.20	0.20	0.20	0.20	0.20	0.20
Spatial Resolution (km)	98x62	60x38	48x28	28x17	16x11	7x4
No. of Feed Horns	1	1	1	1	2	4

Orbit:	Sun-synchronous, 650 km altitude
Number of Channels:	11
Effective Antenna Aperture:	1.26 m
Antenna Beam Efficiency:	90% all frequencies
Sidelobe Isolation:	-26dB at all frequencies
Polarisation Isolation:	-24dB at all frequencies
Swath Width:	700 km (IMR, OCM, SAR) 1000 km (IMR, OCM, ALT)
Incidence Angle on Earth:	50°, Conical Scan
Radiometric Dynamic Range:	$0–313^{o}$ K
Mass:	108 kg without support structure (145 kg with structure)
Power:	197 W (265 W[3] peak); 32 W standby)
Data Rate:	52 kbs^{-1} (maximum)

NOTES:

1. Integration time of 2.5 ms
2. Equivalent integration time for a 98 km x 62 km cell
3. Peak power requirement occurs during spin-up

Table 1: IMR Specification

The swath width is quoted for two of the possible complements of instruments, which cause a different degree of blockage in the available field of view. The major impact is that of the SAR antenna; the Ocean Colour Monitor and Altimeter have negligible effect on the field-of-view.

ANTENNA

The IMR antenna is an offset paraboloid reflector with a cluster of single frequency corrugated feed horns at the prime focus. There are 10 feed horns, one associated with each dual receiver unit and four 90 GHz feeds. The geometry of the reflector is shown in Figure 3; the feed-plane tilt of 4° ensures that the offset angle is sufficiently small to give good polarisation isolation.

Beam diameter: 1.26 m
Cross-polar isolation: -25 dB
Focal length: 1 m
Offset: 40°
$\theta^{*} = 31^{o}$
Feed plane tilt: 4°
Antenna dimensions
1.26m x 1.35 x 10cm

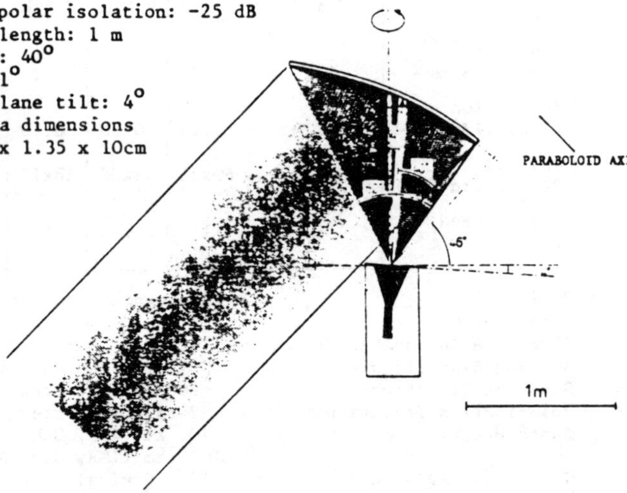

Figure 3: Geometry of the IMR Reflector

The co-polar pattern generated by the feed cluster on the ground is shown in Figure 4. The contours are at -3, -6, -10 and -20 dB with respect to the peaks. The spatial resolution of the instrument is defined by the -6 dB contours, following work at the Technical University of Denmark (TUD) on the physical significance of the definition of spatial resolution.

The reflector is made from aluminium honeycomb, 25 mm deep, with 0.2 mm thick CFRP facing sheets. An aluminised reflecting surface is applied to the reflector in order to ensure adequate reflectivity at the higher frequencies and this surface requires a high standard of finish. Analysis has shown that, for satisfactory performance at 90 GHz, surface imperfections must have amplitude of less than 0.1 mm rms, over small correlation lengths.

The reflecting surface of the antenna is coated with white paint to ensure good thermal coupling with the Earth and, therefore, a tendency to even temperatures. The rear surface is covered with multi-layer insulation, to isolate this surface from solar illumination and radiation to space.

The antenna temperature variation, during one orbit, is from -40°C to -5°C. The resulting thermal distortions slightly degrade the antenna electrical performance; but analysis has shown that this degradation is insignificant.

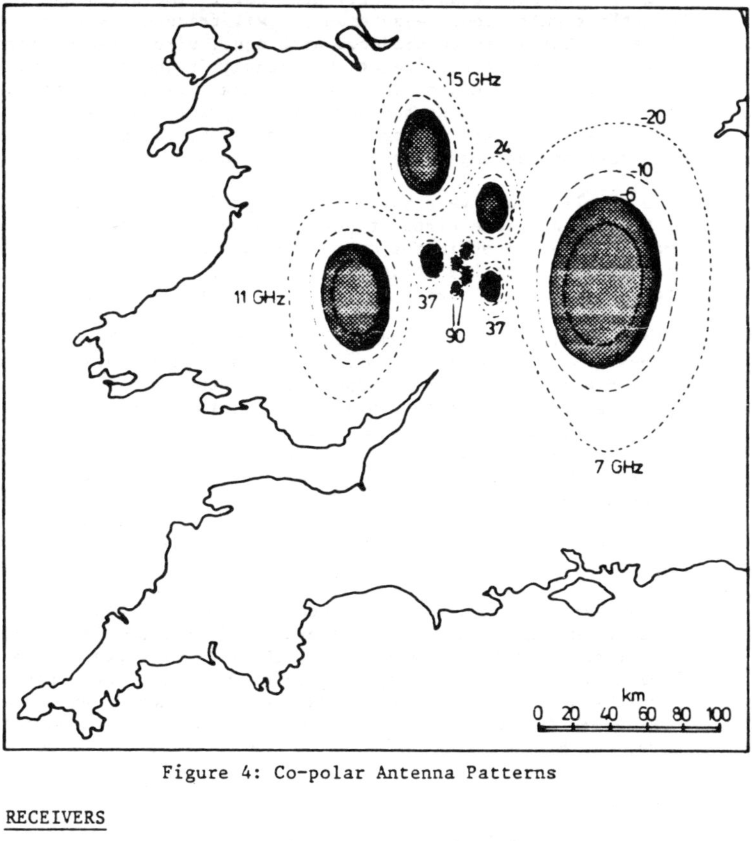

Figure 4: Co-polar Antenna Patterns

RECEIVERS

The radiometer receivers are of the Noise Injection type, though they may also be operated as Dicke radiometers.

A Dicke radiometer makes two measurements (the input signal and an internal reference source) and switches between them at a frequency of about 1 or 2 kHz. This is done to compensate for gain variations in the receiver. The Noise Injection Radiometer takes this principle further, in that the difference between the two signals drives a pulsed noise source which injects noise pulses into the input to bring the two measurements to the same level. The number of noise pulses injected then provides the measurement of input power.

An integration and sampling time of 5 ms is used for all frequencies (except 90 GHz where it is 2.5 ms). This value gives adequate protection against alising errors at 36.5 GHz and represents considerable over-sampling at the lower frequencies and, therefore, a higher data rate than strictly necessary. The advantage of having exactly simultaneous measurement at all frequencies is great, however, and an image produced with equal pixel size - corresponding to the footprint size at the highest frequency - contains much more structural information than an image with conventional pixel sizes. Integration to the correct footprint size, with correspondingly improved radiometric resolution, may be carried out at the data processing stage without any loss of information.

The baseline receivers have been designed with a bandwidth of 1 GHz. Under some circumstances, however, external interference may be experienced in the design so that the bandwidth may be reduced, by switched steps on telecommand. This increases immunity to external interference at the expense of radiometric sensitivity.

The noise injection radiometer, in common with other radiometer types, is very sensitive to its physical temperature. It has been possible to maintain the temperature of the receiver boxes to within a degree of a constant value, during each orbit. This is achieved by a passive thermal control system, an important feature of which is thin films of Mylar over the feed horn and sky horn apertures. In addition, the temperatures of the critical RF components, including the feed horns and other items outside the receiver boxes, will be measured to an accuracy of a few tenths of a degree by a digital thermometer. This will enable some temperature compensation to be applied.

Long term calibration of the receivers is based upon two known reference temperatures. These are separate from the temperature reference within the receiver, which is at the receiver ambient temperature. The two long term references are a cold reference (cold space at 3 K, viewed by dedicated sky-horns) and a warm reference (a section of terminated waveguide at a known temperature, which must be 10-20° K below the internal reference).

MECHANISMS

The mechanisms subsystem has three units:
- Rotation mechanism
- Drive electronics

- Momentum wheel

The rotation mechanism is derived from the Medium Speed Mechanism which has been developed under ESA contract at Dornier. The motor is a brushless DC torquer, which is commutated by a brushless resolver. The motor-windings are on the non-rotating inner section, while the outer section carries the permanent magnets.

Signals and power are transferred from the rotating to the fixed sections via slip-rings - 5 power channels with 4 brushes per channel and 24 signal channels.

Momentum compensation is provided by an off-the-shelf momentum wheel. Dimensional and performance requirements could have been fulfilled by at least 3 different wheels, manufactured by B.Ae, Teldix and Aerospatiale. The Teldix wheel was assumed in the baseline design.

DATA HANDLING

The data handling subsystem of the IMR employs a relatively new ESA-standard concept. The radiometric measurements from the recievers are processed by a set of digital data processing units and this information is multiplexed on to a single serial bus. Similarly, internal housekeeping measurements, in particular, measurements of the physical temperature of critical components, are digitised and this also is multiplexed on to the serial bus. This is achieved by a number of devices distributed through the instrument, known as Remote Terminal Units, or RTU's.

All the data emanating from the IMR can, therefore, be transferred across the rotating joint by two slip rings; though four will be used for redundancy. A Central Terminal Unit in the payload data handling subsystem or the fixed section of the IMR, will interface with the signals from the slip rings.

The data rate from the instrument will be at 52 kbs^{-1}. Since only a part of each scan is used to take measurements, however, buffering and slowing down of the data stream would be possible.

POWER

Raw d.c. power from the spacecraft bus is supplied direct to the mechanism drive electronics. Raw d.c. power is separately supplied, via redundant slip rings, to the power conditioning unit which is on the rotating part of the IMR, from where it is distributed to the remaining subsystems. Minimum use of the power transfer slip rings is therefore made.

In the standby mode, rotation of the instrument is maintained while the other subsystems are switched off. This has a number of advantages:

- For intermittent operation within an orbit, the 5 minute

start-up and slow-down period (at peak power) is avoided.

- Heat is dissipated in the mechanism, the unit most sensitive to cold, without the need for heating elements.

- The even illumination of the rotating portion of the instrument is maintained.

CONCLUSIONS

Some of the details of the design of an Imaging Microwave Radiometer have been presented. The instrument was designed to form part of the payload of the ESA spacecraft ERS-1, due to be launched in 1986, and it should provide useful contributions to the disciplines of oceanography, meteorology and glaciology.

ESTIMATION OF SUGAR BEET PRODUCTIVITY FROM REFLECTION IN THE RED AND INFRARED SPECTRAL BANDS

M.D. Steven
University of Nottingham School of Agriculture,
Sutton Bonington, Loughborough LE12 5RD

and

P.V. Biscoe and K.W. Jaggard
Broom's Barn Experimental Station, Higham,
Bury St Edmunds, Suffolk IP28 6NP

Abstract: The growth of healthy green crops is proportional to the solar radiation intercepted and the fraction of light intercepted by the crop f is the major source of variation in crop yields. A two band spectrophotometer was used to measure solar radiation reflected from crops of sugar beet in the red (600-660 nm) and the near-infrared (780-940 nm) bands. The ratio of infrared/red reflected fluxes increased with the percentage of the ground covered by healthy green leaves and was used to estimate f. On the assumption that the efficiency with which crops convert radiant energy to biomass ε is constant, such measurements were used to determine future rates of growth and predicted the yield of an independently grown crop to within 6%. Preliminary evidence suggests that spectral ratios may also be sensitive to changes in ε and the extension of these techniques to crops limited by disease or environmental stress is discussed.

1. REFLECTION SPECTRA OF CROPS

Many of the applications of remote sensing in agriculture exploit the fact that sunlight reflected from green leaves has a characteristic spectrum, distinctly different from that of soil (Figure 1). The near-infrared band is reflected strongly by leaves due to discontinuities of refractive index in the mesophyll structure (Gaussman, 1974). Reflectivity is low in the visible band because of absorption by pigments within the cell, especially chlorophyll (Heath, 1969). In contrast, soils show a more gradual increase of reflectivity across the spectrum. The spectral signature of a crop changes as it grows, from that of soil to a spectrum more characteristic of a leaf.

Simple ratios or other combinations of the red (R) and infrared (IR) bands have been used to estimate vegetation density (Tucker, 1979; Curran, 1980). However, while there is general agreement on the appropriate wavebands, there is no consensus on what feature of the vegetation should serve as the "ground truth". Strong correlations have been found with leaf area index, leaf cover, biomass, leaf water content, chlorophyll

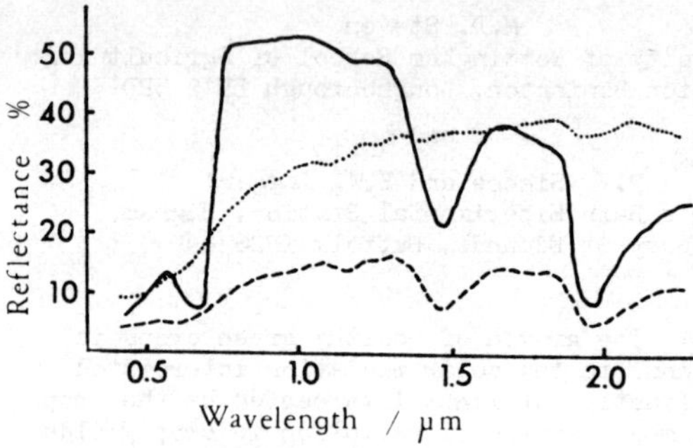

Fig. 1. Spectral reflectance of a green wheat leaf (solid curve) compared with the reflectance of a dry (dotted curve) and a moist (dashed curve) sandy loam soil : From Bunnik (1978).

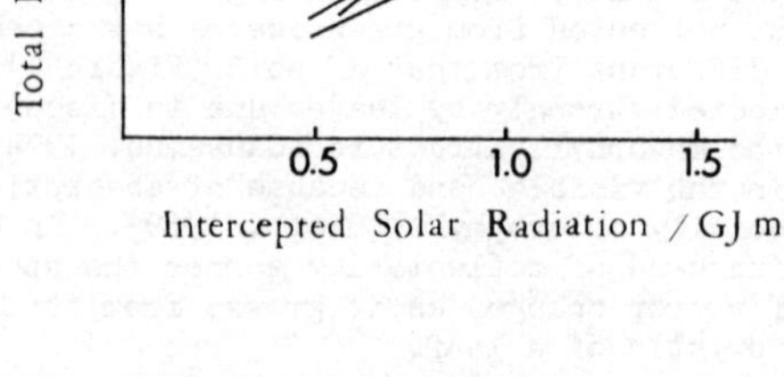

Fig. 2. The relation between total dry matter at harvest and radiation intercepted by foliage throughout the growing season for various crops: a - apples; b - barley; p - potatoes; sb - sugar beet. From Monteith (1977).

content and other related factors (Colwell, 1974; Tucker, 1977; Berg et al, 1978; Tucker et al, 1979; Holben et al, 1980; Curran, 1980). These variables tend to be highly correlated in the field and the relationships found with reflectance spectra are dependent on the form of the correlations. Kumar and Monteith (1981) related the IR/R spectral ratio to measurements of the fraction of the available light intercepted by crops, and showed on theoretical grounds that the relationship is approximately linear. However this linearity depended critically on the reflection and transmission coefficients of leaves in the two spectral bands. In crops that are limited by disease or environmental stress these coefficients may differ and change the relationships between reflection spectra and the various indices of vegetation. None of the indices that have been used as ground truth effectively combines leaf area (however expressed), which determines the infrared reflection, with chlorophyll concentration which determines the red reflection. This paper describes a simple model of crop growth in which IR/R spectral ratios have been used to predict the growth and yield of a crop of sugar beet. An extension of these techniques, with an appropriate choice of ground truth indices relating both to the red and the infrared properties of foliage, is proposed for the investigation of crops limited by stress or disease.

2. SOLAR RADIATION AND PRODUCTIVITY

The basis of all crop growth is photosynthesis - the process of conversion of radiant to chemical energy in plants. A number of recent studies have related the crop productivity - the rate of increase with time t of biomass w per unit ground area - to the fraction f of the incident solar irradiance S intercepted by the crop. (Milford et al, 1980; Gallagher and Biscoe, 1978a). Denoting productivity by $\frac{dw}{dt}$ the relationship may be expressed in the form

$$\frac{dw}{dt} = \epsilon f S \qquad (1)$$

This equation defines ϵ as the net efficiency of conversion of solar energy to biomass. There is now substantial experimental evidence from these studies to confirm that ϵ is relatively stable and that in healthy green crops not limited by disease or water stress it has a value in the range 1.5 to 2 g MJ^{-1}. Indeed, when taken over the whole growing season, Figure 2 shows that the net production per unit intercepted radiation is remarkably similar for a variety of crops in the U.K. (Monteith, 1977). Differences from year to year may be due either to variation in S or variation in f. Fourteen years of solar radiation records at Broom's Barn (52° 16'N, 0° 35'E) show that irradiation totals over the growing season for sugar beet (April-October) varied at most by $\pm$ 12.5% from the mean. However, the major variable in the productivity equation is the fraction of the available light intercepted by leaves. This fraction depends on the history of leaf production, expansion and death and is governed by a complex of environmental, management and genetic factors. The chief environmental factor is temperature which determines the initiation and expansion of leaves (Milford and Riley, 1980; University of Nottingham, 1981), but the final size may be limited by nutrient shortages or water stress (Bunce, 1977; Boyer; 1968; Ma and Hunt, 1975; Loomis and Nevins, 1963). Much of the benefit obtained from the use of fertilisers is a result of increased leaf growth enhancing the interception of radiant (Scott and Jaggard, 1978). Consequently, the development of f is difficult to predict but it may be measured using tube solarimeters (Szeicz, 1974) or derived from measurements of leaf area (Varlet-Grancher and Bonhomme, 1979;

Whitfield and Connor, 1980). Kumar and Monteith (1981) showed that f may also be determined remotely from the IR/R reflection ratio and with this information equation 1 can be integrated with time to predict the future growth of the crop and the ultimate dry weight. In many crops, while yield varies widely, the harvest index - the economically useful fraction of total dry matter - varies less than 10% of its mean value (Milford et al 1980; Gallagher and Biscoe, 1978b), allowing an accurate estimation of the final yield.

3. TECHNIQUES AND CROPS

A Ratio Spectrophotometer built by Macam Photometrics Ltd was used in this study. A fibre-optic probe receives the light input and splits it into two pathways, feeding the light through filters to a pair of silicon diode detectors. The half-power bandwidths for each filter-detector combination are 600-660 nm and 780-940 nm respectively. An electronic unit amplifies the signals and computes their ratio. Two interchangeable probes limit the field-of-view (f.o.v.) of the instrument to either 45° or 10°. The spectrophotometer is powered by rechargeable batteries and is fully portable.

In aerial use, the instrument was fixed in a multi-camera mount and positioned over a hole in the floor of the survey aircraft, a Cessna 206 provided by the Aerial Photography Unit, ADAS. The cameras and the printing of spectral ratios were triggered simultaneously by a timer unit. Flight passes were made at about 150 m altitude using the 10° f.o.v. device to restrict the diameter of the detected area to about 26 m. The spectral ratio measurements were normalised with respect to a Kodak grey reference card to compensate for changes in the spectral quality of the incident light and to permit comparisons with other instruments (Milton, 1980). Measurements were made in 1980 and 1981 over crops of sugar beet (*Beta vulgaris* L) in the vicinity of Broom's Barn.

The same fields were studied from the ground with the instrument mounted on the arm of a portable gallows at a height of about 2 m. The 45° f.o.v. device was used to give an average reading over a circular area of about 1.5 m diameter. Similar measurements made over a variety of other field crops will be reported elsewhere.

The ground truth for this study was based on measurements of light interception. Sets of three tube solarimeters in series were installed at the base of the crop early in the season. Two replicate sets were used in 1981, but only one set in 1980. The solarimeters were arranged across the rows of the crop so that their combined length completely traversed 6 rows. A single tube solarimeter above the crop recorded the incident solar irradiation. The output from the solarimeters was continuously accumulated on an electronic integrator (Delta-T Devices) which was read daily. From these measurements the transmitted fraction of radiation τ was calculated, and f was assumed to be equal to $1-\tau$. Measurements of light interception were supplemented by fractional leaf cover determined from photographs taken looking vertically downwards over the crop. We have evidence to show that light interception is proportional to leaf cover and that, at least in sugar beet, the relationship is almost 1 to 1.

The crops grown in the vicinity of Broom's Barn were on the research station's farm in 1980 and, in 1981, on a commercial farm at Kennett, 8 km West of Broom's Barn. Both crops were sown in early April with seeds spaced 16 cm apart in rows separated by 51 cm. The resulting seedlings were not thinned, but were left to establish a satisfactory stand of 75,000 to 85,000 plants/ha. In 1980 the variety was 'Nomo' and in 1981 'Amono'; the recommended amounts of fertiliser were applied to both crops and neither showed many symptoms of deficiencies or water stress. Aqueous suspensions of sulphur were sprayed at the recommended dose to control the fungal disease powdery mildew (Erysiphe betae).

To test our methods of yield prediction a crop was deliberately sown late in 1980 at Sutton Bonington (52° 50'N, 1° 15'W) to provide a very different pattern of leaf growth and biomass production during the summer. The sowing was made with the variety 'Vytomo' on 4 June. Seeds were spaced 7 cm apart in rows separated by 51 cm and the seedlings were thinned by hand on 2 July to produce a regularly spaced stand of 80,000 plants/ha. The crop was fertilised to ensure an adequate supply of nutrients.

Throughout the summer and autumn, growth of the crops at Sutton Bonington and Kennett was measured by making frequent harvests. Four replicate samples, five at Kennett, each of approximately 10 plants, were taken at every harvest, and were oven-dried and weighed to determine total biomass. On 7 December 1980 replicate samples, each of 12 m^2 were harvested and the late-sown crop at Sutton Bonington produced 11.9 t/ha of dry matter of which 6.8 t/ha was storage root and 5.15 t/ha was sugar. In 1981 the crop at Kennett produced 23.5 t/ha of dry matter of which 12.9 t/ha was sugar when lifted on 26 October.

4. RESULTS

Measurements of spectral ratios of light reflected from two crops of sugar beet are shown in Figure 3 as a function of light interception. The spectral ratios are averages of 20 values measured in a traverse across the field at ground level. The flight data are very similar but rather more scattered. The lines join consecutive measurements except for the first point in 1980 which was a measurement made later in the season over bare local soil. The dates of measurement are included with these data in Table 1. In both years spectral ratios initially increased with light interception but tended to fall later in the season after the maximum level of light interception had been reached. In the 1980 crop, this reversal occurred in August but it was delayed until October in 1981. The two points from July 1980 lie above the general trend of Figure 3, but the relationship with leaf cover measurements (Figure 4) is more consistent with 1981 and suggests that there may have been a systematic error in the light interception data of 1980, obtained from only one set of tube solarimeters. The 1981 data are well fitted by the equation:

$$f = 36.9 \ln r - 3.53 \qquad (2)$$

Where r is the IR/R spectral ratio. This equation was derived by linear regression using the first 8 points of 1981 and accounts for over 98% of the variance. Neglecting for the moment the systematic change in relation between f and r in the late autumn, f can be derived from spectral data with equation 2 and integrated in equation 1 to estimate the production of the crop. This exercise was done retrospectively for the crop of sugar beet grown at Sutton Bonington in 1980. The value chosen for ε was 1.63 g MJ^{-1} which was the value measured in the 1981 Kennett crop from which equation 2 is derived. Daily records of the incident solar irradiance were obtained from the meteorological station at Sutton Bonington and values of f were calculated from measured spectral ratios. Spectral ratio

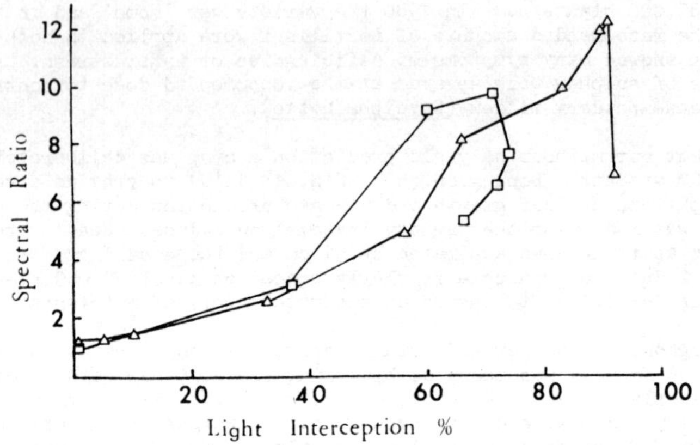

Fig. 3. The relation between IR/R spectral ratios and the fraction of light intercepted in two crops of sugar beet: 1980 (squares) and 1981 (triangles).

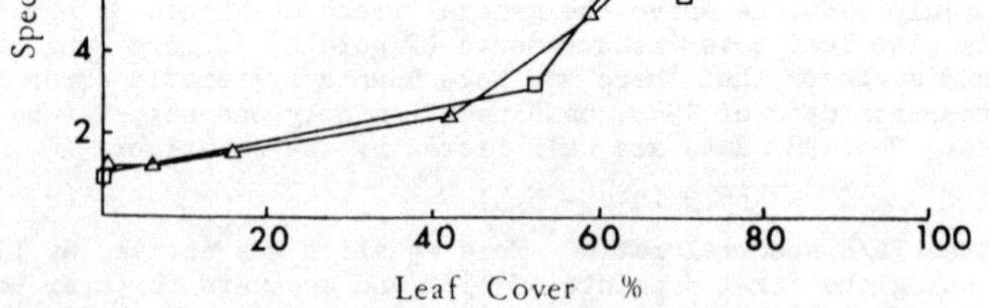

Fig. 4. The relation between IR/R spectral ratios and the fraction of the ground covered by leaves of sugar beet: 1980 (squares) and 1981 (triangles).

Table 1

Spectral ratios and ground truth data in sugar beet: 1980 and 1981

Date	Spectral ratio IR/R	Light interception %	Leaf cover %
1980			
23 June	3.17	37	52
4 July	9.17	60	74
24 July	9.62	71	79
12 August	7.58	74	86
27 August	6.41	72	87
27 August	1.00	0	0 (soil)
30 October	5.26	66	70
1981			
19 May	1.24	0.5	0.6
3 June	1.23	5	6
10 June	1.43	10	16
17 June	2.36	33	42
29 June	4.83	56	59
9 July	8.13	66	75
27 July	9.90	83	84
20 August	11.76	90	90
24 September	12.0*	91	90
23 October	6.8+	92	96

* estimated from flight data

\+ normalised with average grey card data from previous dates measurements in the first 6 weeks of growth were affected by weeds and tended to overestimate light interception by the sugar beet. However, application of a correction for the weed leaf cover, measured from photographs, reduced the predicted dry matter by a trivial 0.15 t ha^{-1}. The data were integrated in equation 1 without the weed correction and the predicted growth of the crop is compared in Figure 5 with the production of dry matter measured at intervals through the growing season. The model overpredicted dry matter at harvest by only 6% (0.7 t ha^{-1}), a figure which is well within the statistical error of the dry matter determinations.

5. DISCUSSION

The success of this technique to predict crop production depends on the availability of incident solar radiation data and on the predictability of the efficiency ϵ of the foliage. Work elsewhere has shown that IR/R spectral ratio measurements made early in the season are highly correlated with the total dry matter accumulated at harvest time in rice (Berg et al, 1978) and in winter wheat (Tucker et al, 1981). In both these studies regressions of accumulated dry matter on spectral ratio were used to explain variations in the yield of neighbouring plots grown during the same season. Thus there were no differences in the incident solar radiation and environmental stresses limiting the efficiency of production would have been very similar in all the plots. Such regression formulae are unlikely to be stable when applied to crops grown in different locations or in different years. Moreover, Tucker et al noted that their regression equations changed with the date of the spectral measurement and to overcome this effect they related the accumulated dry matter at harvest to the integral with time of the measured spectral ratios. Such an approach is superficially similar to that adopted in this study and would be directly comparable if two assumptions were valid:

(a) That spectral ratios are directly proportional to the light interception in the crop, i.e. $f = k\,r$ where k is a constant;

(b) That the incident solar irradiance was constant throughout the growing season.

Our analysis predicts dry matter production from the integral of Equation 1, i.e.

$$w = \int \epsilon\, f\, S\, dt$$

$$\text{or} \qquad w = \epsilon \int AS\, dt$$

if ϵ is constant as we assume.

Assumptions (a) and (b) would allow us to continue the simplification to

$$w = \epsilon\, S\, k \int r\, dt$$

and the slope of the regression line in the work of Tucker et al would be equal to $\epsilon\, S\, k$. However these assumptions are not valid. The most serious error in (a) occurs when f is small because of the positive intercept on the r axis shown in Figure 3. Beyond about 50% interception the assumption is reasonable. Assumption (b) is seriously in error because the crop was grown from winter to mid summer when the incident solar radiation is generally increasing. Both errors imply that the integral approach of Tucker et al places excessive weight on early measurements of spectral ratios. This argument helps to explain why the correlation coefficient with the integral approach was lower than the coefficients

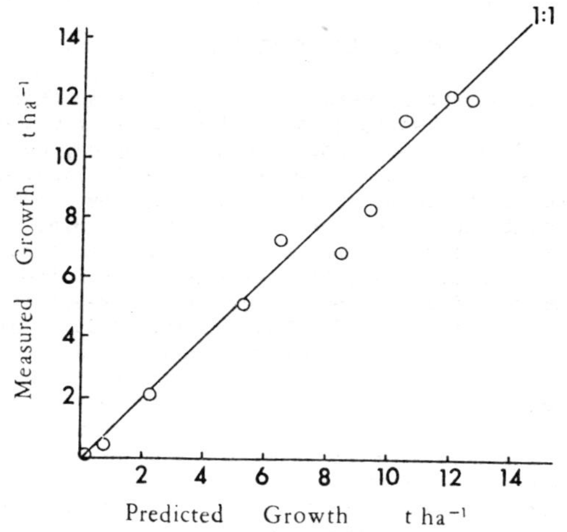

Fig. 5. Measured versus predicted growth for a crop of sugar beet.

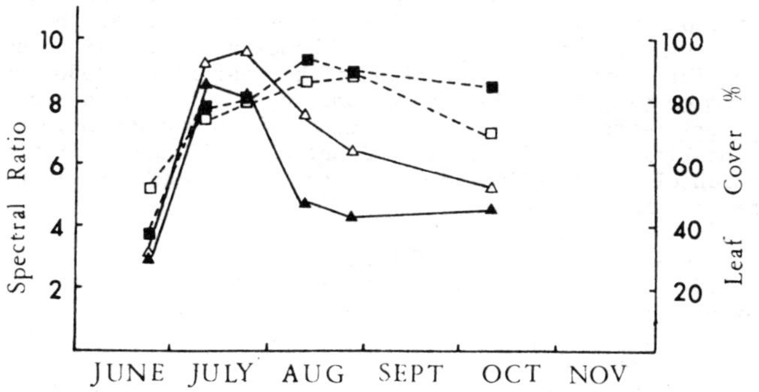

Fig. 6. The change with time of measured spectral ratios (triangles - solid lines) and leaf cover (squares - dotted lines) in sugar beet in 1980: healthy areas (open points) and areas infected with Yellow Virus (black points).

derived from single spectral ratios measured in mid-season when most of the dry matter was produced.

6. FUTURE APPLICATIONS

The techniques in this study were developed on crops that were maintained free of disease and serious water stress. In terms of the present analysis, the effect of such stresses on a crop may be to reduce leaf area and light interception or to decrease the efficiency of conversion of solar energy to biomass, or both. If only the first effect applies, then the present techniques are directly applicable, but changes in the efficiency would require a different approach. Short term changes in ε may result from reversible physiological reactions of the plants such as closure of the stomata which frequently occurs as a response to temporary shortage of water (Turner, 1974; Ishihara et al, 1979) or atmospheric pollution (Unsworth and Black, 1981). Other stresses or disease may cause longer-term changes in efficiency through loss of chlorophyll from the leaves and a consequent reduction in their ability to use the available light (Tsunoda, 1972; Takano and Tsunoda, 1971). Measurements of efficiency of the crop are presently derived over several weeks by regression of the rate of production of dry matter on accumulated intercepted solar radiation and incorporate both long and short term effects. Existing remote sensing techniques are not well suited to study the short term changes, but it may be possible to measure the more serious longer-term variations of efficiency. The red reflectance of foliage is largely determined by chlorophyll content and other studies have shown that red and near-infrared band combinations are well correlated with the percentage of chlorotic leaves in the crop canopy (Tucker et al, 1979). If the efficiency of photosynthesising leaves is directly related to their chlorophyll content and if chlorophyll loss is the main source of variation in ε during a season, then the present IR/R spectral ratios may in fact be more closely related to the product $f\,\varepsilon$ than to f alone. This may be the explanation for the change in the relation between f and r shown in Figure 3 which occurred at about the same time as other crops of sugar beet have shown a decrease of ε (Biscoe et al, 1980). This hypothesis has yet to be tested directly but the accuracy of prediction of biomass in the 1980 crop at Sutton Bonington (Figure 5) does give some support to the idea since there was no _a priori_ reason why the efficiency of two crops grown in different years and localities should have been identical. It may be that errors in the prediction of interception were compensated by differences in the efficiencies of the two crops.

Further evidence is seen in Figure 6 which shows the time course of spectral ratios and leaf cover in two areas of the 1980 sugar beet crop. The open symbols show the trends in a healthy area of the crop while the filled symbols denote a plot infected with Virus Yellows, a disease which causes extensive leaf chlorosis. Figure 6 shows that the spectral ratios in the infected area fell considerably below the values measured in the healthy crop, in spite of similar or slightly higher leaf cover. The original data indicate that greater reflection in the red band was the main cause of the lower spectral ratios in the diseased crop.

In future work, attempts will be made to relate measurements of red and infrared band reflection both to light interception and to the efficiency of the foliage. However our present technique for measuring efficiency requires a long period of growth so that substantial increases in dry matter can accrue. A different ground truth index is required for the investigation of shorter term changes. A suitable surrogate measurement may be the chlorophyll content of leaves if its relationship with ε can be established. Future refinements in remote sensing and in ground truth techniques may allow us to independently determine both f and ε from spectral data. Indeed, with an appropriate choice of spectral bands it may even be possible to use multispectral reflection measurements to locate the occurrence of stress, to identify the cause and to estimate the resultant loss of yield.

7. ACKNOWLEDGEMENTS

The authors thank John Monteith, Mahendra Kumar and Keith Scott for their advice, Chris Clark, Graham Moore, Donald Monteith and Margaret Green for technical assistance and the Aerial Photography Unit, ADAS Cambridge for supplying the aircraft and for aerial photography. This work was supported by a grant from the ARC.

8. REFERENCES

BERG, A., FLOUZAT, G. and DE PARATESI, S.G. (Ed.) (1978) Agricultural resources investigations in Northern Italy and Southern France. AGRESTE Project Final Report, Commission of the European Communities, Ispra.

BISCOE, P.V., DRAYCOTT, A., EDWARDS, P., GLAUERT, W., JAGGARD, K., MESSEM, A., LAST, P. and MILFORD, G. (1980) Rothamsted Experimental Station, Annual Report 1980, Part I, pp. 71-72.

BOYER, J.S. (1968) Relationship of water potential to growth of leaves. Plant Physiol. 43, pp. 1056-1062.

BUNCE, J.A. (1977) Leaf elongation in relation to leaf water potential in soybean. J. exp. Bot. 28, pp. 156-161.

COLWELL, J.E. (1974) Grass canopy bidirectional spectral reflectance. Proc. 9th Intern. Symp. on remote sensing of the environment, Vol. II, pp. 1061-1085, Univ. of Michigan, Ann. Arbor.

CURRAN, P. (1980) Multispectral remote sensing of vegetation amount. Progr. phys. Geogr. 4, pp. 315-341.

GALLAGHER, J.N. and BISCOE, P.V. (1978a) Radiation absorption, growth and yield of cereals. J. Agric. Sci., Camb. 91, pp. 47-60.

GALLAGHER, J.N. and BISCOE, P.V. (1978b) A physiological analysis of cereal yield II, partitioning of dry matter. Agric. Progr. 53, pp. 51-70.

GAUSSMAN, H.W. (1974) Leaf reflectance of near-infrared, Photogramm. Engng. 40, pp. 183-191.

HEATH, O.V.S. (1969) The physiological aspects of photosynthesis. Heinemann, London.

HOLBEN, B.N., TUCKER, C.J. and FAN, C.J. (1980) Spectral assessment of soybean leaf area and leaf biomass. Photogramm. Engng. and Remote Sens. 46, pp. 651-656.

ISHIHARA, K., IIDA, O., HIRASAWA, T. and OGURA, T. (1979) Relationship between nitrogen content in leaf blades and photosynthetic rate of rice plants with reference to stomatal aperture and conductance. Japan. J. Crop Sci. 48, pp. 543-550.

KUMAR, M. and MONTEITH, J.L. (1981) Remote sensing of crop growth. Presented at the British Photobiology Society Symposium, Leicester University, January 1981, Proceedings in preparation.

LOOMIS, R.S. and NEVINS, D.J. (1963) Interrupted nitrogen nutrition effects on growth, sucrose accumulation and foliar development of the sugar beet plant. J. Am. Soc. Sug. Beet Technol. 12, pp. 309-322.

MA, P. and HUNT, L.A. (1975) Photosynthesis of newly matured leaves during the ontogeny of barley grown at different nutrient levels. Can. J. Bot. 53, pp. 2389-2398.

MILFORD, G.F.J. and RILEY, J. (1980) The effects of temperature on leaf growth of sugar beet varieties. Ann. appl. Biol. 94, pp. 431-443.

MILFORD, G.F.J., BISCOE, P.V., JAGGARD, K.W., SCOTT, R.K. and DRAYCOTT, A.P. (1980) Physiological potential for increasing yields of sugar beet. In 'Opportunities for increasing crop yields', Ed. R.G. Hurd, P.V. Biscoe and C. Dennis. Pitman, London, pp. 71-85.

MILTON, E.J. (1980) A portable four-band "Landsat" radiometer for ground data collection in remote sensing. Int. J. Remote Sens. 1, pp. 153-165.

MONTEITH, J.L. (1977) Climate and the efficiency of crop production in Britain. Phil. Trans. R. Soc. Lond. B 281, pp. 277-294.

SCOTT, R.K. and JAGGARD, K.W. (1978) How the crop grows - from seed to sugar. Brit. Sug. Beet Rev. 46, pp. 19-22.

SZEICZ, G. (1974) Solar radiation in crop canopies. J. Appl. Ecol. 11, pp. 1117-1156.

TAKANO, Y. and TSUNODA, S. (1971) Curvilinear regression of the leaf photosynthetic rate on leaf nitrogen content among strains of Oryza species. Japan. J. Breed. 21, pp. 69-76.

TSUNODA, S. (1972) Photosynthetic efficiency in rice and wheat. In 'Rice Breeding', International Rice Research Institute, Los Banos, Philippines, pp. 471-482.

TUCKER, C.J. (1977) Spectral estimation of grass canopy variables. Remote Sens. Environ. 6, pp. 11-26.

TUCKER, C.J. (1979) Red and photographic infrared linear combinations for monitoring vegetation. Remote Sens. Environ. 8, pp. 127-150.

TUCKER, C.J., ELGIN, J.H. and McMURTREY, J.E. (1979) Temporal spectral measurements of corn and soybean crops. Photogramm. Engng. and Remote Sens. 45, pp. 643-653.

TUCKER, C.J., HOLBEN, B.N., ELGIN, J.H. and McMURTREY, J.E. (1981) Remote sensing of total dry matter accumulation in winter wheat. Remote Sens. Environ. 11, pp. 171-189.

TURNER, N.C. (1974) Stomatal response to light and water under field conditions. R. Soc. N.Z. Bull. 12, pp. 423-432.

UNIVERSITY OF NOTTINGHAM (1981) Effects of temperature and atmospheric humidity on the development and growth of Pearl Millet. Report No. 5 (R 3208), Overseas Development Administration Microclimatology Unit, University of Nottingham.

UNSWORTH, M.H. and BLACK, V.J. (1981) Stomatal responses to pollutants. In 'Stomatal Physiology', Ed. P. Jarvis and T. Mansfield, Cambridge University Press, pp. 187-204.

VARLET-GRANCHER, C. and BONHOMME, R. (1979) Application aux couverts végétaux des lois de rayonnement en milieu diffusant II, Interception de l'energie solaire par une culture. Annls Agron. 30, pp. 1-26.

WHITFIELD, D.M. and CONNOR, D.J. (1980) Penetration of photosynthetically active radiation into tobacco crops. Aust. J. Plant Physiol. 7, pp. 449-461.

BUNNIK, N.J.J. (1978) The multispectral reflectance of short wave radiation by agricultural crops in relation to their morphological and optical properties. Communications Agricultural University, Wageningen 78-1.

MICROWAVE RADIOMETERS FOR REMOTE SENSING IN AGRICULTURE

P. Pampaloni
Institute of Environmental Analysis and
Remote Sensing for Agriculture-IATA-CNR
Firenze - Italy

Abstract: The use of microwave radiometry for remote sensing is increasing in many countries. As far as Agriculture is concerned, the first and the most important application has so far been moisture measurement of bare and vegetated soil. Soil moisture is in fact a very useful parameter to predict agricultural productivity. Nevertheless a direct knowledge of other physical parameters of the air-soil interface,with or without vegetation is very important in view of evapotranspiration prediction. Different layers of soil and vegetation can be studied by using radiometers operating at different frequencies, polarizations and observation angles. In order to investigate this problem and after some preliminary experiments with microwave radiometers for radioastronomy, two new instruments at 9.8 and 36.6 GHz have been designed to operate specifically for ground-based remote sensing. These radiometers were first used in June-July 1981 during the ESA-JRC Campaign SAR 580. From a 10 m tower on a corn field in order to study the growth of vegetation. Many physical parameters of air, soil and vegetation were directly measured to compare the radiometric signals with the"ground truth". The first results of the measurement are summarized in the paper.

1. INTRODUCTION

Microwave remote sensing (M.R.S.) is an important tool for the study of the Earth's environment. Its applications range from oceanography and geology to meteorology and agriculture (Ulaby 1976, Moore 1978, Staelin 1981).Microwave sensors are either active or passive; the first ones provide their own illumination source and measure the back-scatter echo while passive systems measure radiation of thermal origin emitted from materials. The main characteristics of the microwave range of the spectrum as opposed to the visible and the infrared ones are its capacity of operating even at night and through clouds and of penetrating the Earth's surface deeper than the infrared radiation.

Moreover microwave sensors can give complementary information on the physical parameters of crops because the emission and reflective properties of observed materials depend heavily on their dielectric constant.

Furthermore, the Synthetic Aperture Radar (SAR) is, at the present , the sensor giving the highest resolving power from satellites.

In this paper only passive systems with particular regard to their applications in Agriculture will be considered.

Radiometric sensors in the 1-100 GHz range have been used since the second half of the 60's from satellites (Tiros, Cosmos, Nimbus, ecc.), aircraft and ground platforms; several other multispectral systems are planned for the near future on satellites Block 5D-2, NOSS, Tiros D.

The most important applications of microwave radiometers are the study of the sea, ice, snow cover, soil moisture, and the measurement of atmospheric characteristics. In agriculture greatest attention has been payed to the moisture of bare and vegetated soil; nevertheless it seems to be interesting to investigate the physical parameters of vegetations. This can be done all the better as much more data is available at different frequencies, polarizations and angles of incidence. Two approaches are possible in using experimental data. One, empirical correlations between the physical parameters of vegetation and radiometric signals are investigated; in the other, theoretical models together with appropriate boundary conditions may be derived by using the transport equation and tested with experimental data.

In both cases remote sensing data must be implemented with "in situ" physical measurements.

2. MICROWAVE RADIOMETRY

Microwave radiometry is based on the measurement of thermal radiation from material objects. Thermal emission can be studied by using the "black body" which is an ideal body in which the maximum conversion of thermal energy into electromagnetic radiation takes place. This body also has the reciprocal property of absorbing all the radiation falling upon it and convert it into thermal energy. The radiation from the black body is described by Planck's well known law which, in the microwave range , may be approximated (with an error of less than 1% at 110 GHz and 300 K) by the Rayleigh-Jeans equation:

$$B(\lambda,T) = \frac{2KT}{\lambda^2} \qquad (1)$$

where:

$B(\lambda T)$ = Brightness, is the power radiated from 1 m^2 of surface per solid angle unit and per wavelength unit $W\,m^{-2}\,sr^{-1}\,m^{-1}$

K = Boltzmann's constant 1.38×10^{-23} J/K

T = Absolute physical temperature (K)

λ = Wavelength (m)

Equation 1 establishes a linear relationship between radiation and physical temperature.

Black bodies don't exist in nature, but natural emission may again be described by a relationship of the same form as 1 by replacing T with the equivalent brightness temperature:

$$T_b = \varepsilon(T, \lambda)\ T \qquad (2)$$

where the emissivity ε (<1) is a coefficient which changes according to the observation angle, frequency, polarization and surface roughness.

The microwave brightness temperature of natural bodies ranges from about 10 K to 310 K. In general radiation is not produced by the uppermost surface of a medium: it is the result of the contribution of several layers at different depths. Depending on the type and on the object of observation these layers may refer to the Earth's surface covered by clouds or to the vegetation over the soil or finally to different levels of bare soil. In any case the emission may be described by the equation of radiative transfer (Chandrasekhar 1960). It is possible to show that the noise power received by the antenna of a radiometer may be expressed by means of a fictitious temperature T_a (Antenna temperature) which represents the temperature, of a resistance equal to the antenna resistance capable of giving same power delivered by the antenna.

When an antenna with a normalized power pattern $P(\theta,\varphi)$ observes a finite source with a brightness temperature $T_b(\theta,\varphi)$, the received power may be expressed by means of T_a as a weighted average of $T_b(\theta,\varphi)$ over the antenna pattern (Moore ,1975) i.e.:

$$T_a = \frac{\int_{4\pi} P(\theta,\varphi)\, T_b(\theta,\varphi)\, d\Omega}{\int_{4\pi} P(\theta,\varphi)\, d\Omega} \qquad (3)$$

If the brightness distribution of the source is constant over the entire antenna recption pattern whe have:

$$T_a = T_b$$

On the other hand, if the source is very small compared to the antenna beam, T_a is reduced according to the ratio of the solid angle of the source to that of the antenna.

3. RADIOMETRIC SYSTEMS

A microwave radiometer is a very sensitive and stable receiver which measures the power delivered by the antenna (Moore et al 1975). Usually the noise signal received is only a small fraction of the noise generated by the system. Both signal and system noises may be expressed in terms of equivalent temperatures T_a and T_s.

There are basically two radiometer configurations: the total power and the switched of Dicke (Dicke, 1946) type. The total power type is the simplest form of radiometer and potentially provides the best sensitivity. It consists basically of a predetection section of bandwidth B and gain G,a square law detector , with output voltage proportional to the input power and an integration section.

The radiometric sensitivity is usually given by the minimum input temperature variation ΔT_{min} which produces a change in the output signal equal to the r m s value of fluctuations. (Minimum detectable temperature).

For the total power radiometer it is:

$$\Delta T_{min} = (T_a + T_s)\sqrt{\frac{1}{B\tau} + \left(\frac{\Delta G}{G}\right)^2} \qquad (4)$$

where T_a and T_s are respectively the equivalent antenna and input system noise temperatures, B is the predetection bandwidth, τ is the integration time and $\Delta G / G$ represents the normalized rms fluctuation of receiver power gain.

In the Dicke type radiometer the receiver input is periodically switche from the antenna to a reference noise source. In this way the gain variatio term of equation (4) may be strongly reduced provided the switching rate b higher than the frequency variation of the gain. The sensitivity of the Dicke radiometer with a duty cycle equal to .5 is:

$$\Delta T_{min} = (T_a + T_{rif} + 2T_s)\sqrt{\frac{1}{B\tau} + \left(\frac{\Delta G}{G}\right)^2 \left(\frac{T_a - T_{rif}}{T_a + T_{rif} + 2T_s}\right)^2} \qquad (5)$$

where T_{rif} is the reference temperature.

We point out that, if $T_a = T_{rif}$, equation 5 becomes:

$$\Delta T_{min} = \frac{2(T_a + T_s)}{\sqrt{B\tau}}$$

i.e. ΔT_{min} is independent of ΔG.

In general ΔT_{min} decreases the nearer T_a gets to T_{rif}.

Microwave radiometers may operate in order to produce images. In this case an antenna movement is necessary. If the radiometer is ground based, a line by line scanning must be used, while with moving platforms (aircraft or satellite) a cross track scanning is usually sufficient. The image may be restored on a black an white or color CRT by using different levels of grey or color intensities.

4. IATA RADIOMETERS

Two radiometers at 9.8 GHz (X band) and 36.6 GHz (Ka band), have been realized by IATA (Institute of Environmental Analysis and Remote Sensing for Agriculture) in cooperation with the Arcetri Astrophysical Observatory to investigate natural surfaces from a ground based platform. Both instruments have basically the same configuration and the same characteristics which are summarized as follows:

Frequency:	9.8GHz - 36.6 GHz
Polarization:	V and H
Antenna HPBW:	20 or 2 degrees
IF Bandwidth:	5 - 500 MHz

Temperature Resolution : .2K (τ = 1 sec)
Absolute Calibration : .5K

The radiometers are weather-proof and temperature stabilized at 25 C $\pm$.3 C (when the external temperature ranges from 0C to 40 C) by means of a Peltier cooling/heating block. The block diagram is shown in fig. 1.

Since the brightness temperature of natural bodies is always less than their physical temperature,the most important range of calibration for land application is between 100 and 300K. In these radiometers calibration is obtained with an internal cold load,the temperature of which,is known with a margin of error of $\pm$ .1K and can be changed from 0 to -20K with regard to the reference temperature.The latter is also known with the same precision.

The calibrator may be connected to the input of the receiver by means of an SPDT remote control switch. Besides,to extend the calibration range, an external known target can be used together with a noise source and a variable attenuator coupled to the antenna arm through a directional coupler. Noise may be added step by step by means of the attenuator. A temperature controlled "Black body" is available for external calibration.

In the Ka band radiometer the two H and V polarizations are obtained by means of a two port Faraday rotator switch with circular waveguide input and rectangular output. The switch is driven in order to select the two components which can be separated in the L.F. section in a sequential manner. In the X band radiometer two rectangular horns driven by an SPDT switch are used. In both cases the insertion loss of the front end waveguide section is less than 1.8 dB.

The receivers are of the superheterodyne type with a Schottky barrier diode balanced mixer,wide band IF section and a square low detector.

The outputs of the radiometers,together with the temperatures of the reference terminations and several housekeeping data are recorded on a digital tape cassette and/or on an analog paper recorder.

Both sensors have been designed to operat from a 10-15 m tower and the angle of observation may be remotely and continuously changed from nadir to zenith . Moreover,since the instruments are very light and compact,they can be easily mounted on an aircraft.

5. APPLICATION AND EXPERIMENTAL RESULTS

The application of passive radiometry to agriculture is relatively recent. Several studies have been devoted to the measurement of soil moisture. The knowledge of the temporal and spatial distribution of soil water is of significant importance in the prediction of agricultural productivity. Research in this field has been carried out in the USA by the NASA and the University of Texas, in Australia by CSIRO and in Switzerland by the University of Berne. Theoretical and experimental work has gone on in URSS since 1973 (Basharinov et al 1974).

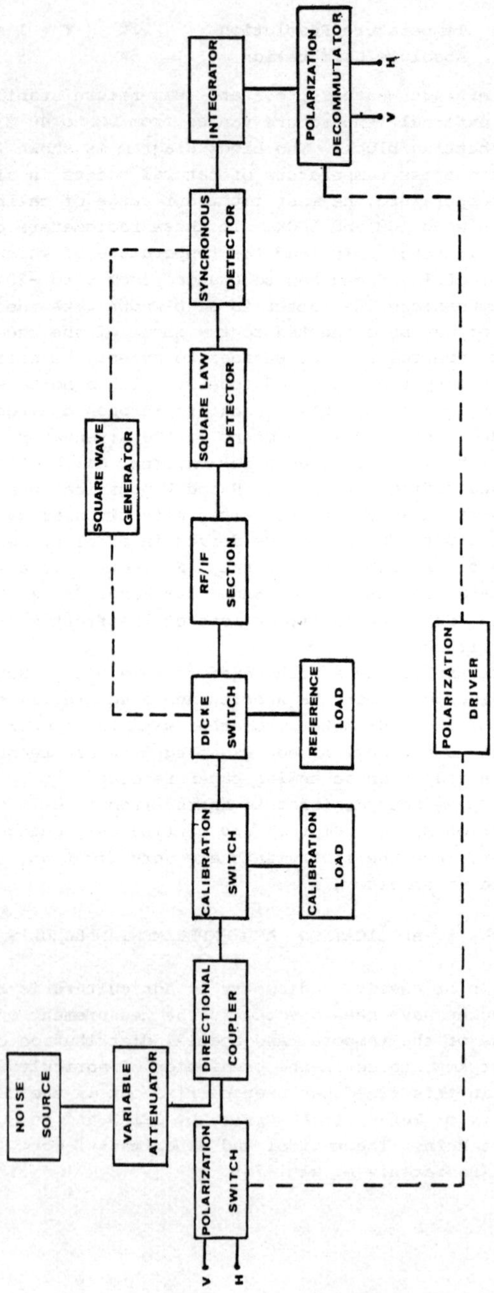

Fig. 1: Block Diagram of Radiometers

The used approaches were both theoretical and experimental .

Since emission from soil is generated within the soil volume, theoretical models basically consider two parameters: the temperature and the absorption coefficient which is greatly influenced by the moisture content.

The objective is to determine the emission properties of the soil from the subsurface physical conditions. This can be done by relating the brightness temperature of the surface to the temperature profile T (h) and to the absorbing coefficient α(h) in the soil by means of the transfer equation:

$$T_b = \varepsilon \int_{-\infty}^{0} T(h)\, \alpha(h)\, \exp\left(-\int_{h}^{0} \alpha(h')dh'\right) dh \qquad (6)$$

where h is the distance from the surface and ε takes into account the loss of energy at the surface.

Prediction of T_b from this equation has been obtained for smooth soil surfaces by Njoku and Kong (1977), Wilheit (1978) and Burke et al (1979).

In order to achieved a better agreement between theoretical and experimental data, models including the roughness effects were considered by Choundhury et al (1979) who compared theoretical results with experimental data. From all these models one can deduce that 1)as regards smooth surfaces pernetration depth is only a small fraction of the wavelength; 2) the roughness effects are important for wet soils where the difference between smooth and rough surfaces can be as great as 50K.

Since the 1974 Schmugge and his collegues (Schmugge et al 1974) demonstrated experimentally that it is possible to monitor soil moisture variations in bare soil by means of airborne radiometers. They carried out measurements at several frequencies from 8 to 21 cm and found the best sensitivity at the longest wavelength. These results were confirmed by Shanda et al (1979) with ground based radiometers and by Barton (1978) who also investigated vegetated surfaces.

Recently an experimental investigation on a 24 hour per day basis for six weeks has been conducted by Newton and Rouse (1980). Measurements at 2.8 cm and 21.4 cm wavelengths have been carried out within a range of angles of incidence from nadir to 50° over bare and vegetated plots. The results confirm that microwave emission, above all at 21 cm, is strongly correlated to the total moisture in the surface soil layer and to the distribution of moisture in depth. In Italy, apart from some preliminary experiments, (Marsiglia et al 1979) a sistematic research on M W R S only began in 1980 when the IATA was founded in Florence by the CNR (National Research Council).

Without neglecting the importance of soil moisture we consider that a remote sensing measurement of the physical parameters of the vegetation could be very useful for the estimate of biomass and water balance. From a theoretical point of view the problem could be restated by means of the transfer equation where the temperature profile and the absorption coefficient are now made to refer to vegetation. Nevertheless, since vegetation is a much less homogeneous medium than soil, only spacially averaged values for the parameters could be considered leading consequently considerable approximations. A more sophisticated approach seems therefore to be required. From the experimental point of view different layers of vegetation can

Fig.2: Brightness Temperature versus angle of incidence

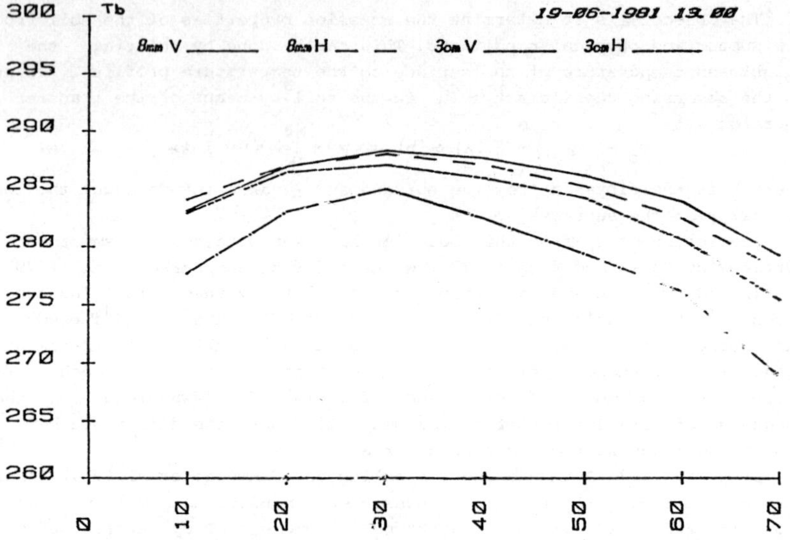

Fig.3:Brightness Temperature versus Soil Moisture (% Weight) at 3 cm λ

■ H polarization ○ V Polarization

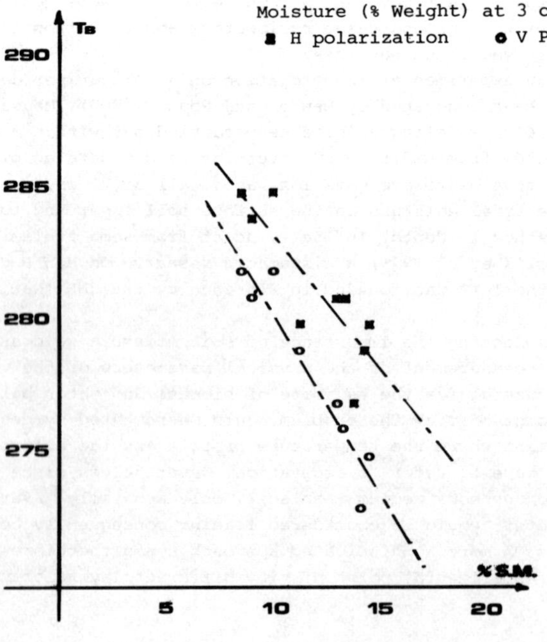

be investigated by using radiometers at different frequencies polarizations and angles of incidence.

The first experimental campaign took place last year in Puglia using the wheat as object (Maracchi et al 1981). Unfortunately research was seriously affected by frequent unusually heavy rains. Nevertheless the possibility of monitoring vegetation temperature variations by means of microwave radiometers was established.

In June-July 1981 the corn growth has been monitored on 24 hour basis for two months. For this experiment, which took place within the ESA-JRC SAR 580 Campaign, the two above described radiometers were used from a ground based platform with angles of incidence between 10 and 60 degrees. (Benincasa et al 1981). In addition to microwave data, many other micrometeorological parameters were collected to compare remote sensing data with ground truth. In all 45 parameters were recorded every 20 minutes during the day and once an hour at night. Though data processing is still in progress some results are already available (Benincasa et al 1981) and may be summarized as follows:

1) The microwave emission of the crop at both frequencies noticeably depends on the polarization and on the angle of incidence (Fig. 2 is one example)
2) For an angle of incidence equal to 10 degrees from nadir a good correlation exists between microwave emission at in the x band and soil moisture measured by weight a depth of 0-10 cm (Fig. 3). On the countrary no appreciable correlation has been found in Ka band.
3) When the angle of incidence is 20 degrees of higher, the microwave emission is related to the physical parameters of the vegetation rather than to soil.

6. CONCLUSIONS

The use of microwave radiometry seems to be of reasonable interest to Agriculture, above all integrated with active sensors. The measurements on a wide scale of soil moisture, evapotranspiration and biomass are realistic targets. Due to the present state of art in microwave components (low loss switches and low noise RF amplifiers and mixers), microwave radiometers with absolute accuracy of about .5K to 100GHz may be realized at relatively low prices . The most critical point seems to be their low spatial resolution; owing to that, aircraft are perhaps better platforms than satellites.

7. AKNOWLEDGEMENTS

Many people participated in the realization of radiometers and in the field measurements; the author wishes to thank Prof. G. Tofani and Dr. M. Catarzi of the Arcetri Astrophysical Observatory and the Staff of the IATA.

8. REFERENCES

Basharinov, A.E., Borodin, L.F., Shutko, A.M., (1974): Passive microwave Sensing of Moist Soils. URSI Meeting Comm. II Berne 1974.

Barton, I.J., (1978): A Case Study Comparison of Microwave Radiometer Measurements over Bare and Vegetated Surfaces. J. Geophys. Res. 83,B7, 3513-3517

Benincasa, F., Maracchi, G., Paloscia, S., Pampaloni, P., Zipoli, G.,(1981): Microwave Remote Sensing over the Corn.¨Conf. on Matching R.S. Tech Londra 1981 - Porter Session .

Burke, W.J., Schmugge, T., and Paris, J.F., (1979): Comparison of 2.8 and 21 cm microwave radiometer observations over soils with emission model calculations. J. Geophys Res 84 287-294 .

Chandrasekhar, S., (1960). Radiative Transfer.Newyork 1960.

Choudhury,B.F., Schmugge,T., Newton,R.W., and Chang,A., (1979):Effect of surface roughness on the microwave emission from soils. J. Geophys Res. 84,5699 -5706.

Dicke, R.H., (1946): The measurement of thermal radiation at microwave frequencies. Rev. Sci. Instr. 17, 268.

Maracchi,G., Pampaloni,P., Zipoli,G., (1981): Ground based microwave radiometry over the wheat. Int. Coll. on Spectral Sign. Avignon 1981 Sept.

Marsiglia,G., Pampaloni,P., Tofani,G., Maracchi,G., (1979): Microwave remote Sensing at the Arcetri Observatory . Earsel Workshop - Paris 1979 80-84.

Moore,R.K., (Editor) (1975): Microwave Remote Sensors - in Manual of Remote Sensing Vol I (by Janza), 499-537.

Moore,R.K., (1978): Active Microwave Sensing of the Earth's Surface-A mini Review. IEEE Trans A.P. A.P.-26, No 6, 843-849.

Morzier,C., Musy,A., (1979): Microwave Emission Measurement on Bare Soil in Switzerland. Earsel Workshop - Paris 1979, 53-63.

Newton,R.W., and Rose,J.W., (1980): Microwave Radiometer Measurement of Soil Moisture Content. IEEE Trans A.P.-28 n5, 680-686.

Njoku,E.G., and Kong, (1977): Theory for passive microwave remote sensing of near-surface soil moisture . J. Geophys. Res. 83 3108-3118.

Schanda,E., Wyssen,D., Hofer,R., Mätzler,C., Meylan,P,.(1979) Microwave Emission Meas. on Bare Soil in Switzerland. Earsel Workshop - Paris 1979,53-63
Schmugge,T.J., Blanchard,B., Anderson,A., and Wang,J., (1974):Remote Sensing of soil moisture with microwave radiometers. J. Gephys Res.79 317,323.

Shmugge,T.J., Jackson,T.J., McKim,H.L., (1980): Survey of Methods for Soil Moisture Determination. Water Researches Res. 16, 6, 961-979.

Staenlin,D.H., (1981): Passive Microwave Techniques for Geophysical Sensing of the Earth from Satellites. IEEE Trans A.P. A.P.29 No 4 683-687.

Ulaby,F.T., (1976): Passive Microwave Remote Sensing of the Earth's Surface. IEEE Trans A.P. A.P. 24 No 1 112-115.

Wilheit,T.J., (1978): Radiative transfer in a plane stratified dielectric. IEEE Trans Geosci Elect. GE-16 138-143.

OPTICAL SCANNERS AND REMOTE SENSING OF ESTUARINE AND COASTAL WATER QUALITY

A.P. Cracknell
Carnegie Laboratory of Physics,
University of Dundee,
DUNDEE DD1 4HN.

Abstract: We have studied data from the following scanners with relevance to the monitoring of water quality: the Ocean Colour Scanner, OCS, the LANDSAT Multi-Spectral Scanner, MSS, the NIMBUS-7 Coastal Zone Colour Scanner, CZCS, and the TIROS-N Advanced Very High Resolution Radiometer, AVHRR. By water quality we refer to suspended sediment arising both from natural causes and from the activities of mankind, to phytoplankton, and to floating oil pollution. The usefulness of each scanner depends on its spatial resolution, its spectral resolution, and the frequency of coverage by the platform on which the scanner is flown.

The published specifications of scanners planned for future satellite missions, LANDSAT-D, and -D', SPOT, ERS-1 and future satellites in the TIROS-N, NOAA-6, -7, etc. series will be discussed and predictions made for the improvements that can be anticipated both in the quality of the remotely-sensed data and in the usefulness of the results.

Finally, a political point will be made. It is no use spending millions of pounds on developing instruments, building satellites, putting them into space and recovering the data generated by them, if all that then happens is that the data are stored in a cupboard. There has to be investment, of finance and human endeavour, in the processing and interpretation of that data.

1. INTRODUCTION

In Dundee University we have considerable interest in marine remote sensing and at the last two conferences in this series I have given talks on some of the detailed aspects of our work. Those two talks were not written up for publication in the proceedings of those two conferences, but the material is mostly published elsewhere or is in press (Singh and Cracknell 1979, Cracknell et al. 1980, Cracknell and Singh 1980, 1981a, 1981b). Given the title of this meeting it seemed reasonable to take a more general view. Therefore, I shall try to give some kind of assessment of what we and other people have been able to achieve in the last few years, with the kinds of remote sensing data that have been available. Then I shall try to indicate what we hope to be able to achieve in the next few years, with the kinds of data that we expect to become available. Although I have used the term "optical scanners" in the title, I shall also include some mention of infrared wavelengths too. Although I shall make some mention of aircraft I shall concentrate more on satellites; this is because relatively few of us are in the position of having an aircraft available or having the funds to charter an aircraft, whereas most of us have some funds to go and buy a few satellite images.

It will be noticed that no images are reproduced in this paper; there are several reasons for this. First, all the relevant images have been published in the papers I have already mentioned. Secondly, and more important, I read in the instructions to authors that I might have to pay the costs of reproduction of such images and that did not seem to me a good idea. The presentation at the conference itself will include some images.

2. MULTI-SPECTRAL SCANNERS

The general principles of the observation of the surface of the Earth using a multi-spectral scanner are assumed to be familiar to people at this conference. The important parameters, for our purposes, are (a) the spectral resolution, (b) the spatial resolution, and (c) the frequency of coverage of a given area. (a) is determined by the construction of the scanner itself, while (b) and (c) are determined both by the optical construction of the instrument, by its power requirements, and by the height at which it is flown above the surface of the Earth.

For many applications it would be useful to have as many spectral channels as possible, i.e. to have as much spectral resolution as possible. However, the more spectral channels there are, the more heavy and expensive the instrument becomes and the more likely there are to be problems of its reliability when operating in space. The numbers and wavelength ranges of some of the instruments used in recent years in the civilian remote sensing programme are given in table 1. The list in table 1 is by no means exhaustive and people have indeed done work on water-quality monitoring using other systems; one example involves the use of data from the Heat Capacity Mapping Mission, HCMM, carried out by Dr. G. Fielder's group at Lancaster University, but I do not wish to mention any details of this because all I have is unpublished reports on their work. In addition to satellite-flown scanners there are several scanners which have been flown on aircraft and which have larger numbers of spectral channels.

There is a trade-off between spatial resolution and the frequency of coverage of a given area. Consider the polar-orbiting satellites in table 1. The linear dimensions of the footprint for the LANDSAT MSS are about 80m but the frequency of coverage is only once, or possibly twice, in 18 days. For the AVHRR on TIROS-N and for the CZCS on NIMBUS-7 the linear dimension of the footprint is an order of magnitude larger, i.e. the spatial resolution is worse, but the frequency of coverage is, from orbit considerations, better than once every day. For the CZCS, the imagery may actually only be obtained rather less frequently; this is because the power requirement of the CZCS is rather large and so the instrument is not necessarily always switched on during every pass over any given area. While a footprint size of about 1 km is fine for oceanographic or meteorological work it is really too big for most estuarine work. Even on a wide estuary such as that of the Tay there are not many pixels between one side of the estuary and the other. On the other hand, for many estuaries the spatial resolution achieved by the LANDSAT MSS is perfectly adequate.

Table 1. Comparison of characteristics of some sensors.

LANDSAT MSS[1]		TIROS-N AVHRR[2]		NIMBUS-7 CZCS[3]		METEOSAT	
Channel	Wavelength (μm)	Channel	Wavelength (μm)	Channel	Wavelength (μm)	Channel	Wavelength (μm)
4	0.5–0.6	1	0.55–0.90	1	0.433–0.453	1	0.4– 1.1
5	0.6–0.7	2	0.725–1.0	2	0.510–0.530	2	10.5–12.5
6	0.7–0.8	3	10.5–11.5	3	0.540–0.560	3	5.7– 7.1
7	0.8–1.1	4	3.55–3.93	4	0.660–0.680	Resolution	
Resolution	~80m	Resolution	~1 km	5	0.700–0.800	Channel 1	~2.4 km
				6	10.5 –12.5	Channels	
				Resolution	~825m	2 & 3	~5 km

1 Multi-Spectral Scanner

2 Advanced Very High Resolution Radiometer

3 Coastal Zone Colour Scanner

3. WATER-QUALITY PARAMETERS AND REMOTE SENSING DATA.

Water-quality parameters that are commonly measured in situ include the temperature and salinity and the concentrations of chlorophyll and of suspended sediment. The suspended sediment may be of natural origin, as it is for instance in the Tay, or it may result from the activities of mankind. Pollutants are also of great importance; these include heavy metals dissolved in the water, oil floating on the surface of the water, and oil-water emulsions which may be on the surface or below the surface.

There are two stages involved in studying the problem of extracting water-quality information from remote-sensing data. The first stage is to ascertain whether the parameter under investigation can, in principle, be extracted from the remotely-sensed data. The second stage, if a positive answer is obtained at the first stage, is to determine how much useful information can actually be obtained at present about the parameter in question, given the current status of the space and ground segment technologies of remote sensing satellite programmes.

It is important to realize that there are some water-quality parameters about which remote sensing, using optical scanners, can provide virtually no information. For example, most heavy metal ions which may be present in sufficient concentration to have quite serious toxic effects on marine life are unlikely, at such concentrations, to make a significant change to the optical properties of the water and are therefore unlikely to be detected using optical scanners. The position with regard to salinity is somewhat similar. There is no doubt that variations in the salinity cause variations in the optical properties of sea water and we are, indeed, doing some theoretical investigations of this ourselves. Nevertheless, the changes in the optical properties are relatively small and are almost certain to be masked by other effects as far as remote sensing from satellites is concerned. On the other hand, oil pollution floating on the surface of the sea should, in principle, be clearly visible in a conventional aerial photograph or in an image from an optical scanner. Although chlorophyll and suspended sediment are not so clearly "visible" as a floating oil film would be, their presence can often be seen in imagery from optical scanners. In principle it should be possible to extract estimates of the concentrations of these materials from remote sensing data, either with the use of simultaneous in situ data for calibration within a "training area" on the scene or, preferably, without the use of any simultaneous in situ data. Finally, sea surface temperatures can be determined using scanners having "thermal" infrared channels, that is corresponding to wavelengths in the vicinity of 10-12 μm.

4. PRESENT STATUS OF WORK ON EXTRACTING WATER QUALITY PARAMETERS.

4.1 Oil pollution

In principle the idea of using an optical sensor flown on a satellite to monitor pollution of the rivers and of the sea is a splendid idea. The satellite is able to observe huge areas with a regularity that is guaranteed as long as the lifetime of the satellite. Moreover, from the point of view of the user, the cost is rather low. The costs of building and launching the satellite, controlling the satellite in orbit, and providing the ground stations to receive the data from the satellite are usually all paid by somebody else (the taxpayer) and the user is only left with having to pay the direct costs of distributing the data. However, having said all that, there are still serious problems to be overcome before we can expect any operational marine pollution monitoring system, based exclusively on the use of satellites, to be introduced. At present those countries that do attempt to run pollution monitoring systems by remote sensing rely very heavily on the use of aircraft and make relatively little use of satellite data.

The main problems that arise are associated with the problems of spatial resolution, frequency of coverage, cloud cover and delay in data delivery. For a major oil spill disaster, such as the IXTOC-1 blow-out in the Gulf of Mexico in 1979, the slick produced is so large that one can expect to see evidence of it in AVHRR imagery or in imagery from the appropriate geostationery satellite, in the case of IXTOC-1 that would be GOES-E instead of METEOSAT. Indeed, we have done some work with data from AVHRR tapes that we got from Washington and have had some success in mapping the oil slick in the early days soon after the IXTOC-1 blowout. This relied on the (panchromatic) visible channel of the AVHRR. We found no evidence of the oil when we looked at the same scenes in the thermal infrared channel; this surprised us slightly but it seems to be consistent with the results of other workers (Hayes 1980). A substantial amount of LANDSAT imagery of the Gulf of Mexico exists giving evidence on the movement of the oil and resulting oil-water emulsions from this blow-out (Hayes 1980, Deutsch et al 1980). The United States Coast Guard did indeed make some use of satellite remote sensing data in monitoring the movement of the oil originating from this spill. But this was an enormous spill which lasted for several months. Most other spills are much smaller and the oil involved disappears much more quickly. We have tried to use the Oil Spill Intelligence Report in conjunction with AVHRR, CZCS and LANDSAT MSS imagery to find evidence in satellite imagery of any of the major spills in European waters in recent years. The problems are greater than with IXTOC-1, both because the slicks produced are smaller and more short-lived and because the cloud-cover is generally greater than in the Gulf of Mexico. We have failed to find any evidence of these smaller spills in the satellite imagery.

The conclusions drawn by Hayes (1980) from the experiences of the United States Coast Guard with IXTOC-1 seem to us to be very fair and worth summarizing here. The meteorological satellites were useful in the early stages, in the absence of all other sources of information, in providing a general sense of the magnitude of the spill and of the general direction of drift of the spilt oil. However, the low resolution and small scale and high distortion of the standard image product were felt to limit these remote sensing systems as reconnaissance tools. As we have already mentioned the oil only seemed to be detectable with the visible sensors and not by the infra-red sensors. During the time of the blowout both the LANDSAT-2 and LANDSAT-3 satellites were operating and a considerable amount of relevant imagery was obtained; it was found to be useful in corroborating the results from aircraft reconnaissance and in helping to establish the overall picture of the situation. But the infrequency of satellite overpasses, the problem of cloud cover, the rather low spatial resolution, and the lack of real-time processing and delivery of imagery prevented the use of LANDSAT for tactical response by the on-scene coordinator of the U.S. Coast Guard. In the event the use of "low altitude visual aircraft reconnaissance backed up by an all-weather airborne remote sensing system (where weather and visibility required it) proved to be the most effective tactical and, along with accurate oil drift computer model forecasting, strategic aid for environmental pollution response planning for the Bay of Campeche oil spill. This will probably be true for some time into the future. The need for frequent, accurate, comprehensive, rapidly processed and analyzed information as input to the on-scene coordinator's decision making process cannot be overemphasised. Until satellite remote sensing systems can deliver useful products in fine enough detail and observational frequency ..... in sufficient time to respond to any likely impact such systems will necessarily occupy a subordinate role in the oil spill support strategy. Developing an all-weather operational satellite oil spill detection and mapping capability is a logical solution to monitor such events which are so unpredictably and widely distributed in time, space and environmental conditions. Orbiting spacecraft could alleviate some of the often immense logistical burdens placed upon conventional reconnaissance methods by the remoteness of the site or hostility of the environment. Whether or not this route shall be chosen will depend most upon the development of capable remote sensors from orbital altitude and the ability to find sufficient multiple applications of the system beyond marine environmental protection to satisfy the requirement for sufficient economic and social benefit to justify the cost."

However, I do not see these conclusions as a counsel of despair. What Hayes says is a much more eloquent version of what I have said in my applications for grants to Research Councils. It is an argument not for giving up hope but an argument for further work and investment, both of money and of human resources in a difficult but worthwhile task. It is only fair to say that the Research Councils have not been entirely deaf to these arguments!

4.2 Sea surface temperatures

The thermal infrared channel of the AVHRR on TIROS-N (see Foote and Draper 1980), or of the CZCS on NIMBUS-7 (see Fusco 1981) can give us good thermal maps of the surface of the open sea. Similar data can also be obtained from geostationary satellites in the METEOSAT programme (see Brimacombe 1981). The important features that need to be considered are (i) spatial resolution, (ii) temporal resolution, and (iii) temperature resolution (or accuracy or sensitivity).

The spatial resolution can be expressed by the size of the footprint of the scanner. For the AVHRR the footprint size is of the order of 1 km square. For the METEOSAT radiometer the spatial resolution is considerably poorer because, being geostationary rather than polar-orbiting, the satellite is very much higher above the surface of the Earth; the footprint size is about 5 km square for the thermal infrared channel of METEOSAT. At the other extreme, of course, for a thermal infrared scanner flown in a light aircraft at a rather low altitude the footprint size may be only about one metre or so; aerial surveys using such scanners are now widely used to monitor heat losses from the roofs of large buildings (see, for example, Anderson and Callison 1981). In areas of open ocean the spatial resolution of the AVHRR, CZCS or METEOSAT radiometer is perfectly adequate for work on sea surface temperatures. Not only is it perfectly adequate - it actually enables the oceanographer to have synoptic maps of sea surface temperatures over enormous areas; these could not previously be obtained on such a scale in any other way before the use of satellites. Such maps are also beginning to find uses in marine resources exploitation and management, e.g. in assisting in the location of fish (see, for example, Domain et al. (1980)).

However, with the AVHRR, CZCS and METEOSAT radiometer, their spatial resolution makes it rather difficult to use data from these instruments for smaller areas such as rivers or estuaries to study, for example, the thermal "pollution" by the cooling water from a power station etc. The width of the river would only correspond to two or three times the linear dimension of the footprint. The scanners on the early satellites in the LANDSAT programme did not have a thermal infrared channel. For good thermal maps of estuaries and coastal regions we await data from new instruments and satellite programmes, such as the thematic mapper on LANDSAT-D, due for launch in 1982 or so.

For the AVHRR satellites in the TIROS-N, NOAA-6,.... series the temporal resolution will be one pass every few hours if there are two satellites in the series in orbit. For METEOSAT, data are transmitted every half hour. However, in each case, the frequency quoted refers to the frequency at which thermal infrared data are generated and transmitted back to Earth. Not all the data will be useful because of the problem of cloud cover and the frequency of reception of useful data on sea surface temperatures will be correspondingly reduced. Moreover, there is also the problem that there may be gaps in a satellite program when one satellite has failed and there is a delay before a replacement satellite can be launched.

As far as temperature resolution and accuracy are concerned it appears that, by using calibration sources observed by the scanner itself, temperature differences in a given scene can be measured to within a few tenths of a degree and absolute temperatures can be measured to within a few degrees. There are two ways in which the uncertainty in the absolute temperatures can be reduced. The first is by attempting to calculate atmospheric corrections (see, for example, Phulpin and Deschamps 1980, Weinreb and Hill 1980). The second is by using some simultaneous in situ measurements of sea surface temperatures, at a few points within the scene, to calibrate the temperatures for the whole scene; we are starting to try to carry out some work along these lines using moored buoys with data recovered via the ARGOS system (Cracknell and Singh 1981a, 1981b). These two approaches, though they are very different, should not be regarded as alternatives; I believe that the best results will be obtained from a combination of both approaches.

4.3 Suspended sediment, chlorophyll, etc.

The problems of spatial resolution and of temporal resolution which we have noted in the previous two sections apply also to the problems of studying suspended sediment, of mineral or human origin, and of biological material of non-human origin. The comments made in the previous two sections about the relative merits of the AVHRR, the CZCS and the METEOSAT radiometer, not to mention scanners flown on aircraft, with regard to spatial resolution, the frequency of satellite overpasses, the problem of cloud cover, and so on will still be relevant to the applications discussed in this section. For projects involving relatively small areas in estuarine or coastal regions there may still be considerable advantages in the use of aircraft rather than satellites; for example, see the discussion by Valerio (1981). However, the question of spectral resolution is rather different. In the case of monitoring oil pollution the panchromatic view of the AVHRR visible channel or the rather small number of visible wavelength channels of the LANDSAT MSS constituted much less of a problem than some of the other difficulties that we noted in the use of data from these instruments. In the case of sea surface temperatures, the optical channels were more or less irrelevant and only the data from the thermal infrared channel, or channels if more than one were available, were relevant. In other words, in those cases, spectral resolution was not a serious problem because the extraction of the required parameters was quite straightforward. For the oil slick monitoring this was principally based on visual interpretation, coupled with geometrical rectification if desired, while for sea surface temperature determination a well-established algorithm, the Planck radiation formula, is available. For suspended sediment and chlorophyll monitoring the extraction of the required parameters from the remotely sensed data is substantially more difficult.

Quantitative information about the concentration of suspended sediment or of biological material is not obtained from the intensity recorded in a single spectral channel of a multispectral scanner, irrespective of whether the channel in question is restricted to a narrow range of visible wavelengths or is panchromatic, i.e. covering the whole range of visible wavelengths. This information has to be gained by using some algorithm for extracting the concentration from data in many spectral channels. There is no single algorithm that is universally accepted for the recovery of suspended sediment or chlorophyll from remote sensing data. However, what is fairly generally accepted is that one needs optical multi-spectral scanning data from a scanner with quite a large number of spectral channels in the visible region of the electromagnetic spectrum. Our own experience (Singh and Cracknell 1979, Cracknell and Singh 1980) has been that the number of visible channels in the LANDSAT MSS is rather too few; essentially only two of the bands (4 and 5) carry any useful data about the contents of the water. With the larger number of channels on the CZCS it is possible to do some reasonable work on algorithm development and testing (Cracknell and Singh 1981b); about four of the six CZCS channels are useful for this purpose. The Ocean Colour Monitor, OCM, planned by ESA for the ERS-1 satellite would provide a little more spectral resolution. At present I think it is fair to say that there is a lot of CZCS data which has been received over the last two or three years and that we and various other people dotted around the world are working quite hard on algorithm development and testing. This is done in the first instance by choosing scenes for which results of simultaneous in situ measurements are available and then, with the experience gained from doing that, moving on to scenes for which no results of simultaneous in situ measurements are available for calibration purposes. One very important aspect of the work is the need to calculate corrections to the intensities recorded at the satellite to allow for the atmosphere that is present between the satellite and the surface of the sea. In this connection we have made extensive use of the work of Sturm (1981). It would, I think, be a fair summary to say that at present there are many different algorithms proposed by various different groups of workers. These algorithms include ratios of the intensities in various pairs of channels, ratios of pairs of combinations of intensities in various channels, simple power laws and exponential laws, general linear transformations of the intensities in all the channels of the scanner, and so on. Although there is no general agreement, as yet, about the algorithms themselves I think it is widely accepted that different algorithms will be needed for different types of water and that it would be extremely naive to expect a single algorithm to be universally applicable. There is still a large amount of CZCS data waiting to be processed; the experience gained and the results produced should assist greatly in handling the data expected to come from the OCM on ERS-1.

5. FUTURE SATELLITE PROGRAMMES.

The discussion in section 4 has been largely in terms of scanners used in existing satellite programmes. Several improvements are planned or expected in the next few years.

In the version of the AVHRR which will be flown on some of the later satellites in the TIROS-N/NOAA-6, -7, ... series it is planned to include an extra channel; this will be in the thermal infra-red and should help to improve the accuracy of sea surface temperature determinations (see Foote and Draper (1980) for details).

The Thematic Mapper that is planned to be carried on the next satellites in the LANDSAT programme (due to be launched in 1982 or later) has already been mentioned. This is a new design of scanner with more spectral channels and with better spectral and spatial resolution than the multispectral scanner carried on the first three satellites in the series, see table 2. The frequency of coverage is expected to be very similar to that of each of the earlier LANDSAT satellites.

Table 2. Thematic Mapper Characteristics

Spectral Bands	Wavelength (μm)	Resolution (m)
Blue	0.45 - 0.52	30
Green	0.52 - 0.60	30
Red	0.63 - 0.69	30
Near infrared	0.76 - 0.90	30
Mid-infrared	1.55 - 1.75	30
Infrared	2.10 - 2.35	30
Thermal infrared	10.4 - 12.5	120

Grey level quantisation: 256 levels (8 bits).

The first satellite in the French SPOT (Système Probatoire d'Observation de la Terre) (due for launch in 1984) is planned to carry two scanners, the MRVIR and the HRV (the Medium Resolution Visible and Infrared Radiometer, and the High Resolution Visible (radiometer), respectively, to give a rather free translation from the French). The spectral resolution of the MRVIR is very similar indeed to that of the Thematic Mapper but its spatial resolution is very much poorer, namely 80m for bands 1 - 6 and 160m for band 7; the swath width is planned to be 240km. The HRV will have fewer channels but a much better spatial resolution (20m) is planned, see table 3; the swath width will be reduced to 60km. The repetitivity of the orbit of the SPOT satellite over a given point on the surface of the Earth is planned to be once in 13 days, i.e. somewhat more frequent than a single LANDSAT satellite. However, it is planned to produce stereoscopic images from different orbits, by tilting the mirror so that the swath scanned need not be vertically beneath the orbit of the satellite. Stereoscopic views of one area will therefore be produced at the expense of lost data for some other area; it remains to be seen how the system will actually be operated in practice and what the consequent frequency of coverage will actually be in the end. The combination of LANDSAT-D, -D' and SPOT should considerably enhance the accuracy that one can expect to obtain

Table 3. SPOT HRV characteristics

Channel	Wavelength (μm)	Resolution (m)
1	0.48 - 0.53	20
2	0.55 - 0.6	20
3	0.62 - 0.7	20
4	0.8 - 1.05	20

in satellite-based remote-sensing work for estuarine and coastal waters in the next few years.

Plans are at the moment being laid by ESA for a mission ERS-1 (previously known as COMSS, see Paci (1980), Duchossois and Plevin (1981)) to include a scanner called the OCM, which has already been mentioned briefly in section 4.3; this instrument can be regarded in many ways as a successor to the CZCS and some of its characteristics are indicated in table 4.

Table 4. Principal OCM requirements

Channel	Wavelength (nm)	3 dB width (nm)	Spatial resolution (m)	Principal physical parameters
1	400	20	800	yellow substance absorption
2	445	20	800	phytoplankton absorption
3	520	20	800	phytoplankton absorption and/or turbidity
4	565	20	800	turbidity
5	620	20	800	turbidity
6	785	30	800	atmospheric correction
7	1020	60	800	atmospheric correction
8	2200 (or 3700)	400	800	sunglint correction
9	8500	500	800	) surface temperature and
10	10800	1000	800	) atmospheric correction
11	12000	1000	800	)

It should have better spectral resolution, and possibly marginally better spatial resolution than the CZCS. Thus, provided this programme goes ahead successfully, we can look forward to ocean-related remote sensing data in the later 1980's capable of giving better results on "water quality" than we are currently able to obtain from the CZCS. Of course, the OCM is just one of several instruments planned for inclusion in the payload of ERS-1, but I have only mentioned the OCM because I restricted the aim of this paper to cover optical and infrared scanners. The whole ERS-1 programme should provide a very powerful package for studying a wide range of ocean parameters and certain meteorological parameters closely related to the behaviour of the coastal waters and oceans (see, for example, table 4.5 of Duchossois and Plevin (1981)).

6. CONCLUSION.

Technologically it is clear, provided remote-sensing satellite programmes that are currently planned actually go ahead successively, that we should see improvements in the supply and quality of remotely-sensed data for marine applications in the next few years. However, this conference is entitled "Matching Remote Sensing Technologies and their Applications - the Past and the Future". I would like to stress that the problems are not entirely technological problems. It is rather pointless investing millions of pounds or dollars or francs or marks or whatever in building a satellite, in launching it, in controlling it in its orbit, in setting up ground stations to recover the data and establishing a data-distribution network, unless people are actually going to make use of the data once it has been generated. As far as ERS-1 is concerned it is the task of ESA to extract funds from its member states to finance the programme right through as far as the data-distribution network. But then it is up to the workers in the individual member states to make use of the data. It is not the job of ESA to follow through with the actual exploitation of the usefulness of the data.

The point we have to stress to Governments in Europe, and to our own Government in particular, is that their responsibility should not end with the payment of their contributions to ESA. If they stop there, then they will not have obtained the full value for the money they have spent. It is necessary to ensure that there are resources and manpower available to process and interpret the data and to try to ensure that the data are used for the benefit of mankind and not just stored in a cupboard. In the end it all comes down to money; our Government must be made to appreciate that we shall need substantial sums of money for facilities, for people and for the education and training of people to work with these data that we expect to be produced by ERS-1. This is not a new problem, Mr. T.F. Smith of the Dept. of the Environment was making this point, albeit in connection with a different satellite, at the "Parliamentary Hearing" of the Council of Europe (1978) in Toulouse: "We have heard the amount of money that is being spent on hardware, the launching of the satellite and the provision of the sensor. But the amount of money in the whole of Europe that is being spent on the end use of the data and making it into useful information is very, very small indeed and I would plead for those responsible to look at the programme as a whole, that is the launch, the sensors and the end users and make sufficient provision to use the data which is so expensive to obtain. If we are committed to these millions of pounds every year on annual programming getting the data, sufficient must be allocated to process the data to be able to use it usefully." There has been some improvement since 1978 but the problem is still there. It is easy enough, as always, to try to blame someone else. But perhaps we should each one look at ourselves and see whether we have been sufficiently "aggressive" in publicising the uses of remote sensing, in trying to improve education and training in remote sensing, in trying to extract funds from the Research Councils and other bodies to finance remote sensing work.

7. REFERENCES.

ANDERSON, J.M. and CALLISON, R., (1981): Interpretation of aerial infra-red survey data in Remote Sensing in Meteorology, Oceanography and Hydrology (Ed. A.P. Cracknell) Ellis Horwood, Chichester, 508-512.

BRIMACOMBE, C.A., (1981): Atlas of METEOSAT Images, ESA, Paris, SP-1030, 494pp.

COUNCIL OF EUROPE (1978): European Parliamentary Hearing on Europe's Specific Needs in the Field of Remote Sensing, Council of Europe, Strasbourg, 158pp.

CRACKNELL, A.P. and SINGH, S.M., (1980): The determination of chlorophyll-a and suspended sediment concentrations for EURASEP test site, during North Sea Ocean Colour Scanner experiment, from an analysis of a LANDSAT scene of 27th June 1977 Proceedings of 14th International Congress of the Society of Photogrammetry, Hamburg, July 1980, International Archives of Photogrammetry 23,B7, 225-34.

CRACKNELL, A.P. and SINGH, S.M. (1981a): Proposed use of System ARGOS with data buoys for calibration of thermal infra-red imagery of North British waters for sea surface temperature maps. Proceedings of ARGOS Workshop, Bergen March 1981. Service ARGOS, Toulouse.

CRACKNELL, A.P. and SINGH, S.M., (1981b): Coastal zone research work at Dundee University using remote sensing techniques. Proceedings of EARSeL-ESA Symposium on Application of Remote Sensing data on the Continental Shelf. Voss, May 1981. ESA, Paris, SP-167, pp157-68.

CRACKNELL, A.P., SINGH, S.M. and MACFARLANE, N., (1980): Remote Sensing of the North Sea using LANDSAT-2 MSS and NIMBUS-7 CZCS DATA, Proceedings of the 14th International Symposium on Remote Sensing of Environment, Costa Rica, April 1980, Environmental Research Institute of Michigan, Ann Arbor, pp1643-51.

DEUTSCH, M., VOLLMERS, R.R. and DEUTSCH, J.P., LANDSAT tracking of oil slicks from the 1979 Gulf of Mexico oil well blowout. Proceedings of the 14th International Symposium on Remote Sensing of Environment, Costa Rica, April 1980, Environmental Research Institute of Michigan, Ann Arbor, pp1197-211.

DOMAIN, F., CITEAU, J. and NOEL, J., (1980): Sea Surface Temperatures Studied by METEOSAT Data along the Coast of Senegal and Mauritania in Coastal and Marine Applications of Remote Sensing, Ed. A.P. Cracknell, Remote Sensing Society, Reading pp59-67.

DUCHOSSOIS, G. and PLEVIN, J., (1981): Remote Sensing from Space: Review of Future Plans and Prospects, in Remote Sensing in Meteorology, Oceanography and Hydrology Ed. A.P. Cracknell, Ellis Horwood, Chichester, pp66-80.

FOOTE, R. and DRAPER, L.T., (1980): TIROS-N Advanced Very High Resolution Radiometer (AVHRR) in Coastal and Marine Applications of Remote Sensing, Ed. A.P. Cracknell, Remote Sensing Society, Reading pp25-35.

FUSCO, L., (1981): NIMBUS-7 Coastal Zone Colour Scanner Data Processing for EARTHNET - Experience to Date ESA Bulletin, No 27, 36-42.

HAYES, R.M., (1980): Operational Use of Remote Sensing during the Campeche Bay Oil Well Blowout, Proceedings of the 14th International Symposium on Remote Sensing of Environment, Costa Rica, April 1980, Environmental Research Institute of Michigan, Ann Arbor, pp1187-96.

PACI, G., (1980): Results of Studies on a Coastal Ocean Monitoring Satellite System, in Coastal and Marine Applications of Remote Sensing, Ed. A.P. Cracknell, Remote Sensing Society, Reading pp105-19.

PHULPIN, T. and DESCHAMPS, P.V., (1980): Estimation of Sea Surface Temperature from AVHRR Infrared Channels Measurements, in Coastal and Marine Applications of Remote Sensing, Ed. A.P. Cracknell, Remote Sensing Society, Reading pp47-68.

SINGH, S.M. and CRACKNELL, A.P., (1979): Analysis of LANDSAT scene of 27th June 1977 for EURASEP test site and correlation with OCS results in Proceedings of Workshop on the EURASEP Ocean Colour Scanner Experiments 1977 Ed. B.M. Sørensen. Joint Research Centre, Ispra, pp171-9.

STURM, B., (1981): The atmospheric correction of remotely sensed data and the quantitative determination of suspended matter in marine water surface layers in Remote Sensing in Meteorology, Oceanography and Hydrology Ed. A.P. Cracknell, Ellis Horwood, Chichester pp163-97.

VALERIO, C., (1981): Airborne remote sensing experiments with a fluorescent tracer in Remote Sensing in Meteorology, Oceanography and Hydrology Ed. A.P. Cracknell, Ellis Horwood, Chichester pp218-48.

WEINREB, M.P. and HILL, M.L., (1980): Calculation of Atmospheric Radiances and Brightness Temperatures in Infrared Window Channels of Satellite Radiometers NOAA Technical Report NESS80 US Dept. of Commerce, 40pp.

Note added in proof: This paper was written and typed before it was announced by ESA that the OCM had been removed from the baseline payload for ERS-1; it is understood, however, that development work on the OCM will continue with a view to having it flown on later satellites. This development affects some matters of detail, but not general principles, outlined in this paper.

Part C

GEOMETRY AND CARTOGRAPHIC ASPECTS

GROUND CONTROL POINTING OF THE UK

M R Gordon
National Remote Sensing Centre
Space Department
Royal Aircraft Establishment
Farnborough
Hants

Abstract: Landsat images as obtained from Earthnet contain residual distortions due to Earth curvature and satellite attitude and orbit variations. For most purposes these errors are not significant, but geometrically corrected images are needed:

1. to overlay scenes of the same area taken at different times and/or with different sensors.

2. to bring adjacent images into registration.

3. to locate common points of interest in different scenes of the same area.

To facilitate this it was decided in April 1980 to produce software to transfer Landsat imagery into the British National Grid. The main part of this task has been to compile a library of ground control points (GCPs) for Great Britain using 1:50000 scale maps.

For each GCP the map and image coordinates are stored and when a scene has been completed (ie about 120 points found) a least squares fit is carried out to calculate the transformation between image and map coordinates. The residuals in the fitting process give an indication of the transformation accuracy and typically produce RMS errors of 50 metres.

Criteria for choosing these GCPs are given together with an explanation of how their image and corresponding map coordinates can be obtained.

Paper presented at a conference on Matching remote sensing technologies and their applications convened in London by the Remote Sensing Society, 16–18 December 1981

1. INTRODUCTION

Most of the data received in the National Remote Sensing Centre at the Royal Aircraft Establishment is from the Landsat Multi Spectral Scanner (MSS) and is supplied by Earthnet. This data has already had corrections applied to remove the worst errors caused by earth curvature and rotation, satellite attitude and sensor non-linearities. However, significant geometric distortions remain with the centre of an image only located to an accuracy of 1 km.

Often these errors do not matter but geometrically correct images are needed:

1. To bring adjacent scenes into register, eg to join two images side by side.

2. To locate common points of interest in different scenes of the same area when the map references of the points are known.

3. To overlay images of the same area:

   a. Taken at different times; to study for example vegetation changes or flood areas.

   b. Taken using different sensors, eg the improved resolution obtained by combining MSS and Return Beam Vidicon (RBV) data.

To facilitate transforming images into a common projection it was decided in April 1980 to compile a library of ground control points of Great Britain based on the British National Grid. This has since been extended to include a set of points for Northern Ireland based on the Irish National Grid.

It is the factors determining the choice of the ground control points (GCPs) and the compilation of these libraries that is the subject of this paper.

2. SELECTION OF GROUND CONTROL POINTS

The first task is to choose a suitable Landsat band 7 image of the area to be ground control pointed, along with a set of maps covering the area. Band 7 images are used as these usually give the clearest single band detail, especially in the case of land/water boundaries. The maps are ideally of 1:50000 scale but smaller scale maps have sometimes had to be used, eg the Irish 1:126720 (half inch) series. They should have a suitable grid; to date maps with some form of Transverse Mercator Grid have always been used. Assuming that several images of the area are available then the chosen image should not only be as free from cloud as possible but checked for haze and also for data faults. The most common data faults are missing or slipped lines, both of which usually occur in multiples of six lines. Images with missing lines should be discarded but those with line slippages may be used provided that care is taken not to choose any GCPs near to the faulty lines.

Having chosen the best available image, some 100 to 120 GCPs are found in a full 185 km by 185 km Landsat scene. (This is a GCP every 15 to 20 kilometres.) The objective is to get an even spread of GCPs but with more points in coastal regions. This increased density of GCPs near coastlines is to ensure that images with only small areas of land on them still contain a substantial number of GCPs. Finding this large a number of GCPs is a lengthy process, but once completed it ensures that subsequent geometric correction exercises can be carried out on images with a considerable amount of cloud or haze. It also gives an insurance against GCPs being lost due to changes of shape, drought, civil engineering, etc. Furthermore, little knowledge of satellite orbit or attitude data is required, unlike other methods which only need a small number of GCPs (Sawada et al, 1981).

The GCPs are of course features that can be identified on both the image and the map; the usual features chosen are as follows:

1. Motorway junctions. These usually show up clearly on images.

2. Bends and junctions of various types in roads, rivers, canals and railways. Obviously these usually show up with a clarity proportional to their width but the contrast in the reflectivity of surrounding areas of the image often modifies this. Narrow features of this type sometimes show up better than expected due to their apparent width being increased by adjacent buildings or woods.

3. Lakes etc. These, in common with most land/water boundaries, make very good GCPs, provided some thought is given to the possible effects of unusually high or low water levels. For this reason the middle of an island (if any) in the lake is ideal. Failing this the centre of a dam at one end of the lake, or a point on the shore against steeply rising ground, should be chosen, as both give only small position changes in the land/water boundary as changes occur in the water level. If none of these is available then the centre of the lake itself is to be preferred to a point near a shallow shoreline.

4. Coastal features make excellent GCPs, which in view of the need for a greater GCP density in their vicinity is fortunate. Ideal locations are small islands, docks that do not dry out at low tide and steep-sided headlands and coves. Areas with a large intertidal zone may be used if the land/intertidal zone boundary is distinct, otherwise they should be avoided if possible.

5. Airfields make very good GCPs if they have intersecting runways or if runway/taxiway junctions are clearly visible. Ends of single runways should not be used as locating the exact end of a runway is almost always impossible.

6. Large areas of woodland sometimes have to be used if nothing better is visible in an image. The problem with using woods is that their boundaries change over the years as trees get cut down or burn down and appear to spread as adjacent areas are afforested.

These are the usual features used as GCPs and most images have a good selection of them. Unfortunately they are not always evenly spread over the image and sometimes quite small ponds and woods have to be used in areas where little else can be made out.

3. COMBINING THE MAP AND IMAGE COORDINATES

Having chosen the GCPs, the next task is to combine the map and image coordinates. This is done using a GEMS display system together with software developed at the National Remote Sensing Centre. The GEMS system is used to rapidly display images and to superimpose a set of manually controlled cross-wires on them. The image coordinates of these cross-wires can be automatically stored.

The band 7 image is displayed after having been sampled so that it fits on to the screen. An area containing a GCP is then chosen using the cross-wires and a sub-image of 257 by 257 pixels centred on the cross-wires is displayed on the screen at full resolution in an edge enhanced form. The cross-wires are then aligned on the approximate GCP position and an area of 65 by 65 pixels centred on the cross-wires is displayed with a four times magnification. This magnification is a two-stage process consisting of a two times software magnification using Hanning interpolation, followed by a two times hardware magnification using pixel replication. This system of magnification gives an acceptable visual result quickly. Using this enlarged part of the image the cross-wires can then be centred on the GCP. The image coordinates are automatically stored as column and line values measured from the top left-hand corner of the complete image. The map grid coordinates are entered in the form of grid eastings and northings, along with the GCP identification, altitude and description, from the keyboard.

After six GCPs have been so recorded, an approximate transformation matrix for the image is calculated. The procedure now becomes faster as the GCP grid coordinates are entered on the keyboard and an enlarged 65 by 65 pixel sub-image, centred on the image coordinates derived from this matrix, is displayed. The cross-wires are then centred on the GCP and the correct image coordinates stored. This process continues, with periodic recalculation of the transformation matrix, until all the chosen GCPs have been entered. As an enlarged portion of the original image has been used to store the image coordinates, they have been recorded to an accuracy of better than one pixel. To assist in the next stage of processing the two sets of coordinates are split into two files, one with the image values and the other with the map values. At this stage the image coordinates are converted to the nearest whole pixel value and the transformation matrix is used to derive the corresponding map coordinates. A linear fit of this set of GCPs usually gives an RMS error in the order of 100 metres but this accuracy can be improved by the following procedure.

4. IMPROVING THE FIT

First of all a set of sub-images, called patches, centred on the image coordinates of each GCP, is prepared as a photographic negative using a Linoscan film writer. These patches have their central pixel marked with a cross. They are made to be as nearly as possible the same scale as the maps being used, and also to be always approximately the same physical size regardless of map scale. For example, using a map scale of 1:50000, a sub-image of 25 by 25 pixels magnified twenty times is used to give a patch scale of 1:50000. Whereas when a map scale of 1:63360 (1" to 1 mile) is used, a sub-image of 31 by 31 pixels magnified sixteen times is produced to give a patch scale of 1:62500, this being the nearest approximation to 1:63360 possible. The size of 31 by 31 pixels at 1:62500 gives approximately the same size patch as 25 by 25 pixels at 1:50000 scale.

As an aid to positioning the patches on the maps, the local orientation angle of the satellite orbit at the centre of the image is calculated by:

$$\text{Cos } c = \text{Cos } i/\text{Cos } l$$

where:

c is the orientation angle at the image centre,

i is the orbit inclination,

l is the latitude at the image centre.

Using this orientation angle as a guide, each patch in turn is aligned over its GCP on the map. A pin is pushed through the cross at the centre of the patch and the map reference of the pin hole recorded. This procedure has to be carried out with care, especially if the patch scale is not quite the same as that of the map. Using these revised map references the coordinates in the previously formed map file are corrected one at a time and a new linear fit of the GCPs carried out after each correction. With each refit the RMS error should decrease. If any "correction" causes an increase in this error the patch of the offending GCP is rechecked against the map to see if the pin hole can legitimately be repositioned. Some corrections never improve the fit no matter how carefully the patch position is checked, this is most probably due to pixel stuffing. Pixel stuffing is part of the Earthnet initial correction procedure whereby extra pixels are inserted into the image to remove some of the larger distortions. If a patch includes any of these extra pixels then it's centre can never be properly aligned with the map.

On completion of this rather lengthy process the RMS error for a linear fit of the GCPs is usually in the order of 50 metres.

5. STORING THE GCPs

For each completed image, three sets of information are stored for future use.

Firstly small sub-images centred on each GCP and called 'chips' are stored in a library for future use with automatic GCP locating programs (Benny, 1981). These chips are 19 by 19 pixels in size and are given file names which correspond to the previously used GCP identification. This GCP identification is based on the map used to find the GCP and consists of the International Registration Letters for the country concerned, the map sheet number and the point number on that map. An allowance of four characters is made for the Registration Letters so that some modification can be made if necessary. For example, points taken from the Northern Ireland one-inch series are named GBNI and those taken from Admiralty charts are named GBAD. The map sheet number has an allowance of six digits and the point number on the map can be a two digit number.

Secondly the image coordinates and GCP identification are stored along with the image path, row and date. The file containing these values is called the Image Master File.

Lastly, for the map values the GCP identification is stored along with the grid eastings and northings, the GCP altitude and its description.

Unlike the image coordinates where only one Master File is used, several Map Master Files are available, based on the various map grids being used.

In the case of Great Britain, four different map grids have been used. These are the British National Grid (BNG), the Irish National Grid (ING) and the Universal Transverse Mercator zones 30 and 31 (UTM30, UTM31). This multiplicity of grids is due to the fact that Landsat scenes spread beyond national boundaries, thus images of Great Britain include parts of both Ireland and France. Similarly, images of Northern Ireland include parts of Great Britain as well as the Irish Republic. All GCPs taken from Great Britain are stored in the BNG Master File, all GCPs taken from Ireland (north or south) are stored in the ING Master File, whilst those taken from France are stored in either the UTM30 or UTM31 Master Files. As mentioned above, images of Great Britain often include parts of Ireland or France and so all GCPs recorded in ING, UTM30 or UTM31 that could possibly appear on a Great Britain image are also stored in the BNG Master File after their grid references have been converted to BNG coordinates. Similar conversions of points from the BNG are carried out to ING, UTM30 or UTM31 as appropriate and stored in the relevant Master Files. There is also an interchange of GCPs between UTM30 and UTM31 to allow for image overlaps between these two grid systems.

6. RESULTS

Using the above procedures a library of some 1500 GCPs has been built up for Great Britain. This quantity is sufficiently large to allow images to have up to 50 percent cloud cover and yet still produce an RMS fitting error of about 50 metres. This error is considerably better than one Landsat pixel and is directly comparable with the errors in the method, these being:

1. An error of up to 20 metres in positioning the patch on the map.

2. An error of up to 20 metres in making the pin hole.

3. Errors in the map scale and mapping techniques, which at a scale of 1:50000 may be up to 50 metres.

These results have been achieved with minimal knowledge of the satellite parameters, the only values required being the ground area covered per pixel, to determine the required patch scales, and the orbit inclination combined with the satellite heading, to get the local orientation angle for the patches.

7. REFERENCES

SAWADA, N., et al (1981): An analytic correction method for satellite MSS geometric distortions. Photogrammetric Engineering and Remote Sensing 47 No. 8 1195-1203.

BENNY, A.H., (1981): Automatic relocation of ground control points in Landsat imagery. This Conference.

AUTOMATIC RELOCATION OF GROUND CONTROL POINTS IN LANDSAT IMAGERY

A H Benny
Space Department
Royal Aircraft Establishment
Farnborough
Hampshire
GU14 6TD

Abstract: Images obtained from the Landsat multispectral scanner are not provided in any normally recognised map projection system, so it is often necessary for a transformation to be made. This is usually done with the aid of ground control points, whose locations on the ground (or on a map) are accurately known and which can be accurately located within Landsat images. The initial selection and location of such ground control points is usually performed manually.

For repeat - ie multitemporal - imagery, the same ground control points can be used. The positions of these points can be relocated within the repeat image either manually or, as described in this paper, by an almost entirely automatic computer process. This process has been found to work speedily and effectively on many repeat images of Britain, usually giving accurate relocation of well over 90% of ground control points.

The process has been developed for use with images of Britain but it could readily be adapted for worldwide use.

1. INTRODUCTION

Each image provided by Landsat MSS (USGS, 1979) covers an area approximately 185 km square, viewed vertically downwards. Even after system correction (ESA, 1976) has been applied, a small amount of spatial distortion is present, and the image lacks a conventional cartographic grid (Colvocoresses, 1974). A convenient treatment is to transform the image to a known cartographic coordinate system, this process making use of recognisable terrestrial features referred to as ground control points (gcps), identifiable both on the satellite image and also on suitable maps. When several gcps have been located, both on the image and on a map, a transformation matrix can be obtained, which establishes a correspondence between all points in the image and the map coordinate system. Then the image may be converted, eg by pixel resampling, this process being referred to as geometric transformation (Shlien, 1979; Williams, 1979).

The identification of gcps, both on an image and on a map, is, in the first instance, an operation which is done manually. However, the Landsat system is a continuing operation, so repeat images are produced for each area, and it is necessary to recognise and relocate known gcps in these repeat images.

Paper presented at a conference on Matching remote sensing technologies and their applications convened in London by the Remote Sensing Society, 16-18 December, 1981.

A distinction should therefore be clearly drawn, between (a) the first identification and location of gcps both on an image and on a map, and (b) the relocation of known gcps in other images. (Relocation on a map is, of course, not needed). (a) has been performed manually (Gordon, 1981) and the results have been entered into data files ready for use by the computer. This paper is concerned with (b) and describes an almost entirely automatic computer process for relocation of specified gcps within a repeat image.

2. AUTOMATIC RELOCATION OF ONE GROUND CONTROL POINT

2.1 Ground control point chips

The task of establishing the location of a gcp in an image is essentially one of pattern recognition. Any gcp is by definition the description of some terrestrial feature of non-zero size. For example, the definition "the intersection of the M4 and M5 motorways" whilst it refers, as a location, to the point where the centre lines of those roads intersect, in fact only has meaning with reference to the existence of the roads themselves for some short distance at, and to either side of, the intersection. The actual location of a gcp therefore, is determined by the description of one particular point in its surrounding context.

For the first location the pattern recognition problem is that of relating a portion of an image with a portion of a map, ie comparing data in digital and non-digital form and this is at present best done using human intelligence.

For the problem of relocating gcps the situation is greatly eased, as it reduces to the matching of portions of one image with another image, that is, both elements are in the same (digital) format.

In readiness for the matching process, small chips of the original image have been prepared, of size 19 by 19 pixels. A library of over 1000 gcp chips is available. The map coordinates of the centres of these chips are recorded in a computer data file.

The problem of relocating gcps is much easier than the general pattern recognition problem since (a) the satellite height varies very little, hence the size of a pixel is almost identical in every image, and (b) the direction of the satellite path varies very little (ie its angle to the local meridian), hence the gcp is always orientated at about the same angle. Thus the pattern recognition problem becomes the manageable one of comparing and relating images of almost identical size and orientation. It is true that the object being imaged may change from one satellite pass to another, eg a field may be ploughed on one occasion and have growing crops on another, and such changes, discussed later, can make the relocation problem more difficult if not impossible in, for example, the case of a snow-covered scene, or of course a cloud-covered gcp.

2.2 Pattern matching by cross correlation

The method of establishing the correspondence of chip and new image location is by means of the correlation coefficient, CC, calculated between the 19 by 19 (=361) pixels of the chip, and a 19 by 19 pixel section of the new image. The correlation coefficient is calculated for any image location by the conventional method:

$$CC = (\Sigma(x_i-x_a)\ (y_i-y_a))/(\Sigma(x_i-x_a)^2\ \Sigma(y_i-y_a)^2)^{\frac{1}{2}} \ \ldots\ldots\ldots\ldots\ldots \ (1)$$

where x_i = i'th pixel value of the chip

y_i = i'th corresponding value for the image

x_a = average of all 361 chip pixel values

y_a = average of all 361 image values

and the summations are conducted for i = 1 to 361.

Expression (1) may be rearranged in the form:

$$CC = (N\Sigma x_i.y_i \ - \Sigma x_i \ \Sigma y_i)/((N\Sigma x_i^2 - (\Sigma x_i)^2)(N\Sigma y_i^2 - (\Sigma y_i)^2))^{\frac{1}{2}} \ \ldots(2)$$

where N = 361, the total number of x_i or y_i values, and the summation is as before.

Expression (2) is more convenient than (1) for computer evaluation because it does not involve the pre-calculation of the average values x_a and y_a. Expression (2) may be written as:

$$CC = R1/(R2.R3)^{\frac{1}{2}} \ldots\ldots\ldots\ldots\ldots\ldots\ldots\ldots\ldots\ldots\ldots \ (3)$$

the meanings of R1 R2 and R3 being clear from inspection of (2). Since R2 contains only x_i values, it may be calculated only once for the particular chip, and will then be available for all further correlation calculations for that chip, whereas the terms R1 and R3 must be recalculated for each position of the chip in relation to the new image. If a portion of the image is uniform (as sometimes occurs) the calculation of CC is indeterminate, so this condition must be recognised and the calculation abandoned.

The calculation of R1 and R3 requires a significant amount of computer effort, as each summation involves 361 values, each of which is the result of a multiplication of two numbers. Further, the expression for CC has to be performed many times for each gcp (see below). In fact, the largest single activity performed by the program is evaluating CC. Any means of reducing this work would therefore be welcome. Barnea and Silverman (1972) discuss a class of algorithms for fast digital image registration. It seems, however, that their method is not applicable to the present problem. It is of interest to note that Orti et al (1979) chose to use the cross-correlation coefficient method for matching purposes.

The numerical value of a correlation coefficient indicates the amount of correlation between the two items being compared. CC values range from +1.0, indicating perfect correlation, or identical patterns, to −1.0, occurring if one pattern is the negative of the other. CC values near to zero imply little correlation. Experience of CC values found with Landsat MSS imagery is mentioned later.

2.3 The correlation coefficient surface

For any particular gcp chip it is possible to calculate CC for many locations on the image. The values can be displayed in three-dimensional form, with the two horizontal axes being the column and row directions of the image and vertical direction being the value of CC, thus forming a CC surface. In practice it is found, as might intuitively be expected, that this surface has the shape of an irregular hill standing out from a surrounding plain having mean value zero, but with local noise leading to individual random values in the range of about -0.25 to +0.25. The peak of the hill corresponds with the position of maximum CC, and this is the best relocation place for that chip on that region of the image.

In practice, the CC surface may show a number of other peaks, these being due to other ground features having a similar general appearance. For example, a chip centred on a small body of water such as a reservoir, will correlate with other such bodies of water of approximately similar shape and size. It is possible therefore, to obtain apparently good relocations which are in fact incorrect.

2.4 Relocating a ground control point

In operation, the exact location of a chip in the new image is not of course known (otherwise there would be no need to find it), but, as will be shown later, the approximate location will be known, together with some measure of the probable maximum distance away from the actual location. These values - approximate column and row number and largest expected distance from the actual location - are passed to a searching subroutine which operates by conducting an outward rectilinear spiral search from the approximate location. The search is stopped either when the gcp has been as positively located (as explained below), or when the specified number of turns of the spiral have been completed.

Due to the method of operation of the main program, it is acceptable to have an occasional mislocation. Experience with many hundreds of gcps in various images of Britain shows that the following algorithm gives an acceptably high proportion of correctly located gcps.

As the spiral search proceeds, a correlation coefficient is calculated for each position. The value of the first CC is stored and each time a higher CC value is calculated, the previous highest value is replaced. If at any time the maximum CC is greater than 0.5 and two complete rings of the spiral have been performed without a higher CC value being found, then the subroutine is brought to a halt, as consideration of the CC surface shows that it is likely that the correct location has been found and passed. In the absence of such a positive relocation, the search continues until the specified number of rings have been examined. The highest CC value found is returned, together with its position. A maximum CC of less than 0.3 is arbitrarily taken as meaning that no relocation can be found.

2.5 Location to a fraction of a pixel

Inspection of CC surfaces revealed that often the largest CC value had an almost equally large adjacent value, implying that the peak lay between the two. This suggested the possibility that a more accurate value could be determined by starting from the integer location of maximum CC and spirallying outwards in rings of 0.1 pixel rather than the usual 1.0 pixel. To perform the calculations of CC, a grid of pixel values at intervals of 0.1 column and row is needed, and these are obtained by linear interpolation from two, or more usually, four, adjacent pixel values. This method enables the location to be obtained to the nearest 0.1 pixel. Linear interpolation is chosen as probably the quickest method, and although clearly not mathematically the most accurate is probably more than sufficient for the present purpose.

3. AUTOMATIC RELOCATION OF ALL GROUND CONTROL POINTS IN AN IMAGE

A computer program was evolved which uses the process described in section 2 to relocate automatically most or all gcps within an entire Landsat image. This program requests information on the expected approximate location of one gcp (chosen by the user) and then automatically conducts its search, first finding the position of the chosen gcp, next relocating the majority of gcps and finally providing a list of their locations. This program will now be described in some detail.

3.1 Relocation of first ground control point

The choice of the first gcp to be located is at the user's discretion. It should preferably be one which stands out clearly from its surroundings, is well away from the edges of the image and is closely surrounded by other gcps.

If only cloud-free images were to be examined, one gcp could be selected as the standard first gcp for each track/frame in the Landsat system. However this program may successfully be used on images which have considerable cloud cover, provided that a proportion of the gcps are visible, and in such cases user discretion in the choice of gcp is needed.

The user is asked for some indication of the position of the first gcp, as the search is only conducted up to 40 pixels in all directions. It is sufficient to provide measurements taken by eye with a ruler from a 180 mm square Linoscan (Neilson, 1979) picture of the Landsat image.

When this information has been provided, all other work is done automatically.

First the map location file is searched for the chosen chip and the relevant map easting and northing are read. An attempt is then made to relocate the chosen gcp chip by conducting a search from the location provided, for up to 40 rings or turns of the spiral. If the gcp is not relocated the user will be requested to supply a different starting position, or choose another gcp. In practice the first gcp chosen is usually relocated, but it may be useful to have chosen a second one for occasional cases of difficulty.

3.2 Estimation of initial transformation matrix

Once a chosen gcp has been relocated within the image, the program can calculate the coefficients of linear transformation equations of the form:

$$X = ax + by + c$$

$$Y = dx + ey + f$$

where:

X, Y = column, row of the new image
x, y = easting, northing of the coordinate system
a, b, c, d, e, f are coefficients

This allows the approximate positions of the second and subsequent gcps to be calculated. It is recognised that these coefficients are approximate, but is only necessary for them to be good enough to enable several nearby gcps to be located.

The calculation of the initial matrix assumes a pixel size of 57 by 79 metres, these being typical values. The approximate expected orientation of the image with respect to the map coordinate system is calculated for that image track and frame number. The rotations of x and y directions are both taken to be this value, which implies that the image is not sheared. The angle of shear depends on the yaw angle of the satellite, ie any deviation of the scan direction from perpendicularity with the direction of satellite travel, combined with earth rotation effects, and it has proved satisfactory for the present purpose to assume zero shear. Using these pixel values and this rotation angle, four of the six matrix coefficients can be calculated. These four values determine the x and y scales and orientation of the image in relation to the map coordinate system. The final two matrix values determine the relative x and y displacements and are provided by the relocation position of the gcp within the image, together with the map coordinates of the gcp.

3.3 Relocation of other gcps

After the first gcp has been successfully relocated and the initial transformation matrix prepared, the relocation of other gcps can commence. Since the transformation matrix is known to be only an approximation, the accuracy of prediction of the position of other gcps will become progressively worse as the distance of the gcp from the first one increases. The next operation, therefore, is a scan through the map location file to list all gcps which seem likely, on the basis of the initial matrix, to be inside or near to the image in use. Thus a batch of, typically, 100 to 150 gcps is formed and these are placed in order of distance from the starting gcp.

Each gcp in the batch is then examined, in ascending order of distance, in a similar manner to the first one. Its approximate location is calculated, using its easting and northing together with the initial transformation matrix, and an attempt is made to relocate it. Since the estimated position of each gcp should now be known much more accurately than for the first gcp, a much smaller spiral search is sufficient, the number of rings specified being dependent upon the distance of the gcp from the initial one. In practice this is found to be suitable, provided that the first gcp has been correctly located: if it has not, then there is little likelihood of any others being correctly found. As each gcp is successfully relocated, details of the position and correlation coeffficient are stored. Information on the progress of the program can optionally be provided at the user's terminal, or the details may be saved for subsequent inspection.

After 10 gcps have been relocated, with CC of 0.6 or more, a new first-order transformation matrix is calculated, based on the actual positions of the 10 gcps, and this should represent an improvement upon the initial transformation matrix, which was not based on actual gcps. (The requirement for a CC of over 0.6 tends to ensure that only confident relocations are used, not doubtful ones.) Following this calculation, subsequent searches only explore up to 12 rings, as the better transformation matrix now available increases the accuracy of the estimated positions.

All the remaining gcps in the batch are examined, with a 12-ring search, calculating a fresh transformation matrix after every ten more have been relocated. When all in the batch have been examined, a final transformation matrix is calculated, together with least-squares residual errors for the position of each gcp.

3.4 Rechecking of doubtful locations

To complete the initial relocation process, poorly located gcps are rechecked. Usually some of the gcps have residuals which are rather larger than might be expected, this limit being arbitrarily specified as 3 pixels. Such gcps are therefore rechecked, by a repeat of the relocation process. It is sometimes found that a gcp can be relocated at a position more consistent with the others, usually because the previous search for that gcp started too far away to find it, or perhaps because a lower peak in the CC surface had been met previously. The number of such better relocations is usually small, perhaps two per image, dependent on the quality of that image. An image with haze is amongst the most troublesome, leading to many mislocations which have apparently acceptably high CC values.

3.5 Completion of the process

The last stage of the gcp relocation process is to perform a search to the nearest 0.1 pixel, on all gcps with residuals less than 7 pixels. Two spiral searches are made in this case: first a recheck of the integer location, followed by a fractional search. Occasionally the integer location is modified during the integer search, and this is usually for the better as judged by the resulting residual.

Finally the program outputs the results in the form of a list of a relocation positions to the nearest 0.1 pixel, together with the CC values. From this list, users can calculate a transformation matrix in any way they choose and use this to transform the image to the map coordinate system.

4. RESULTS AND CONCLUSIONS

The product of the process is a list of gcps together with their locations within the given image. Ideally, all gcps within the image will be relocated to within 0.1 pixel, but in practice, due to several causes, an accuracy of about 30 metres (0.4 pixels) is to be expected.

An alternative to using the process is to perform the operation manually, visually relocating each gcp in the new image (Method A). During program development, an interim process (B) was used, with the spiral search algorithm, but each gcp and its approximate position had to be selected by the user.

The success of the automatic process is shown by comparing it with the previous methods. For a typical image containing 100 gcps.

Method	Computer CPU Time	User man-hours
(A) Visual relocation	1 minute	80
(B) Semi-automatic	20 minutes	8
(C) Automatic	20 minutes	0.1

The human involvement, therefore, is reduced from about two weeks work (for about 100 gcps) to about five minutes, with similar success rate in number and accuracy of location of gcps. The automatic process takes typically 20 minutes to relocate 100 gcps, using a Prime 750 computer, and can be left to run unattended.

The process is therefore very much more convenient to use than manual methods. Automatic relocation has now been performed on many different images (considerably in excess of 100) by several users and no user would now contemplate manual relocation, although method B might be used for an image which was so cloud-covered that only a very small number of widely scattered gcps could be discerned. The process can certainly be said to be very much better than manual methods.

The proportion of gcps relocated within an image varies, depending on such things as the season of the year compared with when the gcp chips were obtained. For similar seasons, the proportion relocated approaches 100%, whilst in the worst cases attempted, about 60% of the gcps were found, when the image was about six months different in season from the original chips, with snow present in some portions of the scene. It may be that the visual method is superior for very difficult images, but in general such images are not of much interest to users.

The accuracy of relocation may be judged from the standard deviations of the gcps. In the best cases the standard deviation may be less than 50 m (about 0.6 to 0.8 pixel) with one or two out of perhaps 150 being badly out of place. In the worst situation, the standard deviation may be 120 m with up to 15-20% badly relocated, mainly because they cannot be correlated at all.

In summary, the process provides a rapid method of relocating gcps, with little user effort, and aids the transformation of large quantities of imagery.

5. REFERENCES

BARNEA, D. I., SILVERMAN, H.F., (1972): A class of algorithms for fast digital image registration. IEEE Transactions on Computers, C-21 179-186.

COLVOCORESSES, A. P., (1974): Space Oblique Mercator. Photogrammetric Engineering, 40 921-926.

ESA, (1976): Format specifications for Landsat MSS system corrected computer compatible tape. European Space Agency DPS-SES-NT/0230-02/200576.

GORDON, M. R., (1981): Ground control pointing of the U.K. (This Conference).

NEILSON, P. J. (1979): Use of the Linoscan 204D image generator in remote sensing applications. The Photogrammetric Record, IX (54) 849-852.

ORTI, F., GARCIA, A., MARTIN, M. A., (1979): Geometric correction of Landsat MSS images using a ground control point library. Proceedings 5 Annual Conference of the Remote Sensing Society, 17-26.

SHLIEN, S., (1979): Geometric correction, registration, and resampling of Landsat imagery. Canadian J Remote Sensing 5 No. 1 74-88.

USGS, (1979): Landsat data users' handbook. U.S. Geological Survey.

WILLIAMS, J. M., (1979): Geometric correction of satellite imagery. RAE Technical Report 79121.

THE USE OF TERRAIN HEIGHT INFORMATION WITH SATELLITE IMAGERY

P A Roberts
Space Department
Royal Aircraft Establishment
Farnborough
Hants
GU14 6TD

Abstract: One potentially important area of work is the use of topographic data in conjunction with satellite imagery to improve classification algorithms. This paper deals with techniques for the use of terrain height data in conjunction with such imagery. An algorithm is described that transforms digitised height contours into an equispaced grid of points. A method is also described which utilises this height matrix to derive the magnitude and direction of the slope of the terrain, information which can in turn be used to calculate the variation in direct solar illumination over an area. Finally, techniques are described for using terrain data to improve the reliability of multispectral classification of Landsat data.

1. INTRODUCTION

Much more information can often be obtained from remotely sensed data by using it in association with existing ground truth information. This paper is concerned with techniques for using topographic information from maps, in conjunction with imagery of land surfaces. An algorithm is described that can be used to transform digitised contour height information into an equispaced grid of points, a format that is suitable for easy manipulation and processing and for comparison with many types of digital imagery. The height matrix can in turn be used to derive the magnitude and direction of the slope of the terrain and the variation in the direct solar illumination arising from it. These quantities can all be displayed as grey tone images for independent interpretation or interpretation in conjunction with other imagery. In addition, the intensity of Landsat pixels can be adjusted so as to compensate for variations in direct solar illumination arising from undulating terrain.

Paper presented at a conference on Matching remote sensing technologies and their applications convened in London by the Remote Sensing Society, 16-18 December, 1981.

In this paper the factors that need to be taken into account when digitising a contour map are first briefly discussed. This is then followed by a description of a technique for producing a terrain height matrix from a map comprised of height contours. Subsequent sections deal with parameters that can be obtained from the terrain height matrix, the display of these quantities as grey tone images and finally some applications of this work to Landsat imagery.

2. DIGITISATION OF CONTOURS

Often the investigator will not have the contour information already digitised, but rather will only have a paper copy of the map thus necessitating digitisation of the contours. It is necessary therefore for the investigator to decide at what interval, around contours, points are to be digitised. The evaluation of this interval involves an estimate being made of the interval along a contour that can be represented by a straight line. This interval is dependent upon the grid spacing of the terrain height matrix and the curvature of the contour in question; the smaller the grid spacing or the greater the curvature of the contour, the smaller this interval must be in order to represent the contour to the required accuracy. More precisely, if d is the digitisation interval, r is the radius of curvature of the contour (which generally is a function of position along the contour) and s is the maximum permissible deviation of a straight line, joining two points on the contour, from the contour, then $d \approx \sqrt{8rs}$, where d is such that the angle the contour turns through is small. The author has found that for his work it is adequate that s be half the grid spacing, but this may not be acceptable for all applications.

If the digitisation interval is greater than d then the contour is under-digitised, and if it is less it is over-digitised. The only consequence of the latter is that the number of data points describing the contour is greater than necessary. Contour fitting algorithms that are considerably better than linear interpolation can be employed and should be adopted when contours are under-digitised. A description of one such algorithm can be found in Roberts (1980).

The digitisation of contours can take either of the following forms (amongst others):

i Constant separation between data points.

ii Variable density system with the interval between data points depending upon the curvature of the contour. This reduces the number of points required to describe a contour.

If it is decided to opt for a constant separation between data points then the map will be correctly digitised if the evaluation of the digitisation interval is based upon the smallest radius of curvature. The variable density method requires a continual evaluation of the curvature of the contour being digitised and is best suited to automated digitisation. If the digitisation is being performed manually using this method then the investigator must make a subjective estimate of the required digitisation interval.

3. TRANSFORMING DIGITISED CONTOURS TO A HEIGHT MATRIX

The basic problem in generating a height matrix is to fit a surface to the contour data such that every contour lies on the surface. Since it is, assumed that there will be no further information available as to the nature of the surface, the intuitively desirable conditions of smoothness and continuity are adopted. This surface can then be sampled at the appropriate points to produce the required grid. It should be remembered that the surface has to be constrained such that between contours it remains within the range of values defined by the contours.

The method adopted does not perform a true two-dimensional fit but instead takes the average of two orthogonal one dimensional fits. This involves two orthogonal sets of cuts being taken across the contour map; the points of intersection between these grid lines define the points at which interpolated data is required. For each grid line the points of intersection of the line with the contours are calculated and then interpolation is carried out along the grid line to obtain values at the grid points lying on the line. Also produced are weighting values reflecting the confidence in the interpolated values. This procedure results in two grids of points being obtained corresponding to the two sets of cuts, together with two grids containing the associated weighting values. These weighting values are used to combine the grids to produce the final height matrix.

It is necessary to take two mutually orthogonal sets of cuts across the map since it may be that some contours run parallel to one of the sets of cuts and thus would be ignored by that set unless they fell exactly on the cuts. Contours missed by one set of cuts are therefore incorporated in the orthogonal set. In this way the orientation of the cuts with respect to the contour map is not important, whilst it may be crucial if only one set of cuts were to be taken.

When converting a contour map into a height matrix it is necessary to define a grid spacing. It should be remembered that the smaller the grid spacing the more accurately the contours can be reconstructed from the height matrix, and in any case it should be no greater than half the scale size of the smallest feature it is required to preserve in the transformation. Figure 1 shows a 5 km square area of Hampshire after digitisation of the contours and spot heights from a 1:25000 Ordnance Survey Map (sheet SU54). This map has been converted into a grid 500 x 500 using a 10 m grid spacing. The accuracy of the height matrix can be ascertained by comparison with the spot heights for the area, although it should be realised that beyond the outermost contours the interpolation technique is not very reliable. Values can be derived from the height matrix at the positions of the spot heights by using the 16 surrounding matrix values. A comparison between the two sets of values is given in Table 1. As is to be expected, there are no differences greater than the contour interval of 25 ft. The mean difference between the two sets of heights is only 0.04, a value that is probably rather fortuitous for a sample of only 21 points. The standard deviation of the differences is 7.1 ft.

A much more detailed description of the algorithm used to generate the height matrix from contours is given in Roberts (1980). This reference also includes a description and a listing of a computer program that can be used for evaluation of the algorithm. Further information and refinements of the algorithm are to be found in Roberts (1981).

Fig 1 A 5km square of Ordnance Survey map SU54 after digitisation of contours and spot heights

(Note: all heights are in feet)

Source: Author

Table 1. Height matrix compared with spot heights for Figure 1

Spot Height (feet)	Grid Height (feet)	Difference (feet)
436.0	433.7	2.3
343.0	346.4	-3.4
365.0	359.3	5.7
313.0	320.6	-7.6
353.0	358.2	-5.2
281.0	294.2	-13.2
335.0	330.7	4.3
318.0	321.8	-3.8
602.0	600.9	1.1
394.0	400.4	-6.4
379.0	378.1	0.9
367.0	366.4	0.6
442.0	428.6	13.4
489.0	481.7	7.3
414.0	419.6	-5.6
453.0	436.4	16.6
381.0	381.8	-0.8
316.0	320.9	-4.9
377.0	376.5	0.5
571.0	566.9	4.1
329.0	335.7	-6.7

Mean difference: -0.04 ft
Standard deviation of the difference = 7.1 ft

Source: Author

4. QUANTITIES DERIVED FROM THE HEIGHT MATRIX

Using the height matrix it is possible to derive the magnitude and direction of the gradient of the terrain. Henceforth in this discussion the magnitude of the gradient will be referred to as slope and the direction as aspect. The slope and aspect are derived from the x and y direction gradients m_x and m_y that are obtained from a least squares fit of a two dimensional second order polynomial over a small $n \times n$ matrix of points. This least squares fitting is considerably simplified by the fact that the points are an equispaced grid and the point of interest is the one at the centre. The least squares fit is performed using either 3 x 3, 5 x 5 or 7 x 7 points depending upon how much smoothing is considered desirable.

The slope and aspect are obtained from the x and y direction gradients m_x and m_y as follows:

$$\text{(slope)} \quad m = \sqrt{m_x^2 + m_y^2}$$

$$\text{(aspect)} \quad a_s = \tan^{-1}\left(\frac{m_x}{m_y}\right) \quad m = 0$$

$$= \text{undefined} \quad m \neq 0$$

The direct illumination from the Sun at a point on the ground can in turn be obtained from the slope and aspect, viz:

$$I_s = \left\{ \cos(\alpha - \gamma) \cos\beta \sin\delta + \sin\beta \cos\delta \right\} I_o$$

where α is the Sun azimuth, β is the Sun elevation, γ the aspect, δ the slope and I_o the solar illumination on a unit area perpendicular to the direction of the illumination; α and γ are measured from north. It must be emphasised that this equation only refers to direct illumination and neglects totally the not insignificant contribution that will arise from backscattered radiation. A quantity that is particularly useful in investigating the variation in brightness arising from the surface topography is given by

$$R_d = \frac{\sin\delta \cos(\alpha - \gamma)}{\tan\beta} + \cos\delta$$

This quantity is unity for flat terrain, greater than unity for terrain inclined towards the Sun, less than unity for terrain inclined away from the Sun and negative for terrain in shadow. Also, the variation in R_d over a given area increases as the Sun elevation decreases.

Neither of these equations takes into account the possibility that some parts of the terrain may be shaded from direct illumination by others. Thus, these equations are only valid for all values of the terrain aspect and Sun azimuth angles when the Sun elevation angle is greater than the steepest slope ie $\beta > \delta$.

5. DISPLAYING TERRAIN PARAMETERS AS GREY TONE IMAGES

Height, slope and variation in direct solar illumination present no problems in being displayed as grey tone images, since lower values can be assigned to lower pixel intensities and higher values to higher pixel intensities. Aspect can also be displayed as an image although the interpretation of such an image is rather more difficult. This is because:

1. The aspect can vary between 0° and 360°, however 0° and 360° represent the same direction.

2. The aspect for flat terrain is undefined.

3. The value of the aspect becomes more unreliable as the slope becomes smaller.

One convention that can be adopted for displaying aspect and the one that is adopted here, although there is some ambiguity in the interpretation of the minimum and maximum permissible image intensities, is to assign an aspect value of 0° to an intensity of zero, 360° to the maximum image intensity, and flat terrain to a zero image intensity. Other conventions can of course be envisaged.

When considering the height matrix as a grey tone image it must be remembered that each value is associated with the centre of a pixel and does not represent the average value over a pixel. Where the terrain is changing slowly compared with the size of a pixel then the values of the height matrix will be a good approximation to the average value over a pixel. However, where the terrain is changing rapidly compared with the pixel size the height estimate will not necessarily be a good approximation to the mean value over a pixel. In this situation the height matrix must first be evaluated using a grid spacing for which the terrain is changing slowly and then a suitable average taken to obtain an estimate of the mean value for the required pixel size. This technique can also be applied to the variation in direct solar illumination but it cannot be applied to the slope and aspect. For these quantities it is necessary to evaluate the mean x and y direction gradients in the manner described and thence to derive the mean slope and aspect.

As an example, the central 4 km square area of the map shown in Figure 1 has been converted to a height matrix using a grid spacing of 10 m. This was then used to derive the height, slope, aspect and variation in direct solar illumination corresponding to 50 m pixels. (The reason for choosing this pixel size will become evident in the next section.) The values associated with the pixels have been obtained by averaging the 25 values of the matrix covering each pixel. The four terrain parameters are displayed as grey tone images in Figure 2.

6. APPLICATION TO LANDSAT IMAGERY

Recently, an increasing amount of interest has been given to the possibility of improving the accuracy of multispectral classification with Landsat imagery by taking into account effects arising from surface topography. These effects arise from the fact that although two areas may have the same cover type, and thus for a given spectral band would be expected to have the same intensity, in practice they will have different intensities if the terrain is such that the two areas receive differing

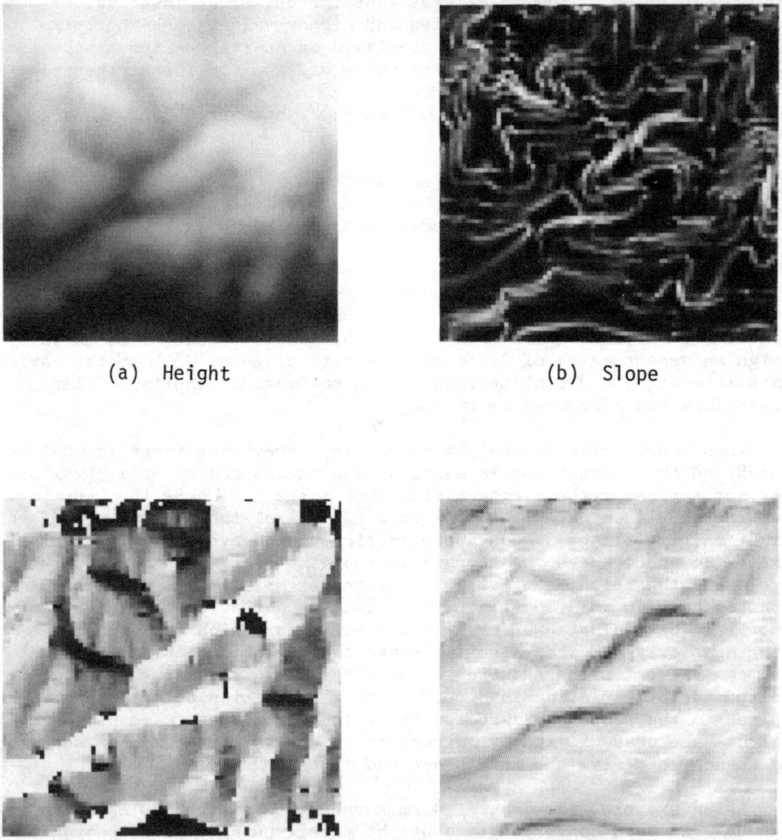

(a) Height

(b) Slope

(c) Aspect

(d) Variation in direct solar illumination

Fig 2 Grey tone image representation of the four terrain parameters for the central 4 km square of the test area. The pixel size is 50 m.

Source: Author

amounts of illumination. If data describing the surface topography is available it should be possible to reduce this topographic effect and thus improve the reliability of classifications. This can be achieved by two methods: one method involves using the terrain parameters as extra image planes and the other involves modifying the pixel intensities in the Landsat image.

In the first method, the four terrain parameters can be considered as four additional image planes to be used in conjunction with the four image planes associated with the four spectral bands of a Landsat scene. All eight image planes can be analysed together either on an interactive image processing system such as the IDP3000 (Balston, 1978) or on a mainframe computer. Thus, in classification, it would be possible to consider only areas exceeding a specified height or only flat areas. When classifying areas of similar cover type the reliability of the classification may be improved by performing a series of classifications each involving pixels which represent areas which received similar amounts of illumination. This can be achieved by using the image plane containing the variation in direct solar illumination R_d in conjunction with the four Landsat image planes.

The second method of improving the reliability of classification is rather more difficult and involves more assumptions. Here, corrections are made to the intensity of pixels so as to account for both the variation in illumination and for the scattering properties of the surface. The total illumination consists of direct solar illumination and a contribution from a diffuse component comprised primarily of scattered skylight, although there is a small contribution from light scattered onto the surface by surrounding ground. The diffuse component has been shown to be a small proportion of the total illumination ie less than 12% of the total for each of the four Landsat bands (Smith et al, 1979), under clear sky non-hazy conditions, conditions which will be satisfied by many Landsat scenes that would generally be considered suitable for classifying surface features. The relative contribution of the diffuse component is however wavelength dependent, being greatest for short wavelengths and smallest for long wavelengths. A further complication arises from the fact that most natural surfaces have preferred directions of scattering and this should be taken into account. This is difficult because the scattering properties of surfaces are dependent upon their composition and roughness and hence are hard to model. Further, the nature of the scattering may be heavily dependent on transient conditions eg the presence of surface water on vegetation after rain or a heavy dew. Because of the complexity of the problem the far from insignificant effect due to the scattering properties of the surface are ignored here. A detailed discussion of the nature of scattering by natural surfaces can be found in Justice and Holben (1979). Thus, assuming that all surfaces scatter light equally in all directions, pixel values can be adjusted to as to take account of variations in the surface illumination by using the formula:

$$I_c = \left(\frac{1 + r}{R + r}\right) I_o \qquad \begin{array}{lll} R = R_d & & R_d \geqslant 0 \\ R = 0 & & R_d < 0 \end{array}$$

where I_o = uncorrected pixel intensity
I_c = corrected pixel intensity
and r = diffuse component expressed as a fraction of the direct illumination for a horizontal surface.

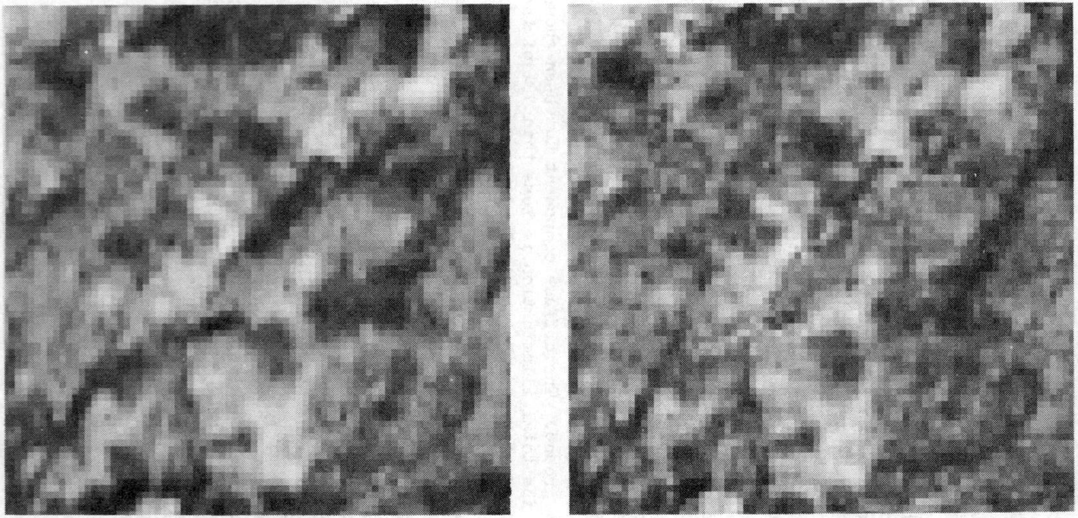

(a) Before correction

(b) After correction

Fig 3 Landsat band 7 for the central 4 km square of the test area before and after correcting for the topographic effect. The pixel size is 50 m.

Source: Author

It must be realised that attempting to correct pixels in shadow will not be expected to produce much improvement in classification because of the small range of pixel values in such areas.

To illustrate the effects of this technique, a Landsat scene covering the area illustrated in Figure 1 has been selected. Each of the four Landsat bands has been destriped, geometrically corrected and brought into registration with the National Grid. A subscene corresponding to the central 4 km square of Figure 1 has been extracted from the main scene to enable direct comparison to be made with the terrain parameters for the area. The pixel size for the Landsat data has been selected to be 50 m square so as to preserve as much as possible the accuracy of the original data. (The Landsat image has to be resampled during the geometrical correction process. For further details see for example Benny (1981).) A winter scene with a Sun elevation of only 15° has been selected so as to highlight the effect of the variation in direct solar illumination over undulating terrain. In comparison, all but 10% of the pixels are associated with slopes less than 5° and the mean slope is only 2.7°. Each Landsat pixel was modified using the above equation with a value of 0.14 being adopted for r. Figure 3 shows the Landsat band 7 image, corresponding to the central 4 km square of Figure 1, both before and after correction. Figure 3a is the original data after a suitable contrast has been performed and Figure 3b is the same image but after the topographic correction has been applied. As can be seen from a comparison of the two images, the correction has quite a marked effect, which is perhaps especially important since the test area is typical of many areas of gently undulating terrain found in southern England. The effect will be expected to be even greater in areas where steeper gradients exist and farther north where lower Sun elevations are encountered.

7. CONCLUSIONS

A technique has been described that can be used to generate a terrain height matrix from a set of contours. The use of this terrain matrix to derive slope, aspect, and the variation in direct solar illumination has also been described. As an example these parameters were derived from data obtained from an Ordnance Survey map and presented in the form of grey tone images. Finally, the terrain data was used to demonstrate how an attempt could be made to reduce the topographic effect on a Landsat image. However, the extent of any improvement can only be ascertained if an area is selected where accurate ground truth data exists. It is expected that the use of terrain data with Landsat imagery will increase as the availability and quality of suitable topographic data bases improves and provision is made for its routine use.

8. REFERENCES

BALSTON, D.M., (1978): IDP3000 - An image analysis system for earth resource surveys. Systems Technology, No 28.

BENNY, A.H., (1981): Automatic relocation of ground control points in Landsat imagery. RAE Technical Report, No 81071.

JUSTICE, C., and HOLBEN, B., (1979): Examination of Lambertian and non-Lambertian models for simulating the topographic effect on remotely sensed data. NASA Technical Memorandum, No 80557.

ROBERTS, P.A., (1980): Computer based technique for converting a contour map into an equispaced grid of points. RAE Technical Report, No 81010.

ROBERTS, P.A., (1981): Techniques for manipulating contour data and applications associated with the use of terrain data with imagery. RAE Technical Report, (To be published).

SMITH, J.A., TZEU, L.L., and RANSON, K.J., (1980): The Lambertian assumption and Landsat data. Photogrammetric Engineering and Remote Sensing 46 No 9 1183-1189.

A COMPOSITE LANDSAT IMAGE OF THE UK

R H Merson
National Remote Sensing Centre
Royal Aircraft Establishment
Farnborough
Hants

Abstract: A composite image of the UK has been constructed from 43 Landsat scenes by a wholly digital procedure, using a PRIME 750 computer. The scenes were first transformed to National Grid coordinates, using 1200 Ground Control Points, the resampling giving 100 m square pixels.

A magnetic tape storage and retrieval system was designed to enable segments of a partially constructed composite image to be transferred to and from a working disk. As each new scene was added, its intensity levels in each of four spectral bands were stretched to match the partially constructed image in the area of overlap.

The whole image has a size of 10,000 x 7,000 pixels and is stored on 17 magnetic tapes.

1. INTRODUCTION

The first aerial photograph of the earth was taken from a balloon over Paris over a century ago. This was regarded merely as a curiosity and it was not until the 1914–18 war period and afterwards that the potential of aerial photography was recognised. As photographic methods developed, the general procedure grew to be more or less routine. The aircraft would fly up and down fixed lines and the resulting strips of positives would be cut and joined together in the form of a mosaic which could then be photographed to give a final picture. The relatively low height of the aircraft resulted in picture distortion and much of the development was directed towards the design of distortion-free cameras. The whole procedure was very costly and the area covered could not be very large. With the advent of artificial satellites, the possibilities were enormously increased. In the 1960s the TIROS satellites provided the means for taking photographs of a complete hemisphere of the earth. These, in fact, go to the opposite extreme, being low-resolution pictures of very large areas. It was not until the Landsat 1 satellite was launched in 1972 to a height of 900 km that a source of relatively cheap, medium-resolution and almost distortion-free images became available.

Paper presented at a conference on Matching remote sensing technologies and their applications convened in London by the Remote Sensing Society, 16–18 December, 1981

By 1975 enough cloud-free images of the USA had been obtained to encourage the National Geographic Society to collaborate with NASA and the Photographic Laboratory of the General Electric company to produce a mosaic of the USA. This picture was made from 569 Landsat scenes but the general procedure, using scissors and paste and a paint-brush, was much as before. An article on 'Portrait USA', as the venture was called, was published in the National Geographic Magazine issue of July 1976. I first saw this article in August 1980, at a time when the ground control pointing of the UK was well under way. It prompted my colleagues and I to consider whether we could produce an image of the UK, not by the scissors and paste method, but entirely by digital methods. At this time software for transforming images to the UK national grid was already available.

2. FIRST STEPS

Two questions were raised in the first instance:

1. Were there enough cloud-free summer scenes of the UK available?

and

2. What pixel size in the transformed images would make the computer storage problem tractable?

Inspection of prints of the available scenes showed that the answer to the first question was, strictly speaking, no. Certain areas in Scotland and Wales appeared to have some cloud on every satellite pass. It was decided, however, that a small amount of cloud would be acceptable.

With regard to pixel size, bearing in mind that the pixels of the original scenes are about 57 m x 79 m, a choice of 50 m x 50 m for the transformed scenes would give a negligible information loss. It was calculated, however, that the working space required would be approximately 360 Megabytes, which is more than could be accommodated on one disk (300 Megabytes). Moreover the computer time required for joining a new scene to the partly-formed composite image might be all of 12 hours. It was decided, therefore, that the transformed images would have 100 m x 100 m pixels.

3. DATA HANDLING

Software for handling the storage and retrieval of large images on disks and magnetic tapes was prepared by A W Odell during September 1980. The program is called GEOBIG and is designed to handle four bands of Landsat data where each pixel is represented by four 8-bit integers. The basic idea is to regard the complete composite image as broken up into blocks, each block consisting of 8 rows, each of 256 pixels in 4 bands. This form was chosen to optimise the disk-to-disk and disk-to-tape timing. The blocks can be regarded as super-pixels of a virtual image and are identified by row and column numbers just like ordinary pixels. At any given time the bulk of the blocks will be held on magnetic tapes and only identification information is held on disk. The information on disk is held in a 'virtual image index file' and consists, for each block, of the number of the magnetic tape where it is to be found, together with its position on the tape. In order to cater for undefined pixels, ie pixels in a transformed image or composite image which are not part of a scene, it is necessary to set aside one of the 256 intensity levels. The particular level can be chosen by the user of GEOBIG. For the UK image, and more generally, for transformed images supplied to customers, the level chosen for such background pixels is 255. If such a level exists in a real image then it is converted to 254 during the transformation process.

When an image is to be picked up and transferred from disk into the 'big image' the required super-pixels are set up and placed at the end of the index file, still on disk.

In order to detail further the scope of the GEOBIG program, here is a list of the 10 main commands, with comments:

1. INITIALIZE This requires the user to define the bottom left and top-right corners of the composite image in grid coordinates, the x and y pixel sizes in metres and the names of the index file and the image.

2. UPDATE TAPES After one or more transformed images have been added to the index file, this command enables the transfer to magnetic tapes.

3. DELETE PART OF DISK Once blocks have been transferred to tape, those not overlapping the next images to be added can be removed from the end of the index file. This can be done in two ways: either by chopping off the last so many blocks or by coordinate specification.

4. UPDATE DISK FROM TAPE This command is used to bring back data from tape to disk by specifying the coordinates of the area required. The computer lists the required tapes to be loaded.

5. INPUT IMAGE After a new segment of the composite image has been prepared this command causes the formation of the appropriate blocks at the end of the index file. The position of the input image in relation to the composite image may be indicated either by manually set coordinates or by reference to a CONTROL file created during the transformation process.

6. OUTPUT IMAGE This enables any rectangular sub-image of the composite image - provided it has been transferred to disk - to be unpacked and stored in normal image format for up to four bands. The sub-image may be specified by manually set coordinates or by reference to a CONTROL file.

7. DISK COVERAGE At any given time, each block will be in one of four possible states:

   a. on disk only,
   b. on tape only,
   c. on disk and tape, or
   d. not yet formed.

   (A formed block is a block with at least one non-background pixel.)

   This information, for all blocks, can be displayed in the form of an image, using four distinct types of shading, on a TEKTRONIX 4014 graphics display terminal.

8. COPY INDEX — The index file is crucial to the recovery of the image from tape. If the index file were lost, the tapes would contain nothing but a meaningless jumble of data. After each occasion on which the command 'UPDATE TAPES' is used it is essential to copy the index to another disk or disks. This command does not copy any blocks following the index.

9. STOP — Run terminator.

10. SUMMARY — This command presents an outline of the storage situation including a list of tapes and how many valid blocks each contains.

4. SECURITY PRECAUTIONS

If anything goes wrong during the active command processes, such as the computer going down or a tape read failure, then no immediate recovery is possible. It is essential, therefore, to create fall-back compatible index files and tapes at frequent intervals, say every fourth input image. In simple terms, never output data to an old tape unless a copy of the tape has been made.

5. THE JOINING PROCESS

5.1 Outline of Operations

The core of the procedure for forming a composite image is the program for joining a new scene to the partially constructed image. This program is called IM.WELD.

Before using IM.WELD, two images of the same size have to be prepared. The new scene is first destriped, then its matrix of transformation to the UK National Grid is determined - using the programs GCP.FIND and GCP.FIT mentioned by A H Benny in an earlier paper of this session. The program GEOMAP is used to set up the control file mentioned in Section 3 above, and the image is then transformed to grid coordinates and resampled to 100 m square pixels using the program GEOTRANS. In the resampling procedure, linear interpolation is normally used. The transformed image will be referred to as image B. Next, if the index file does not already hold the existing blocks for the area required, the UPDATE DISK FROM TAPE procedure is carried out, followed by the OUTPUT IMAGE procedure using the CONTROL file to obtain an image of the same size and covering the same geographical area as image B. This segment of the partial composite image will be referred to as image A. The program IM.WELD is then run - separately for each of the four bands - and hopefully produces a resultant image which combines A and B without any joins showing. After passing a visual inspection test, the new segment is finally put into the composite image using the INPUT IMAGE command.

5.2 The Program IM.WELD

Looking at images A and B together, each pixel can be in one of four different states:

1. 255 in A and B, ie purely background.
2. 255 in A but not in B, called a B-only pixel.

3. 255 in B but not in A, called an A-only pixel.
4. Not 255 in either image, called an overlap pixel.

The overlap pixels may be imagined to join up to form one or more overlap areas. For simplicity, each interior pixel is assumed to have four neighbours only, the four orthogonal adjacent pixels. In the normal case there will be a single overlap area, but the program will handle the general situation.

It is clear that purely background pixels will remain the same in the output image, as will A-only pixels. For overlap pixels, two operations are called for, matching and merging. Matching is the mapping of B image intensity levels to new values which match as closely as possible those of the A image. Merging is the adding together in some way of the intensity level in the A image to the mapped intensity level in the B image for each pixel in the overlap.

In IM.WELD each B image intensity level I(b) is mapped to a new level I'(b) by the linear operation $I'(b)=c*I(b)+d$, where c and d are chosen to equalise the means and standard deviations of the A and B intensity levels in the overlap region. The B-only pixels then simply take on their mapped levels for the output image. Each pixel in the overlap region, however, now has two levels, I(a) and I'(b), from which to form an output value. To avoid any semblance of a join in the composite image, it is clearly desirable that in an area of overlap close to an area of A-only pixels, the value I(a) should be weighted more highly than I'(b), whilst the converse holds in an area close to B-only pixels. In the event, the following weighting has been adopted to give the output intensity I(o):

$$I(o) = w*I(a) + (1-w)*I'(b)$$

where $w = x*x*(x+x-3)+1$

$x = da/(da + db)$

and da,db are the 'taxicab distances' to the nearest A-only and B-only pixels, respectively, adjacent to the particular overlap region in which the pixel lies. Here, the taxicab distance between points with cartesian coordinates (x,y) and (p,q) is $|x-p|+|y-q|$. This definition of distance is the cheapest as regards computer time. It will be noted that the weighting function is the simplest one which satisfies the obvious criteria: $w(0)=1$, $w(1)=0$, $w'(0)=0$, $w'(1)=0$ and the symmetry relation $w(1-x)=1-w(x)$, with $w'(x)=dw/dx$.

In order to determine the distances da and db, it is necessary to scan the A and B images twice. In the first scan - pixels left to right, rows top to bottom - the histograms in the overlap region can be set up and pixel distances from left and top edges of overlap regions can be developed. This positional information is stored in two temporary image files of type 'I2', where each entry is allocated 2 bytes and therefore has a range -32768 to 32767.

After the first scan is completed, the histogram matching is carried out and the intensity map file for the B image is created. During the second scan (pixels right to left, rows bottom to top), as each pixel is taken, its final distances da and db are determined where applicable and the output image is formed.

5.3 Problem Areas

Towards the end of September 1980, two midland scenes (numbers 219/23 of 27 May 1977 and 218/23 of 27 June 1976) were transformed to 100 m pixels and successfully joined in all four bands, as was seen by viewing the composite scene on the IDP 3000 visual display unit. Five further scenes were added - and then the problems started.

The difficulties that were encountered during the formation of the composite image can be separated into three distinct types:

1. Poor scene matching,
2. Missing and corrupted lines,
3. Clouds and cloud shadows.

These will be discussed in turn.

5.3.1 Poor scene matching

In order that two overlapping scenes can be satisfactorily merged together, it is clear that, apart from a matching contrast stretch, the two scenes should not be very different. Now in the case of land areas there is a variation with season but, by and large, there is not much change for a given month from one year to the next, snow and floods excepted. The sea, however, is a different proposition altogether. Using the algorithm described in Section 5.2 above, no two segments of sea have been satisfactorily joined. That is why the sea around the UK has been given a uniform tone in pictures produced from the composite image.

The whole or part of 43 snaps were used in making the composite image, where the term 'snap' is used to describe a given scene on a given date. 41 scenes were required and two snaps of each of the scenes 217/25 and 218/24 were used. The distribution of the snaps by month was as follows:

Month	Number of Snaps
February	1
April	8
May	10
June	8
July	5
Autust	7
September	4

This represents a much larger seasonal scatter than one would like, but apart from the February snap - used to fill a small gap north of Edinburgh - the matching was generally better than might be expected.

5.3.2 Missing and corrupted lines

The first problem that arose concerned the snap of scene 218/24 dated 27 June 1976. When running the GCP.FIT program for performing a linear fit to the ground control point data, the north-south residuals were $\pm$300 metres as against the usual $\pm$60 metres. Those in the north were of one sign and those in the south of the opposite sign. This is a direct indicator of missing lines. A close inspection of the image showed that 12 successive lines were corrupted and the 6 following lines were completely missing. What to do? In this instance it was decided to replace the corrupt lines with the background level 255 and add in 6 extra background lines, after which the normal procedure was followed. This is a tedious business, however, and subsequently, when missing or corrupted lines were found, the maximum possible good sub-image was taken. In the case of scene 218/24, a second snap with cloud at the top, but otherwise clear, was chopped in half and the lower half used to patch in the missing segment.

5.3.3 Clouds

Clouds cannot be merged with ground features. If, therefore, there are two overlapping scenes with some clouds in one but not in the other, it is best to cut out the clouds and replace them with background pixels. Some attempts were made to do this automatically by noting the range of intensity levels covered by the clouds. This method is not generally applicable, however, for two reasons. First, there are sometimes genuine ground features, such as beaches, which have the same intensity level as clouds and second, even if one gets rid of the clouds, their shadows remain to spoil the image. For the composite image the clouds were eliminated by displaying the image on the Tektronix unit and using the cursor to determine the corners of rectangular areas containing the clouds. These areas were then replaced by 255s using the program IM.COPY, a general purpose processing program.

6. THE STORED COMPOSITE IMAGE

The complete composite image - 7,000 pixels * 10,000 lines - is stored on 17 magnetic tapes. For convenience, the Northern Isles (Orkneys and Shetlands) are positioned in the North Sea area. The image is still in its raw state, that is to say, little cleaning up of the sea regions has been carried out.

7. THE QUARTER-SCALE IMAGE

In order to be able to prepare some primary products such as negatives, prints and outline drawings, a version of the image that could be stored on one tape was prepared. This was done by taking every fourth pixel of every fourth row of the composite image to produce a quarter-scale image at a resolution of 400 m and having a size of 1706 pixels * 2500 rows * 4 bands. This image was cleaned up using the IDP 3000 digital interactive image display facility to produce a final version which has been used to provide various products, including the false-colour poster of the UK.

8. FUTURE IMPROVEMENTS

In the production of further composite images, currently being contemplated, it is likely that some improvements both in the software and in general procedures will be adopted. As it stands at the moment, the UK composite image is basically unstretched, and bands 4 and 5, at least in some parts of the image, only span four or five intensity levels. So the first suggested improvement is to apply some contrast stretch to the first scene of a composite image before anything else is done. This will provide more room for manoeuvre in subsequent matching processes. It is proposed next to try out another histogram matching procedure in place of the two-parameter stretch. Finally, a semi-automatic method of cloud elimination is to be introduced. In this method the cloudy part of an image will be shown on a display unit and the individual clouds and shadows will be traced using a cursor and the areas traced out will then automatically be filled with 'white', ie background pixels.

REMOTE SENSING AND DIGITAL CARTOGRAPHY - MERGING OF TECHNOLOGIES AT THE US GEOLOGICAL SURVEY

R B McEwen, Office of Cartographic Research, USGS, Reston, Virginia

Rapid developments in digital cartography and digital image processing are producing a symbiotic merger of technologies and the result will have a profound effect on remote sensing, photogrammetry and the profession of cartography. The National Mapping Division of the US Geological Survey is deeply involved in this development. Planning, research, and development are underway to create a digital cartographic data base that will be topologically structured to provide maximum advantage for automated analysis of data acquired from aircraft and satellite sensors. Typical examples include the merging of digital elevation models with multi-spectral radiance values to improve the classification of phenomena that are sensitive to elevation, slope, or aspect. Some of the same computer programs developed for image processing are now suitable for processing raster-formatted digital cartographic files. It is certain that the use of remote sensing data for mapping will increase and the production of maps will be highly automated.

A COMPARISON OF SENSORS FOR CARTOGRAPHIC REQUIREMENTS

G Konecny, Institut fur Photogrammetrie, Universitat Hannover

Preparations are being made for the metric camera experiment of the first European Spacelab mission. The paper outlines why camera technology is still of interest for later operational uses from space. An attempt is made to compare photographic techniques with the potential of scanners and push-broom devices as well as tadar with the aim to derive cartographic products. An analysis of possible costs is also made.

MULTIPURPOSE CADASTRE AND REMOTE SENSING

R K Bullard
Division of Cadastre, Land Information and Remote Sensing
Department of Land Surveying
North East London Polytechnic
Forest Road
London
E17 4JB

Abstract: A cadastre is an information system which shows the extent, value, and ownership of land, and is used for the registration of land with particular attention to the boundaries of the land being registered. The object of a multipurpose cadastre is to group the basic information together with the other specialized information to form an information data bank which can be used for more than the registration of land.

Remote sensing imagery can be used to collect the information for the multipurpose cadastre. In the developed world, most of the existing boundaries and services are well established, though not necessarily well known. The evidence of boundaries and services requires the detection of both dividing lines between properties (boundaries), and the detection of services; water, gas, electricity, telephone, oil, and sewage.

As boundaries have been established over the centuries, and as the services have been provided by many Nationalised Industries, together with private organisations, without much liason with local authorities, and probably even less liason with the mapping authorities, there is therefore a need to demarcate these services and boundaries because of their increasing complexity, and their continuous need for repair and even replacement.

This paper sets out the ways in which boundaries and services can be established using remote sensing imagery, suggesting which imagery is the most suitable for each detecting task.

Paper presented at a conference on Matching remote sensing technologies and their applications convened in London by the Remote Sensing Society, 16 - 18 December 1981

1. INTRODUCTION

The cadastre is usually concerned with the boundaries of properties, their value, and ownership. Some countries, particularly those which have been under the influence of Roman Law have adopted a numerical cadastre, which involves the establishment of geometrical values for the boundaries. Other countries, like England, have opted for a 'general boundary' system whereby the actual position of the boundary has not been precisely determined.

The multipurpose cadastre will contain a variety of information in addition to that required for boundary demarcation, and will assist in the efficient administration of land.

According to Petersohn (1981) the reason for the acceptance of the multipurpose cadastre in the United States of America are as follows :-

1. Frustration with the archaity of the title registration and transfer system in use in North America.
2. The exponential rise in real estate prices, most prominently felt in urban areas.
3. Since the adoption of strict environmental control legislation during the sixties and seventies, an increase in information demands by the planning professions in urban, regional, and national resource planning efforts.
4. The duplicity and lack of coordination among government and private agencies in the handling of land data.
5. Quantum strides in the sensing equipment available to the surveying disciplines, and similar advances in computing machinery.
6. The rapid development and increasing acceptance of automated information storage and retrieval systems.

The reasons listed above could apply to many countries, not least the United Kingdom, especially items 2, 3 and 5.

McLaughlin (1981) is of the opinion that the evolution of the multipurpose cadastre concept could be viewed as an outgrowth of three developments of the early 1960s in North America :-

1. The introduction of the integrated survey concept.
2. The renewed interest in active land registration systems.
3. The development of geographic information systems concepts and technologies.

The objective in presenting this paper, is to show how remote sensing techniques can be used to collect some of the information that is required for and contained in a multipurpose cadastre.

2. THE MAPPING PROCESS

In most countries surveys are undertaken for a variety of reasons. These surveys result in the production of maps and plans which are generally intended for single purpose usage. Examples of the uses for which maps and plans are constructed are listed below :-

1. Maps to show topographic details as in National Series.
2. Maps for tourism.
3. Military maps.
4. Planning maps as included in Structure Plans.
5. Land Reform plans.
6. Large scale engineering plans.

From the above list it can be seen that mapping is for a specific task, and what may be lost sight of is that surveying itself is being carried out for each of the maps and plans, while except for the change in scale of the surveys it is mainly the mapping exercise that is changing.

The type of map required will also very largely depend on the economy of the country being mapped. Table 1 below gives examples of the needs for mapping in three different economies.

Table 1. Types of Economy and Need for Mapping

Type of Economy	Predominant Production Type	Need for Mapping	
		Small Scale	Large Scale
Pre-industrial	Agriculture	Transport 1:250 000; Defence 1:50 000 where required	Tax Cadastre where desired
Industrial	Consumer Goods	Planning of natural resources 1:50 000	Construction; Management of natural resources
Post-industrial	Services	Derived from large scale plans	Planning and management 1:5 000 to 1:1 000 data banks

Source: G.Fochler-Hauke (1974)

The maps and plans referred to above, and those outlined in Table 1 are analogue records of data supplied from the survey, and interpolated from the results as happens in a photogrammetric survey. Too often the survey records are unavailable or have been lost or destroyed, and only maps and plans remain. The ideal for a multipurpose cadastre is to collect and store the data in digital form.

Historically the mapping and revision task has been considered a purpose in itself, and therefore by todays standards the justification for making and revising a map and plan must lie in its use as a data base for a multipurpose environmental information system (Konecny 1976).

The most suitable method for collecting data for the production of maps and plans will be in digital form, and this can now be achieved by both ground and aerial surveys. Ground surveys using electronic tacheometers, and aerial surveys using scanners will result in the data being recorded on tape in digital form, ideal for multipurpose information systems.

3. MULTIPURPOSE CADASTRE

The designation, multipurpose cadastre indicates that the cadastre can be used in a diversified manner. The multipurpose cadastre is a geographical basis of localisation and of definition of technical, legal, fiscal and economic information relating to land (Gerhard 1977).

Figure 1. The Multipurpose Cadastre in an Integrated Land Information System

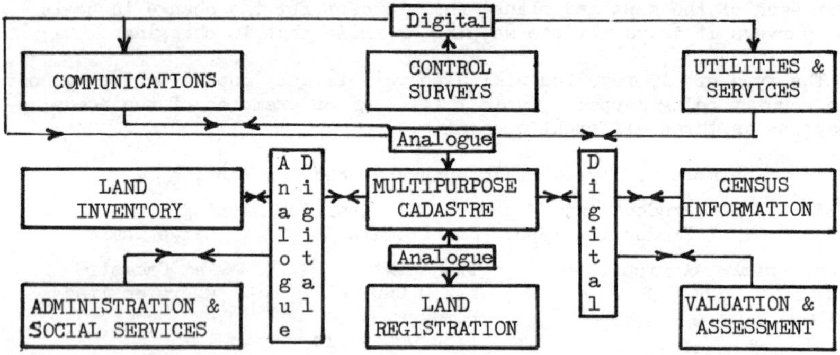

Digital - Computer Records   Analogue - Maps, plans, Documents

Source - After Hamilton, A.C (1970) and Dale, P.F (1976)

Figure 1 shows the multipurpose cadastre at the centre of an integrated information system. The multipurpose cadastre will have records in both digital and analogue form, while some of the other elements will be available in both or either digital or analogue form. The ideal situation would be to bring all forms of data into compatible form, which means digital for ease of integration, speed of recall of information and accessability. Storage in digital form will also simplify the updating of records, the security and the confidentiality of the information.

The multipurpose cadastre will require either coordinated property boundaries, or values for centres of properties, together with unique land parcel identifiers. This information will be essential for the smooth running of a land information system to ensure that every property is recorded and adequately labelled.

4. BOUNDARIES AND SERVICES

When considering the boundaries and services in Figure 1 they have been considered in the relationship that their position and presence can partly be detected by remote sensing technology. Remote sensing methods together with photogrammetric techniques will be the major collecting approach, with the two systems overlapping in their instrumentation. The ground control and ground data for photogrammetric mapping and remote sensing respectively can be collected at the same time thereby reducing the costs and the time taken to collect the data.

The boundaries will consist of the following :-

1. Property boundaries.
2. Transport corridors - Highways, Roads, Railways, Waterways, Airports, Pipelines.
3. High water boundaries.
4. Land use boundaries.

The above can be partly collected with the aid of remote sensing techniques, there will be other boundaries that require other sources namely, Regional, County, District, and Parish, and others which cannot be collected other than from existing records in Local and Central Government Offices.

The services that will be recorded, and therefore available for the information system are as follows :-

1. Gas.
2. Electricity.
3. Telephone.
4. Water.
5. Sewerage.
6. Drainage.
7. Pipelines.

These services, with the exception of electricity and telephone are found below the surface of the ground.

5. DETECTION USING REMOTE SENSING IMAGERY

The detection of boundaries and services will require information to be collected from above and below the ground surface.

As the data required for the multipurpose cadastre will be more manageable in digital form, it follows that collection directly onto tape, or the transference indirectly onto tape should be a priority. With regard to remote sensing imagery it would be ideal to collect data by means of a scanner so that data is directly stored onto tape. Where this cannot be achieved and where aerial photographs are used for collecting the data, they should be converted into digital form, this will result in a degraded result however, because of the additional processing involved.

The digital map can be produced from orthophotographs (Hesler 1977). When information is collected with the aid of a scanner it must still be geometrically corrected before further processing.

In the National Land Survey of Sweden (Asplund 1978) it is suggested that remote sensing imagery can speed up and decrease the cost of map revision. The extent of the use of aerial survey in Sweden can be appreciated when it is known that 750 000 aerial photographs are contained in the library of the National Land Survey in Gävle. Few areas in the country are covered with imagery older than 7 years, and even in the mountainous areas imagery is repeated at 10 year intervals.

The present satellite, Landsat, will have a small though complimentary role to perform in the collection of data for a multipurpose cadastre, but as the resolution of the imagery improves with future satellites, Spot and others, then the role will be increased and extended. The thermal scanner to be incorporated in Landsat D will extend the function for the data collecting process.

The primary source of data for a multipurpose cadastre could be obtained from an airborne multichannel multispectral scanner which has a range of detection from the ultraviolet to the thermal infrared portion of the electromagnetic spectrum.

Synthetic Aperture Radar (SAR) and Side Looking Aperture Radar (SLAR) will have an important application for data collection in areas where there is continuous or nearly complete cloud coverage throughout the year. Radar will be at present limited because of its resolution, though like satellite imagery, as this improves so will its applications be extended. Radar will be largely restricted to surface data. Thermal imagery will be of importance in the detection of underground and surface services.

It is important that remote sensing imagery is collected for the specific task and not as so often happens that existing imagery which is not entirely suitable is made to fit unintended circumstances, with the resultant loss in detail, or temporal changes due to outdated imagery.

Kirby (1980) puts forward the case for remote sensing imagery in the collection of data for land assessment, that remote sensing is only a surrogate for direct ground observations, but results are more often immediately available, more economically gathered in terms of both time and labour costs, and may yield a better result. All of these advantages are required for the collection of data for a multipurpose cadastre.

When the task of collecting data for the multipurpose cadastre has been completed, this should not be seen as the end of the task. There will be an increasing need to update the cadastre, not only to keep the records up to date, but because of the decreasing resources and the shortage of land suitable for the multiplicity of uses, not least of these being the land required for agricultural purposes. A monitoring process will have to be established (Bullard 1981) and remote sensing would be the technique most suitable to provide the revision mechanism to update the multipurpose cadastre.

6. CONCLUSIONS

The collection of data for a multipurpose cadastre can be readily obtained with the aid of remote sensing imagery. Where the imagery can be collected directly onto tape, and where the scale is sufficiently large to obtain a resolution to enable the boundaries and services to be detected, then remote sensing imagery can be justified. Because the information for the multipurpose cadastre would ideally be stored on tape, the scale at which an area is surveyed is only of importance when it relates to resolution, unlike the analogue system where the results are plotted onto a plan or map and the scale becomes of major importance. With digital information, scale can be altered to suit the display required, while with analogue information the scale cannot be altered in the same way, so readily, or without loss in detail and accuracy.

There is a need to persuade Governments that multipurpose cadastres and land information systems are essential, and that they can be justified financially over a period of time. There is also a need to convince those involved with the setting up of multipurpose cadastres of the important role that remote sensing can fulfil in the collection of data. In the meantime it is important to continue to educate the population in the basic concepts of remote sensing so that they are familiar with the applications. This education in remote sensing should start at school level so that in time a broad section of the community can be aware of the applications.

To show the magnitude of the task in constructing a multipurpose cadastre in the United States of America, a Panel of the National Research Council (1980), in its publication 'Need for a Multipurpose Cadastre', sets out the need for a nationwide cadastre and then sets out the involvement that Federal Government should have in such a scheme. It is suggested that at the least, over a period of 10 years, that a sum of $1.2 Billion should be allocated, and each State should have an Office of Land Information Systems. It is suggested that this expenditure would save the country tens of billions of dollars in the next decade, the problem is to encourage the treasury to supply the funds, particularly in the present economic climate. Other countries should take note of the Council's findings and prepare themselves for the advent of a multipurpose cadastre in their country, and in the meantime establish suitable technologies for the collection of and storage of data, and the monitoring that follows the initial survey.

Where systems of multipurpose cadastre and land information systems are being established, they have enabled a multiplicity of uses to be tested, and Kanakubo and Nonomura (1981) list land use planning, urban planning, road planning, water resource planning, environment impact assessment, determination of gravitational correction factors, and the analysis of wave transmission characteristics as a few examples attempted in Japan.

In his paper 'Exploring the Cadastre', Greulich (1981) concludes that the loss of prime farmland and pollution in the United States of America may not be stopped by the implementation of a multipurpose cadastre, but it can provide a reliable inventory of all land, it can assist in the planning and proper allocation of this life-giving resource called earth. Greulich (1981) also considers that the intelligent application of the surveyors' (remote sensors') know-how in the development of a multipurpose cadastre will be evermore essential for the welfare of mankind.

More attention should be made of the importance of the data that has and will be obtained to provide the data base on which the multipurpose cadastre is constructed.

The remote sensing community should prepare itself for the potential expansion of its activities when multipurpose cadastres are introduced, and establish methodology for the collection, storage and retrieval of data in digital form. It should also establish monitoring techniques for the updating of the system.

7. REFERENCES

Asplund, L., (1978): Vigade uppgifter inom den tekniska versamheten. Lantäteriverket (LMV) informerar on sin verksamhet, Nr 6 1978

Bullard, R.K., (1981): Abandoned land detection, reclamation proposals, and monitoring aided by remote sensing. Unpublished Ph.D. thesis, Faculty of Pure Science, University of Sheffield.

Dale, P.F., (1976): Cadastral surveys within the Commonwealth. H.M.S.O., London.

Fochler-Hauke, G., (1974): Der fischer weltalmanch 1975. Fischer Taschenbuch.

Gerhard, L., (1977): The evolution of the existing cadastres towards the multipurpose cadastre. XVth International Congress of Surveyors, Stockholm, June, 1977. Paper 702.2.

Greulich, G.H., (1981): Exploring the cadastre. XVIth International Congress of Surveyors, Montreux, August, 1981. Paper 708.1.

Hamilton, A.C., (1970): Data banks on the basis of integrated surveys. Integrated Surveying and Mapping, Papers of the 1970 Annual Meeting of the Canadian Institute of Surveying. Paper 10.

Hesler, A. von, (1978): Information-und planungssystem. Information-und Planungssystem, Regionale Planungsgemeinschaft Untermain, Erstellt im Rahmen des Erdwissenschaftlichen FlugzeugmeBprogramms.

Kanakubo, T., and K.Nonomura, (1981): The digital national land information its contents, procedures and uses. XVIth International Congress of Surveyors, Montreux, August, 1981. Paper 306.7.

Kirby, R.P., (1980): The role of remote sensing in land assessment. Chapter 2 in, Land Assessment in Scotland. Editors M.F.Thomas and J.T.Coppock. Proceedings of the Royal Scottish Geographical Symposium. Aberdeen University Press, 1980.

Konecny, G., (1976): Methods of data acquisition for the establishment of a multipurpose land registration system in urban and rural areas with special emphasis on photogrammetry. World Cartography 14 1976 pp 97-102

McLaughlin, J., (1981): North American contributions of the development of the multipurpose cadastre concept. XVIth International Congress of of Surveyors, Montreux, August, 1981. Paper 704.5

National Research Council, (1980): Need for a multipurpose cadastre. National Academy Press, Washington, D.C. 1980.

Petersohn, C., (1981): Origins and recent activities of selected private organisations in the U.S. and Canada in the multipurpose cadastral field. XVIth International Congress of Surveyors, Montreux, August, 1981. Paper L II 4.

PHOTOINTERPRETATION AND COMPUTER AIDED CARTOGRAPHY OF BROADLAND VEGETATION

R M FULLER, NATURAL ENVIRONMENT
RESEARCH COUNCIL - INSTITUTE OF
TERRESTRIAL ECOLOGY, AND
J E DRUMMOND, NATURAL ENVIRONMENT
RESEARCH COUNCIL - EXPERIMENTAL
CARTOGRAPHY UNIT

Abstract: The aim of this Natural Environment Research Council project is the production of data and a series of computer draughted maps at 1:10,000 scale, characterising the land use types of Broadland. The input maps have been produced at the Institute of Terrestrial Ecology (ITE) by photointerpreting 1:10,000 panchromatic 1980 aerial photographs. Details were transferred to Ordnance Survey 1:10,000 base maps using a Stereo Zoom Transfer Scope.

Following ground-truthing by ITE the input maps are being digitized at the Experimental Cartography Unit (ECU) where the data, after interactive editing, will be used to generate monochromatic maps on the Unit's flat bed projection plotter.

Sample areas of the output maps will be verified quantitatively. Areas of habitats and lengths of linear features will be calculated for all of Broadlands.

The digitized outlines will provide a readily accessible data base for the management of a changing Broadland environment by the Broads Authority who have funded the work. The data will also form part of ITE's Terrestrial Environment Information System. In addition it will provide a basis for assessing other methodologies (especially the satellite methods as they are being developed by ECU) for monitoring the Broadland environment.

1. INTRODUCTION

The Broadland area of Norfolk and Suffolk, England, consists of the floodplains of four major rivers - the Ant, Bure, Yare, and Waveney. Within these floodplains are freshwater lakes, called Broads, which are the flooded remains of medieval peat excavations (Lambert 1953). The banks of Broads and rivers have been colonised by reedswamp which, in time, has trapped sediments and deposited new peat, to raise the ground surface in many places above water level. Fen communities of mixed reeds, herbs, and grasses have developed on these peat soils and, in turn the fens may have been colonised by trees, such as alder and willow, to give a wet woodland called "carr". Oak woodland established itself at sites as the soil surface built up with plant litter, and the water table dropped due to plant transpiration. Lower in the valleys silt and clay soils have supported mixed reed marshes which, themselves, may have gone through similar successions. The vegetation of Broadland has been described in detail by Lambert (1951) and Wheeler (1978).

Management by man has long been part of Broadland and has included the artificial drainage of the peaty fens and silt based marshes. According to soil type and the degree of drainage, reed beds have been cut for thatching, or grassland used as summer grazing for cattle.

But more recently the character of Broadland has been radically altered (O'Riordan 1979; Caufield 1981). Drains have been cut to about two metres below the soil surface of some marshes and pumped nearly dry, allowing grazing marsh to be turned into arable land. Pleasure boats, have eroded the waterways, damaged vegetation along banks and destroyed the tranquility of much of the area. Run-off of fertilizers, applied to agricultural land, plus the release of phosphates and nitrates in treated sewage effluent have been blamed for the nutrient enrichment of many of the waterways (Moss 1979). This favoured the growth of micro-algae turning the water green and murky, and prevented the growth of water weeds. The introduction of alien geese and coypu has also prevented reedswamp growth and thus hindered the natural succession (Boorman and Fuller 1981).

The management of this changing landscape is now co-ordinated by the newly formed Broads Authority (Anon 1979). In order to make their management more objectively based they wish to take stock of the present status of the Broadland environment. By being able to detect and measure changes, appropriate corrective measures can be applied as necessary. A series of maps would provide the basic information they need; but a more adaptable set of data files could be interrogated to highlight information relevant at any one time. The main aims of this project were thus to provide both ecological maps and data files for the Broadland area. The data files would also provide local, detailed information for the Institute of Terrestrial Ecology (ITE), Terrestrial Environment Information System (TEIS) which is a 'catalogue' of ecological data based on an indicator species analysis of 10 x 10 Km squares of environmental data extracted from existing Ordnance Survey (OS) maps, and also for the Natural Environment Research Council (NERC) ECOBASE.

2. PHOTOINTERPRETATION

2.1 Methods

Semi-natural habitats and other land-use categories were interpreted from 1:10,000 panchromatic aerial photographs, commissioned by the Broads Authority and flown in January 1980 by an aircraft of the Cambridge University Committee for Aerial Photography. Outline maps were made by transferring details from the photographs onto 1:10,000 or 1:10,560 OS Sheets using a Bausch and Lomb Stereo Zoom Transferscope. Using this instrument a view of the stereoimage was seen superimposed over the relevant section of a map. For this particular study the view of the map was enlarged by a 4x map objective lens. Photographic scale was then matched, by means of zoom controls, to that of the base maps. Tilts on the photographs, where present, were counteracted by an equal and opposite distortion of the photographic image, using a 'stretch' facility provided on the instrument. As Broadland is a flat area, relief distortions were negligible.

The outlines of land classes were then transferred onto tracing film overlays placed on each OS sheet. The instrument was set in the mono viewing mode for transfer of outlines but stereo was selected, at will, to assist in interpretation. Photointerpretation was backed by extensive field checks to ensure that approximately 90% accuracy was achieved for each habitat.

2.2 Results

An area of 300 Km^2 was mapped at either 1:10,000 or 1:10,560 scale depending on the availability of revised OS maps. Sixteen semi-natural habitats and fifteen other land-use categories were recognised during interpretation (TABLE 1).

The January photographs proved good for separating most habitats, since reeds, at that time of year, and sedges were straw coloured, rushes were brown, grassland green, and arable land mostly bare. Differences may well have shown better than they would have done in summer months. Low sun-angle, occasional partial cloud cover and consequent shadows did provide some problems which may effect the final accuracy. However it is felt that 90% accuracy was achieved overall, since ground checks were made more intensively where such problems arose (see below, 5.1).

3. DATA INPUT

3.1 Digitisation

The data files were generated by digital encoding of the land classification units (land-use parcels) as represented on the tracing film overlays, at the Experimental Cartography Unit's (ECU's) ARISTO work station. These are driven by a PDP 11-45 Central Processing Unit and each consists of:

i) an ARISTOGRID digitising surface plus cursor for line following digitising;

ii) a TEKTRONIX 4014 or 4010 for displaying the vector data as it is digitised;

iii) a DECSCOPE VT52 for command input, status display etc, but especially feature coding the line segments as they are digitised.

The document to be digitised is attached to the ARISTOGRID digitising surface. The surface incorporates a wire grid for coarse and fine measurements. The grid emits impulses which an induction coil in the cursor receives. These are converted into table co-ordinates for subsequent transformation into a more useful co-ordinate system which in the case of this particular project was the British National Grid. These table co-ordinates can be recorded at the rate of 160 co-ordinate pairs per second, but this was unnecessarily high for this project, and was reduced by ECU personnel selecting a digitising resolution of 200µm, and a minimum distance between recordings of 300µm. (To produce 160 co-ordinate pair records per second would require digitisation personnel to trace almost 5 cm of line work/sec. The complexity of the Broadland land-use parcels precluded this.)

The TEKTRONIX display screen allowed digitisation personnel to ensure all lines were completely digitised and their positions adequate, to highlight any spurious lines, and to determine whether junctions met the required standard for this project.

TABLE 1

LANDUSE	HABITAT
<u>Constructions</u>	
a) Communications	Railways
	Roads - A, B, Unclassified
	Tracks and Footpaths
b) Structures	Residential and Services
	Commercial
	Agricultural
	Boat Yards
c) Water Management	Embankments
	Piled Water Margin
	Old Limit of Flood Plain
<u>Agriculture and Forestry</u>	
	Arable/Short Term Ley
	Improved Permanent Pasture
	Plantation
<u>Disturbance</u>	
	Bare Ground
<u>Natural and Semi Natural</u>	
a) Woody	Woodland
	Carr
	Scrub
	Carr/Fen
	Bracken
b) Grassland	Rough Grassland
	Marsh/Fen Grassland
	Rough Grazing
	Marsh/Fen Grazing
c) Fen	Mown Fen
	Mixed Herbaceous Fen
	Marsh/Fen - Reed-Cut
	Marsh/Fen - Reed-Uncut
c) Swamp	Reed Swamp
	Intertidal Saltmarsh
<u>Water</u>	
	Drainage Ditches
	Broads, Rivers, and Ponds

Display was continual during digitisation. The whole document or selected parts of it could be displayed as it was digitised.

In the context of this project the DECSCOPE VT52 allowed the attachment of feature codes to the line segments as they were digitised. Line segments formed the land parcel boundaries. Each line segment could have up to two lineal functions and two areal functions. For example a farm track whose centre line coincided with the Broads Authority Boundary and also separated improved grazing from arable land would have had the following lineal functions:

* track
* administration boundary

and the following areal functions

* improved grazing edge
* arable land edge

In this example feature codes for tracks, the Broads Authority Boundary, improved grazing, and arable land, would be 18, 78, 63, and 62 respectively, and such a line would therefore be given these codes at the DECSCOPE prior to digitising on the ARISTOGRID surface.

3.2 Interactive Editing

Once digitisation was completed a 30μm line weight film based check plot was produced on ECU's Flatbed plotter. The scale of the check plot was the same as the original photointerpretation document, and the check plot superimposed on the original could be used to check the completeness and line quality of the digitisation. Part of ECU's mapping software package is its GRAPHICS 300 program which allows interactive editing of digitised files at a graphics screen. Any line quality errors appearing in a check plot can therefore be corrected. Feature code errors can be demonstrated by displaying one feature code's lines at a time - any open polygons which result indicate miscoded line segments.

3.3 ECOBASE

ECOBASE is a series of data files of environmental variables generated by NERC funding, at the Experimental Cartography Unit. It consists primarily of digitised cartographic information on the British hydrological and communications networks, urban and forested areas.

Some of the data has been derived from digitising 1:250,000 OS maps, and some from digitising 1:50,000 OS maps. However it is hoped to gradually upgrade ECOBASE by adding data from larger scale sources such as the data generated by this project. If this is to be done in the future, feature coding must be compatible between ECOBASE and the Broadland data, and amendments were made to achieve this compatibility.

3.4 ECU Data Structure

ECU has two data structure, one compact and used for storage or rapid display, and the other suitable for interrogation and manipulation. Considering the second structure only, each item in an ECU file contains two parts. The first (known at the Image Attributes) consists of header information, ie:

* item type (whether point, line, or polygon);
* item start and end point co-ordinates;
* item feature codes.

The second (known as the Segment Generator) consists of an item's co-ordinate string, or pointers to items generating its co-ordinate string. Linkages between the two parts of an item's record exist. Item attribute corrections are made rapidly to the first part of the data. Geometric corrections are made to the Segment Generator.

4. DATA INTERROGATION

4.1 Polygon Generation

Polygons represnting land-use parcels are generated by ECU software from the digitised lines. For example polygon E in FIG 1 is formed by parts of the lines which were initially digitised as A B C and D. Digitised lines are split at junction points by ECU's software (eg at 1, 2, 3, and 4). The double arrows represent the direction of digitising which was selected by the operator on purely ergonomic grounds, and the areal feature codes attributed to the lines in this example are as follows:

Line	Left Areal Code	Right Areal Code
A	5	1
B	1	3
C	1	4
D	2	1

where u = feature code (FC) 1, v = FC 2, w = FC 3, x = FC 4, and y = FC 5

A software redirecting of the lines takes place, allowing the digitising direction around a land parcel to be consistent. Thus if the section 1-4 of line segment C is the first item taken to generate the polygon, then lines D and A need redirecting. The coding on the constituent lines can thus temporarily be considered as:

Line	Left Areal Code	Right Areal Code
C	1	4
D	1	2
B	1	5
A	1	5

The internal code for the polygon is the common temporary code of 1 assigned to the left side and polygon E acquires this code.

FIG 1

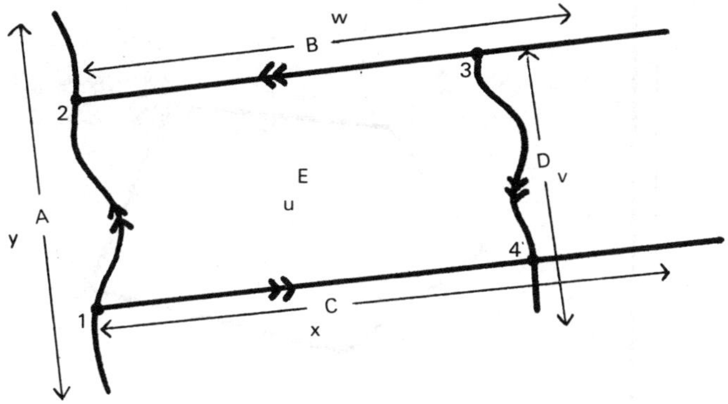

4.2 Polygon Dimension Computation

Once the polygon has been generated its dimensions have to be determined. This is done within ECU's software by an integration method summing the trapezia generated by each increment (consecutive pair of co-ordinate pairs) and the x-axis.

In FIG 2 there is a Polygon A formed by five increments (1-2, 2-3, 3-4, 4-5, 5-1). The area generated by each increment with respect to the x-axis is:

$$\tfrac{1}{2}(x_{n+1} - x_n)(y_n + y_{n+1})$$

where n is the sequence number of the co-ordinate pair starting the increment under consideration. Areas will have negative values for the two trapezia formed by the increments 4-5 and 5-1 with the x-axis, the remaining three increments will generate trapezia of positive area. Thus summing the trapezia will produce the dimension of Polygon A, and the general formula

$$\tfrac{1}{2} \sum_{n=1}^{p} (x_{n+1} - x_n)(y_n + y_{n+1})$$

(where p is the total number of increments in the line segments used to generate the polygon, and n is as define above) is used in ECU's software.

FIG 2

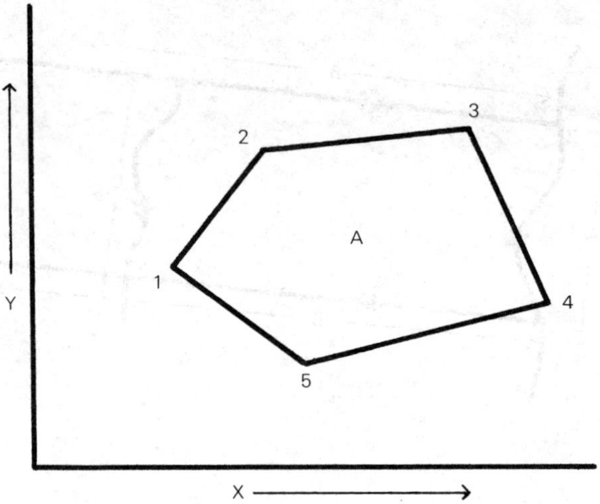

4.3 Line Lengths of Land-Use Interfaces

In FIG 3 the interfaces between reedswamp and river are shown by the lines a, b, and c. Variations in these interface lengths with time are indicators of Broadland environmental change. This particular parameter can be determined in less computer time than polygon dimensions and it is therefore an efficient parameter to use to monitor changes. The Broadland data files can be accessed and the interface lengths of any two selected land use combinations (as they existed at the time of photography in Jan 1980) determined, and these lengths compared with similar measurements taken from new imagery as it is obtained.

FIG 3

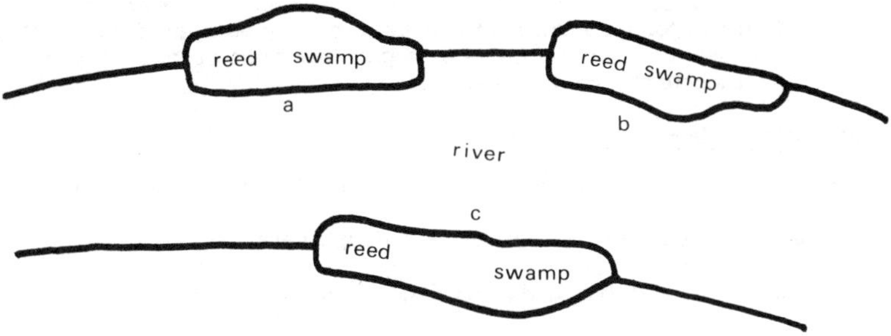

4.4 Data Display

The Broadland data is currently displayed in three ways:

i) as vector plotted maps at 1:10,000 with selected line styles and alphanumeric symbols to depict land-use;

ii) as vector displays on ECU's TEKTRONIX screens;

iii) as a raster display on ECU's I^2S Image Analysis System.

Approach i) above is essentially for field use. Approach ii) allows great flexibility, permitting a user to select those land-use types to be displayed, and the symbols in which they are to be displayed. Approach iii) will allow comparison of the land parcel geometry with LANDSAT imagery.

5. QUALITY CONTROL

5.1 Photointerpretation - Methods and Results

Field checks have been made to ensure that approximately 90% accuracy could be achieved during photointerpretation. However if real changes are to be measured in the future, it is essential that errors be quantified objectively. To this end quadrat points were selected on a stratified random basis, using alternate intersections of the OS 1Km grid. About 250 of these are to be mapped in detail to compare results with the original survey. Because much of the Broadland area is difficult, and in some parts, almost impossible to reach, vertical colour transparencies were taken from the Wallis Autogyro at 1:1,000 scale using a Vinten 70mm Aerial Camera. It was considered that the extra resolution and detail of these photographs would provide a standard by which the accuracy of the 1:10,000 scale photointerpretation could be judged.

The photoquadrats have yet to be mapped in detail but preliminary observations confirm that the overall standard of accuracy is likely to be well within the prescribed limits. The final statistics concerning areas of habitats will incorporate statements of reliability so that future resurveys will allow changes to be detected and separated from errors of photointerpretation and inaccurate cartography.

5.2 Digitisation Quality Control

Errors in the digitisation process are introduced by (amongst others):

i) the resolution of the ARISTOGRID digitisers (ΔE_1);

ii) manual line following during digitising (ΔE_2);

iii) software rounding up errors (ΔE_3);

iv) scaling up errors affecting those maps originally digitised at 1:10,560, but plotted at 1:10,000 scale (ΔE_4).

The above four errors will contribute to planimetric errors in the final digital product, and obviously plotter errors would also have to be included if the accuracy of the graphic product was to be considered. These four errors will also contribute to the polygon dimension computation results, and in addition the further error introduced by the area computation algorithm (ΔE_5).

Approximate magnitudes associated with these errors are:

$\Delta E_1 \simeq$ 0.1mm (related to the tolerance selected by digitisation personnel);

$\Delta E_2 \simeq$ 0.1mm (half the typical line width on the original document);

$\Delta E_3 =$ 0.005mm

$\Delta E_4 =$ 0.005mm

$\Delta E_5 =$ 0.05mm^2

Thus if the planimetric error is represented by ΔP, then the approximate relationship:

$$\Delta P = \sqrt{\Delta E_1{}^2 + \Delta E_2{}^2 + \Delta E_3{}^2 + \Delta E_4{}^2}$$

exists, and can be used to determine the maximum planimetric error in the digital data files.

As well as the above system errors, blunders can occur. These arise because of miscoding, undigitised lines, polygon misgeneration, etc. ECU's quality control is essentially directed towards blunder removal. The methods used to correct miscoded or undigitised lines are described in Section 3.2 above. Polygon misgeneration occurs when land parcel boundary junctions have been badly digitised and gaps or overshoots exist. Poor junctions can be identified in the checkplots. Alternatively the resulting misgenerated polygons can be detected after polygon generation when all lines are displayed in one symbol on a TEKTRONIX screen, and superimposed on this all polygons are displayed sequentially in another symbol.

6. FUTURE APPROACHES

This project has produced reference data suitable for future monitoring of changes in Broadland land-use. Satellites are a possible source of imagery for this and vegetation monitoring using LANDSAT imagery is well established. Image analysis techniques have been developed at ECU (Baker, 1981) which may be suitable for change monitoring once the relationships between image characteristics and the environment have been established by ecologists.

7. SUMMARY AND CONCLUSIONS

The method chosen has produced basic maps to a standard that will allow the Broads Authority to plan its management strategy using objective information about the distribution of plant communities and other land-uses.

Future changes will also be measureable within the constraints imposed by the limits of accuracy of the survey. The exact figures for accuracy still remain to be worked out, but it is believed that 90% of each land-use type has been correctly interpreted.

The use of computer mapping techniques has not provided maps that are really different from those which a draughtsman might have produced. But the strength of the technique lies in the fact that, although a map is little more than raw data, in this study we have created versatile data files which can be used to provide the specific information that we and other users need. This can be done without resorting to further manual methods of abstracting detail. We have also provided a product which can be added to or updated more simply than the conventional map sheet. In achieving these results we have also provided local detail for ECOBASE and TEIS, of which the latter is a growing systm within ITE and will help tell ecologists much more about the way natural and semi-natural communities are distributed, and how they respond to environmental factors which change in space and time.

8. ACKNOWLEDGEMENTS

We thank the Broads Authority who commissioned this project with grant aid from the Countryside Commission: also Davina Salter of ECU for assistance in the preparation of this paper.

9. REFERENCES

ANON, (1979): The Broads: comment and decision. Countryside Commission, Cheltenham.

BAKER, J R, (1981): The Norfolk Broads: Monitoring Changes in vegetation cover using LANDSAT data. Remote Sensing Society, International Conference; LONDON DEC 1981

BOORMAN, L A, and FULLER, R M, (1981): The changing status of reedswamp in the Norfolk Broads. Journal of Applied Ecology, 18, 241-269.

CAUFIELD, C, (1981): What can we save of the Broadlands? New Scientist, 89 (1234), 28-31

LAMBERT, J M, (1951): Alluvial stratigraphy and vegetational succession in the region of the Bure Valley broads. III Classification, status and distribution of communities. Journal of Ecology 39, 149-170.

LAMBERT, J M, (1953): The past, present and future of the Norfolk Broads. Transactions of the Norfolk and Norwich Naturalists Society, 17, 223-258.

Moss, B, (1979): Alarm call for the Broads - an ecosystem out of phase. Geographical Magazine, 52, 47-49.

O'RIORDAN, T, (1979): Alarm call for the Broads - signs of disaster and policy for survival. Geographical Magazine, 52, 50-53.

WHEELER, B D, (1978): The wetland communities of the River Ant Valley, Norfolk. Transactions of the Norfolk and Norwich Naturalists Soceity, 24 (4), 153-187.

A SAMPLE SURVEY OF URBAN OPEN SPACE (USING COLOUR INFRA-RED AERIAL PHOTOGRAPHS) IN MERSEYSIDE METROPOLITAN COUNTY

Dr. Stamatis Sekliziotis
Ministry of Co-ordination and Planning
Athens
Greece

Dr. W. Gordon Collins
Remote Sensing Unit
Department of Civil Engineering
University of Aston In Birmingham
England, B4 7ET.

ABSTRACT: The Merseyside County Planning Authority needed to acquire information relating to its urban open space. A notation was compiled and urban open space maps were produced separately from both black and white and colour infra-red aerial photographs. The results were compared and eleven sample areas were selected to produce, from colour infra-red photography, a detailed inventory of urban open space. Digital computer maps were produced of each sample area and statistics were compiled to give the type and amount of urban open space which exists. The County Authority provided details of the socio-economic class of each sample area and these were compared with the type and amount of urban open space to reveal some interesting relationships between population class and one aspect of the environmental quality of their residential areas.

Paper presented at a conference on Matching remote sensing technologies and their applications convened in London by the Remote Sensing Society, 16 - 18 December, 1981

1. INTRODUCTION

This paper outlines the first three stages of an air photo based study of urban open space carried out in close co-operation with the Planning Authority of Merseyside Metropolitan County Council, England.

With the change in the U.K. government policy - from building new towns on the dwindling supply of agricultural land, to the redevelopment of the declining city centres - came the urgent need for a systematic and comprehensive inventory of the land utilisation of the city centres.

One important aspect of urban land is open space, its type, extent and distribution, and it is this particular aspect which is the main subject oi this research project.

The problem was defined as the need to acquire in the most rapid, accurate and cost effective manner, details of the urban open space, and present this information in both statistical and map form in a way which would be of direct and immediate value to the planning authority. The proposed solution was to investigate the extent to which aerial photographs might provide the prime source material to supply this data, and to develop a suitable method of producing computer maps and graphs. Vertical aerial photographs were available at a scale of 1:10,000 in both black and white and colour infra-red, and comparisons were made of their relative value for open space surveys. An ICL 1904S computer and Calcomp plotter were available and the methodology was based on the assumption that most County Authorities had access to such equipment.

The information needs of the Planning Authority dictated the type and amount of open space information to be collected: hence an essential first step was to define and classify "open space" in relation to these information needs. The first task in compiling a notation is to clearly define what is meant by the term "open space".

Bearing in mind the applied nature of this study the authors worked in close co-operation with the County Planners to ensure that the end product of the work would provide the information required by them for their particular uses. "Open space" was defined as: "That land within the urban framework which is not covered by buildings".

2. STAGE ONE

A notation was developed for the first stage of this study (Figure 1). This notation contained fourteen major units which have been subdivided, mostly on the basis of vegetation, into a total of fifty eight sub units.

In most instances four categories of open space were distinguished in terms of vegetation coverage and maintenance:

Category 1: Unvegetated (Hardly landscaped open spaces)
Category 2: Selfmaintained vegetation growth. Unmanaged or semi-managed (Limited management)
Category 3: Managed vegetation (High grade maintenance)
Category 4: Poorly maintained vegetation (Poor vegetation coverage)

Familiar types of open space, such as a golf course or a soccer pitch, where the vegetation cover varies in terms of maintenance, have been used as a key to identify and classify other types of open space management.

Figure 1 Urban Open Space Notation

PRIVATE GARDENS

1 PRIVATE GARDENS - UNVEGETATED
2 PRIVATE GARDENS - SELF-MAINTAINED VEGETATION
3 PRIVATE GARDENS - MANAGED VEGETATION
4 PRIVATE GARDENS - POORLY MAINTAINED VEGETATION

PUBLIC GARDENS

6 PARKS - UNVEGETATED
7 PARKS - SELF-MAINTAINED VEGETATION
8 PARKS - MANAGED VEGETATION
9 PARKS - POORLY MAINTAINED VEGETATION
10 AMENITY OPEN SPACE - UNVEGETATED
11 AMENITY OPEN SPACE - SELF-MAINTAINED VEGETATION
12 AMENITY OPEN SPACE - MANAGED VEGETATION
13 AMENITY OPEN SPACE - POORLY MAINTAINED VEGETATION

LANDSCAPED OPEN SPACE FOR CIRCULATION

15 STREETS LINED WITH TREES

PLAYSPACE, RECREATION, SPORTS

19 CHILDREN'S PLAYGROUNDS - UNVEGETATED
20 CHILDREN'S PLAYGROUNDS - SELF-MAINTAINED VEGETATION
21 CHILDREN'S PLAYGROUNDS - MANAGED VEGETATION
22 CHILDREN'S PLAYGROUNDS - POORLY MAINTAINED VEGETATION
23 PLAYFIELDS - UNVEGETATED
24 PLAYFIELDS - SELF-MAINTAINED VEGETATION
25 PLAYFIELDS - MANAGED VEGETATION
26 PLAYFIELDS - POORLY MAINTAINED VEGETATION
27 GOLF COURSES - UNVEGETATED
28 GOLF COURSES - SELF-MAINTAINED VEGETATION
29 GOLF COURSES - MANAGED VEGETATION
30 GOLF COURSES - POORLY MAINTAINED VEGETATION
31 EDUCATIONAL PLAY SPACES - UNVEGETATED
32 EDUCATIONAL PLAY SPACES - SELF-MAINTAINED VEGETATION
33 EDUCATIONAL PLAY SPACES - MANAGED VEGETATION
34 EDUCATIONAL PLAY SPACES - POORLY MAINTAINED VEGETATION
35 OTHER RECREATIONAL SPACE

WOODLAND

37 WOODLAND OR CLUMPS OF TREES - HEALTHY
38 WOODLAND OR CLUMPS OF TREES - DAMAGED
40 WOODLAND - UNVEGETATED AREAS

ALLOTMENT GARDENS

41 ALLOTMENTS - AUXILIARY AND UNUSED SPACE
42 ALLOTMENTS - IN USE
43 ALLOTMENTS - NOT IN USE

AGRICULTURE

46 AGRICULTURAL GRASSLAND AND FALLOW
47 AGRICULTURAL CROPLAND, HORTICULTURE, ORCHARDS

UNDISTURBED OPEN SITES

48 ESTABLISHED HEATH
50 SHRUBS - HEALTHY
52 SHRUBS - DAMAGED OR DEFICIENT FOLIAGE
54 SCRUB - HEALTHY
55 SCRUB - DAMAGED OR DEFICIENT FOLIAGE
56 MOSSLAND
57 SAND DUNES
69 MARSHLAND

DERELICT AND SPOILED LAND

58 DERELICT AND SPOILED LAND - UNVEGETATED AREAS
59 DERELICT AND SPOILED LAND - VEGETATION WELL ESTABLISHED
60 DERELICT AND SPOILED LAND - VEGETATION NOT WELL ESTABLISHED
61 DERELICT AND SPOILED LAND - SCATTERED VEGETATION

NEGLECTED WASTE LAND

62 NEGLECTED WASTE LAND - NO VEGETATION
63 NEGLECTED WASTE LAND - VEGETATION WELL ESTABLISHED
64 NEGLECTED WASTE LAND - VEGETATION NOT WELL ESTABLISHED
65 NEGLECTED WASTE LAND - SCATTERED VEGETATION

WATER BODIES

68 OPEN WATER (RIVERS, CANALS, LAKES - NOT RESERVOIRS ETC.)

CEMETERIES

70 CEMETERIES

OTHER OPEN SPACES

71 OTHER OPEN SPACE

OTHER LAND USES (NOT OPEN SPACE)

99 OTHER LAND USES

On the colour infra-red photography well cut areas of grass, as the "greens" in the golf courses, expose a purple hue variation. The "sands" expose a white or white bluish variation. "Hummocks" expose a blue variation changing to reddish purple or purple towards the periphery. This is because of the appearance of undisturbed organic top soil in the middle with some scattered overused grass cover, changing to a uniform well managed grass cover towards the periphery.

The "gorse" areas expose red, light red, and dark green variations as they are covered with self-maintained vegetation growth as rough grass, shrubs and heaths.

On the black and white photography the range of colour (tonal) variation is very much less, and textural differences become more important in identifying the surface appearance. The effect of this lack of surface distinction on the black and white photography confirms that it is a much less sensitive indicator of vegetation differences.

Three enumeration districts were selected as the sample area for the first stage of this study. These three districts vary widely in many respects, particularly in housing type, land use and cover, and socio-economic class of the population.

3. DATA EXTRACTION AND RECORDING

Using a Carl Zeiss Interpretoscope the black and white aerial photographs were stereoscopically examined to identify those units of open space as recorded in the notation (Figure 1). A transparent grid overlay was placed over the photographs, first the black and white, then the colour infra-red, so that the data could be recorded on a grid square basis. The grid represented a 20m x 20m square at ground scale, and that use which fell under the cross at the centre of the grid was the use recorded for the entire cell. The location of each grid cell was recorded on the coding form so that spatial information could be retrieved for mapping purposes.

The two sets of survey information - one from the black and white photography, the other from the colour infra-red photography - were then fed into a computer data bank. Computer programmes were developed to provide a variety of statistical data from the information fed into the system.

4. SURVEY RESULTS

In order to make useful and easily understandable comparisons (as a source of "open space" data) between the value of black and white and colour infra-red aerial photographs the information was compiled in various ways.

The first comparison was a statistical output showing the total amounts of open space recorded in the major categories only (Figure 2). This did not include details of the vegetation, surface cover or condition. The results clearly show that, at this level of information, there is virtually no difference in the type or amount of open space information which can be derived from the two sources. However, when details of the vegetation cover and its condition are included, then some significant differences do emerge.

Figure 2 Open Space Statistics

E.D.68

URBAN OPEN SPACE GROUPS	PHOTOGRAPHY USED: BLACK & WHITE (Ha)	PHOTOGRAPHY USED: FALSE COLOUR (Truth) (Ha)	DIFFERENCES ± (Ha)
Private gardens	4.16	4.84	0.68
Public gardens	0.32	0.28	-0.04
Landscaped open space for circulation	0.80	0.88	0.08
Playspace, Recreation, Sports	8.04	7.76	-0.28
Woodland	0.04	0.00	-0.04
Allotment gardens	1.36	1.24	-0.12
Agriculture	0.04	0.12	0.08
Undisturbed open space	0.00	0.00	0.00
Derelict and spoiled land	0.00	0.00	0.00
Neglected waste land	4.20	3.64	-0.56
Water bodies	0.00	0.00	0.00
Cemeteries	0.00	0.00	0.00
Other open spaces	0.32	0.40	0.08
TOTAL OPEN SPACE	19.28	19.16	-00.12
Other land uses	3.52	3.52	0.00
SURVEYED TOTAL AREA	22.80	22.68	-00.12

\+ = more area recorded with false colour
\- = less area recorded with false colour

Calcomp computer maps of the open space were derived from both the black and white and colour infra-red photography (Figure 3). For each of the three enumeration districts, maps were compiled to show the gross (real and apparent) differences between that information derived from the black and white photography and that derived from the colour infra-red photography.

Some of these differences of interpretation were not real, but were due to slight geometric distortions in the photography which resulted in the same grid dot not being on identical ground points on the two sets of photography. This affected points both within the area and also along the perimeter. Other differences were due to changes having taken place in the field between the black and white 1974 photographs and the colour infra-red 1975 photographs.

These apparent differences were identified and deleted. A new map was compiled which showed only the "real" differences of photo-interpretation at both group and sub-group levels (Figure 4).

The differences in the information obtained were presented as a statistical list of two major types (containing eight minor types) of differences.

5. RESULTS OF PHOTO INTERPRETATION

Analysis of Differences (Figure 5)

1. Apparent Differences (A)

a) Both decisions possible: This occurred when the cross landed on a certain point on the one photograph whilst the identical cross did not land on the same point on the other photograph - due to orientation difference, slight scale difference etc. There were 176 cases of this recorded for the three enumeration districts.

b) Changes in land use: Disagreements due to land use changes were recorded. The B+W aerial photography was 1974 cover and the CIR was 1975 cover: both taken in summer. Such changes in land use are likely to occur in a period of one year. There were eight cases recorded.

c) Boundary differences: The perimeters of the three enumeration districts had to be delineated twice, one on the grid superimposed on the B+W and one on the grid superimposed on the CIR photography. The same line had to be traced twice by hand which means that different grid cells were left out or included in the two cases. Such disagreements occurred along the perimeters. Twenty seven cases of this were recorded.

2. Real Differences (B)

a) Within groups: B+W assessed as correct and CIR as in error. These were cases where the cross was representing a land use value "beyond doubt" on the B+W photography on a "unit" or "sub-unit level". but the identical cross on the CIR photography was recorded as showing a different land use. In these cases the B+W had to be accepted as correct. No such case was recorded.

b) Within groups: CIR assessed as correct and B+W in error. These were cases where the cross represented a land use value "beyond doubt" on the CIR photography on a "unit" or "sub-unit level" according to the criteria established, but the identical cross on the B+W photography was recorded as showing a different land use. In these cases the CIR was accepted as being correct. Three hundred and seventy cases of this were recorded.

Figure 3 Open Space Map
Derived from colour infra-
red aerial photographs

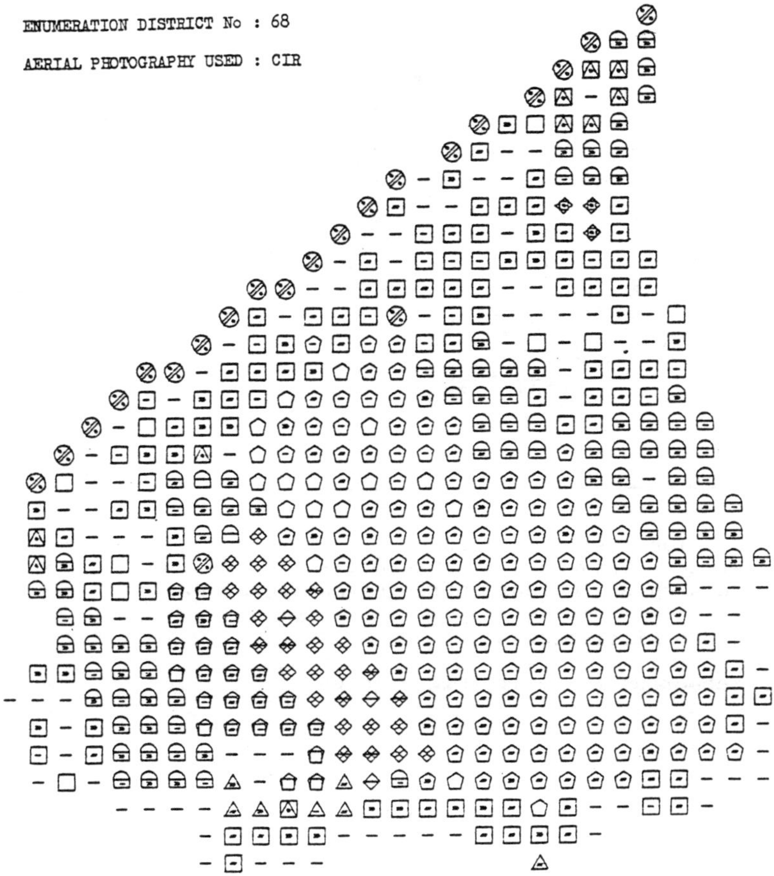

Figure 4 Real Differences in Interpretation

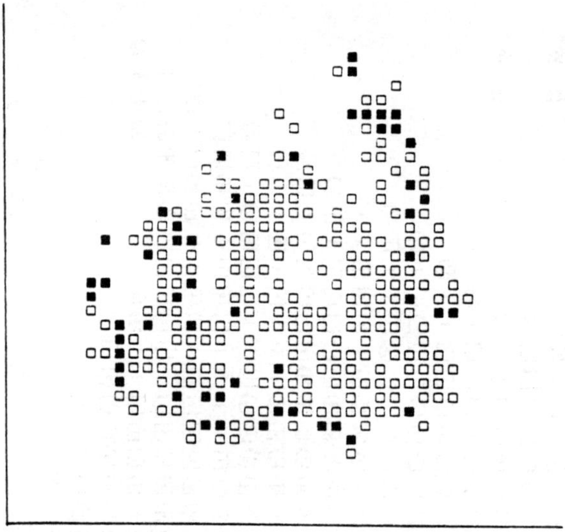

Enumeration district 68/1046
Differences between photointerpretation of black and white photography and colour infra-red photography.

□ Group Differences

■ Sub-group (or unit) Differences

Figure 5 Results of Photo-interpretation. Analysis of Differences.

ENUMERATION DISTRICT	APPARENT DIFFERENCES (A)			REAL DIFFERENCES (B)				FIELD CHECK NEEDED	AGREEMENTS	DISAGREEMENTS	TOTAL OBSERVATIONS PER ENUMERATION DISTRICT
	BOTH DECISIONS POSSIBLE (a)	CHANGES IN LAND USE (b)	BOUNDARY DIFFERENCES (c)	DIFF. WITHIN GROUPS (B+W CORRECT) (a)	DIFF. WITHIN GROUPS (CIR CORRECT) (b)	DIFF. BETWEEN GROUPS (B+W CORRECT) (c)	DIFF. BETWEEN GROUPS (CIR CORRECT) (d)	UNCLEAR IN BOTH IMAGES (e)			
68/1046	72	7	7	0	239	0	18	2	225	345	570
64/1046	32	0	10	0	40	0	0	0	108	82	190
44/1054	72	1	10	0	91	2	13	0	194	189	383
TOTALS	176	8	27	0	370	2	31	2	527	616	1143
	TOTAL OF APPAR. DIFF: 211			TOTAL OF REAL DIFF : 403							
				TOTAL OF CIR CORRECT : 401							
				TOTAL OF B+W CORRECT : 2							

B+W : Black and white photography
CIR : Colour infrared photography
DIFF : Differences

c) Between groups: B+W assessed as correct and CIR as in error. In this instance the open space units were easy to describe but not so easy to allocate to a particular unit in the notation. The results were that a mis-classification occurred when using the CIR photography. Two cases of this were recorded.

d) Between groups: CIR assessed as correct and B+W as in error. These are very significant differences in which the open space was correctly identified on the CIR and incorrectly identified on the B+W. Thirty one cases of this were recorded.

e) Field check necessary: In only two cases was it found that the type of open space could not be identified from either of the two types of photography.

The results of the open space statistics derived from the B+W and CIR aerial photographs confirm the advantages of the CIR when the notation includes information relating to vegetation cover and condition.

6. SUBSEQUENT STAGES

The CIR was used exclusively to digitise, map and record the data of eleven $1/4km^2$ sample areas selected by the County Planning Department. These sample areas contained a wide range of land use, land cover and socio-economic class of population. Six hunded and twenty five (625) sets of digitised data were recorded for each of the eleven (11) $1/4km^2$ sample areas.

In the third stage some of the results of a socio-economic survey carried out by the County Planners were used to identify the socio-economic class of the population of the eleven sample areas. Each sample area was allocated, on a four point scale from 1 (high) to 4 (very low), the socio-economic class of its population. The open space data was re-organised to relate to these four categories of socio-economic class.

The fourth and final stage involved a comparison of the amount and type of open space with each socio-economic class. Because all the data was fed into the computer store, it was possible to make this comparison for any of the ten groups or thirty nine units of the Open Space Notation (Figure 6). Details of the surface cover, type, condition and management were also included.

As examples the two groups "Neglected Land" and "Private Gardens" are selected (Figures 7 and 8 respectively).

7. OPEN SPACE AND SOCIO-ECONOMIC CLASS

Neglected Land

A variety of different units occur under this group: these include among others rough grassland, derelict land, mineral extraction, waste disposal sites and cleared land.

Figure 7 shows a steady increase in the extent of neglected land (7%, 9%, 13% and 23%), with respect to decreasing social category. This is to be expected for the quality of the environment, both natural and man made is closely linked to the visual quality of the landscape, and this is always very strongly related to the socio-economic class of the population.

Figure 6 Urban Open Space Use Notation.

CODE NUMBER	URBAN OPEN SPACE UNITS	
01	Heathland	1. Semi-natural Environments
02	Woodland	
03	Moss-land	
04	Sand Dunes	
05	Beaches	
06	Marshland	
07	Streams and Rivers	2. Water bodies
08	Canals	
09	Lakes and Ponds	
10	Reservoirs	
11	Oceans	
12	Private Gardens	3. Private Gardens
13	Parks	4. Amenity Open Space
14	Amenity Open Space-general access.	
15	Amenity Open Space-limited access, Institutional etc.	
16	Amenity Open Space - Industrial, Commercial.	
17	Streets lined with trees	5. Space for Transportation
18	Streets not lined with trees	
19	Railways	
20	Motorways	
21	Open air car parks	
22	Airfields	
23	Children's Playgrounds	6. Play and Recreation
24	Sportsfields and Stadia	
25	Golf Courses	
26	Educational Playspace	
27	Industrial Sports Facilities	
28	Other Open Air Playspaces	
29	Allotment Gardens	7. Agriculture and Horticulture
30	Agriculture & Horticulture	
31	Industrial/Commercial Ancillary Open Space (other)	8. Neglected land
32	Rough grassland	
33	Scrubland	
34	Derelict land	
35	Mineral Extraction	
36	Waste Disposal Sites	
37	Cleared Land	
38	Cemeteries	9. Cemeteries
39	Other Open Spaces	10. Other
00	Built Environment	

Figure 7

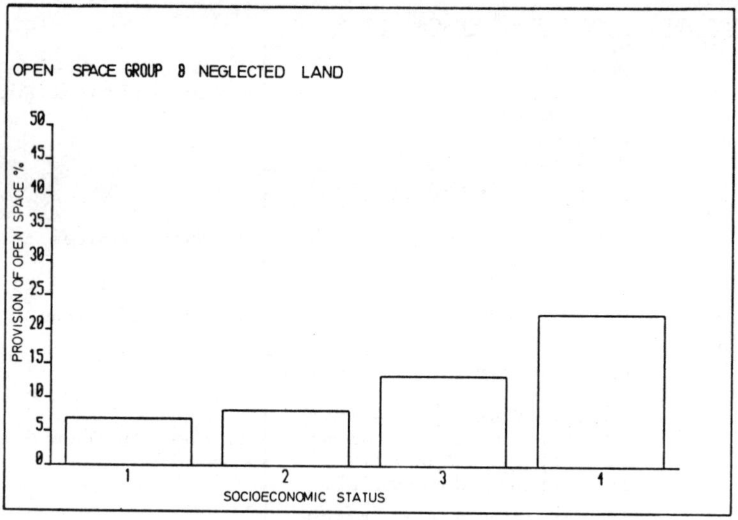
OPEN SPACE GROUP 8 NEGLECTED LAND
PROVISION OF OPEN SPACE %
50
45
40
35
30
25
20
15
10
5
0
1
2
3
4
SOCIOECONOMIC STATUS

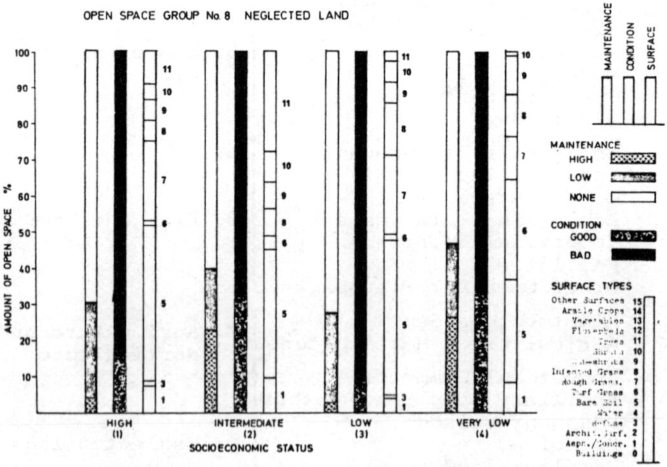
OPEN SPACE GROUP No. 8 NEGLECTED LAND
AMOUNT OF OPEN SPACE %
HIGH (1)
INTERMEDIATE (2)
LOW (3)
VERY LOW (4)
SOCIOECONOMIC STATUS
MAINTENANCE
CONDITION
SURFACE
MAINTENANCE
HIGH
LOW
NONE
CONDITION
GOOD
BAD
SURFACE TYPES
Other Surfaces 15
Bare Soil 5

Figure 8

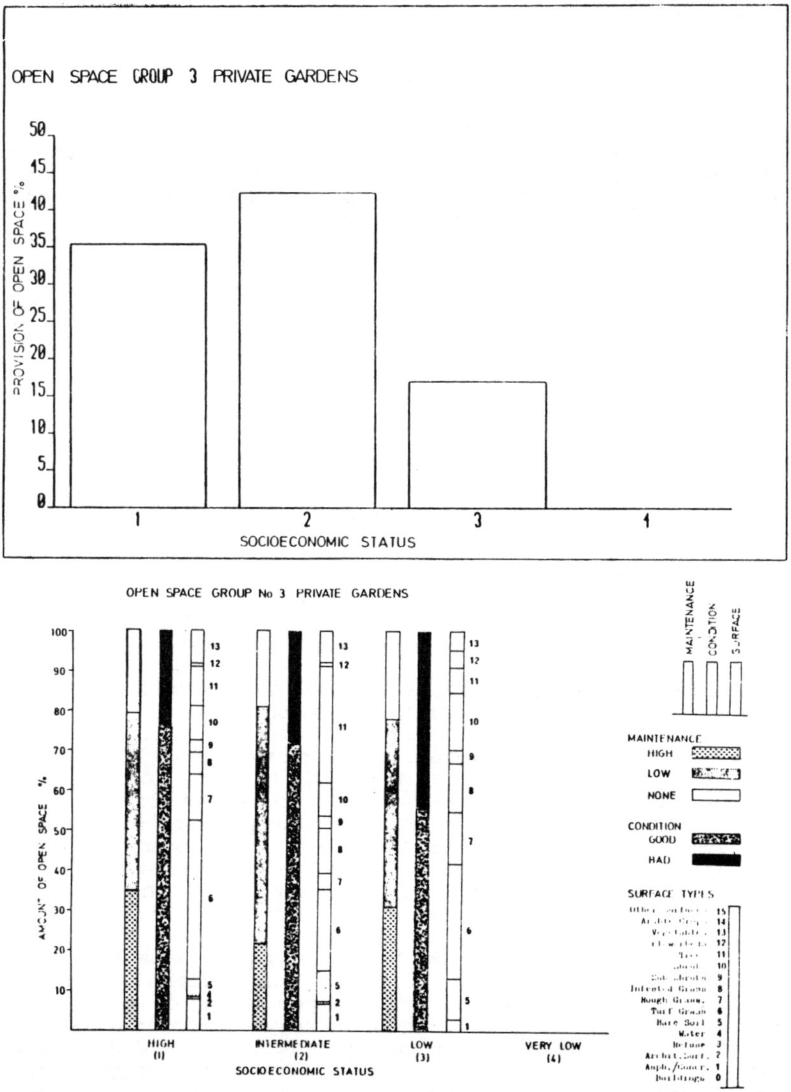
OPEN SPACE GROUP 3 PRIVATE GARDENS
PROVISION OF OPEN SPACE %
50
45
40
35
30
25
20
15
10
5
0
1
2
3
4
SOCIOECONOMIC STATUS
OPEN SPACE GROUP No 3 PRIVATE GARDENS
AMOUNT OF OPEN SPACE %
100
90
80
70
60
50
40
30
20
10
HIGH (1)
INTERMEDIATE (2)
LOW (3)
VERY LOW (4)
SOCIOECONOMIC STATUS
MAINTENANCE
CONDITION
SURFACE
MAINTENANCE
HIGH
LOW
NONE
CONDITION
GOOD
BAD
SURFACE TYPES
Rough Grass 7
Turf Grass 6
Bare Soil 5
Water 4
Refuse 3
Buildings 0

Different types of neglected land predominate in the four social categories. In Category 1 scrubland, cleared land and rough grassland are most prominent, with only a very small amount of derelict sites. In Category 2, cleared land and scrubland are most prominent, with little derelict land and industrial open areas. Category 3 is also predominently scrubland with a small amount of cleared land. Associated with Category 4, cleared land was most common, and derelict sites and industrial open areas were also quite significant. In all categories there was very little hard surface cover of asphalt or concrete.

In terms of management, condition and surface cover, much of the findings could be anticipated. The management of neglected land is mainly non-existent, or poor: the condition dominantly bad, and the surface cover includes a large amount of bare soil and poor grassland. However, what might appear to be a very negative aspect of the urban environment, has a potentially very positive part to play in urban renewal and improvement programmes. For it is in these neglected land areas - dominantly soft covered, distributed throughout the county but with increasing concentration in the socially deprived areas - that the opportunity to improve the environment is greatest. Already associated work has been carried out to map, from aerial photography, all the derelict land in the Merseyside County.

Private Gardens

Figure 8 shows the relationship between the distribution of private gardens and socio-economic status. This reveals that in relation to socio-economic status (Category 1, high to 4, low) the percentages of open space devoted to private gardens are 36%, 42%, 17% and zero respectively.

It is hard to accept that no private gardens occur in all the other Category 4 areas in the County, but this finding reflects a very poor provision of private gardens in inner areas, where the housing stock consists of old terraces of very poor quality, or fairly recently built blocks of flats with open landscaped land around them.

There is usually a direct relationship between urban form in general (and housing in particular) with the extent of the area which is devoted to private gardens. Figure 8 shows that Category 2, which may appear to be out of step, does in fact accord to this relationship. This is because of the age of the residential dwellings in Category 2. This was formerly a high quality residential area, built at a time when planning control was not very strong, and when land was relatively cheap. Large houses with very large gardens were usual in these middle class areas, in contrast with the situation now, when both buildings and land are substantially more expensive, and planning controls more rigorous. This area is now in decline with the large houses being split into flats, or becoming derelict.

The urban form is different in the Category 3 areas where the common type of housing is in the form of tower blocks, relatively recent council estates, semi-detached houses built between the 1930's and the immediate post war period, and many Victorian terraces of intermediate and poor quality. Such areas are mostly covered by buildings so that the space left has to provide such open space facilities as private gardens, street layouts and recreation. Here the average amount of open space devoted to private gardens is very small (17%).

In the Category 4 sample area the residential buildings are tall blocks of flats, some quite high rise. In such circumstances no provision is made for private gardens: instead there are modestly sized, open, landscaped - mostly grass - areas surrounding the buildings.

Figure 8 shows the relationship between maintenance, condition, and type of surface of private gardens, and socio-economic status. In terms of maintenance the percentages of private gardens in high maintenance in relation to Categories 1, 2, 3, are 34%, 23% and 33% respectively. Categories 1 and 3 are very similar in relation to the maintenance situation, though the size of the gardens is very much smaller in Category 3.

In terms of surface condition there is a steady decline with decreasing social category of 77%, 66% and 55% respectively of surfaces considered to be in a good condition.

In terms of surface cover, Category 1 has a greater amount of good quality turf grass than Categories 2 and 3. A fine lawn appears to be mandatory in middle class housing areas! Category 1 is alone in having some water surfaces, these are mostly swimming pools, with a few ornamental ponds. The amount of asphalt/concrete cover increases with social category; the forecourts and drives of Category 1 have quite extensive areas of hard surface which are covered with two or three cars, plus caravans and boats. The amount of tree cover varies considerably: 11%, 29% and 6% (Categories 1, 2 and 3 respectively). The main reason for the large extent in Category 2 is related to the size of the building plots and the date and purpose for which the estate was originally developed.

The percentage of the gardens in a good condition decreases with status: 77%, 66% and 54% respectively.

The standard of maintenance is not similarly reflected, with status grades 1 and 3 showing strikingly similar patterns in spite of their marked differences in other respects. The low status grade areas have a high proportion of council housing, and some control seems to be evident in garden maintenance. The intermediate status group consists mainly of older rented property, where less control or provision of garden maintenance is evident.

A large amount of useful information has been acquired in this study of urban open space, and its relation to the socio-economic class of the population. This information and the methodology developed in this work provide a valuable input for those planning authorities concerned with urban redevelopment.

APPLICATION OF REMOTE SENSING TO PLANETARY TOPOGRAPHIC MAPPING

S S C Wu, United States Geological Survey Branch of Astrogeologic Studies, Flagstaff, Arizona

Remote sensing plays an important role in planetary topographic mapping. Topographic maps of planetary bodies are compiled by the synthesis of remotely sensed data, imaging or nonimaging, transmitted by spacecraft or received from various sensors.

A global topographic map of Mars was compiled by the synthesis of remote sensing data which includes S-band radio-occultation ultraviolet spectrometer (UVS), infrared radiometer (IRR), infrared interferometer spectrometer (IRIS) and images from vidicon cameras on board the Mariner 9 spacecraft, as well as earth-based radar oppositions of Mars. With almost 60,000 Mars pictures transmitted back to Earth by the two Viking spacecraft, Mars is currently being systematically mapped in great detail.

By using topographic data derived from Apollo and Lunar Orbiter photographs and radar altimetry as well as Earth observations, a global topographic map of the Moom is being compiled. Using radar altimetry data transmitted from the Pioneer Venus spacecraft, a preliminary global topographic map of Venus has been compiled. For the compilation of more detailed maps of Venus, the technology of using side-looking radar stereo images from the proposed VOIR mission (Venus Orbiter Imaging Radar) is under development.

Techniques of remote sensing data processing and methods of map compilation are discussed along with the presentation of cartographic products; global maps of Mars, the Moon and Venus.

COMPUTER ANALYSIS OF MULTITEMPORAL LANDSAT DATA FOR MAPPING OF LAND-USE FOREST CLEARCUTS AND MIRES - METHODOLOGICAL STUDIES

Leif Wastenson, Wolter Arnberg,
Laine Boresjö and Margareta Ihse
Remote Sensing Laboratory
Department of Physical Geography
University of Stockholm

Abstract: A brief summary is given of experiments with land-use mapping from LANDSAT-2 data and from airborne MSS in the Linköping region, Sweden. The results from a single-date analysis have been compared with a multi-date analysis. Methods for monitoring forest clear cutting during the period 1975-1979 using digital multi-temporal LANDSAT imageries are also presented. Another study deals with methods to use digitized map information in the multi-date LANDSAT classification process and methods to select the most informative combination of wavelength bands and registration dates. In the first application of these methods we have tried to classify mires.

INTRODUCTION

Since 1973 - one year after the first Earth Resources Technology Satellite was launched - computer mapping experiments using LANDSAT computer compatible tapes have been run at the Remote Sensing Laboratory, Department of Physical Geography, University of Stockholm. The experiments started during implementation, at the Stockholm Computer Center, of a general image analysis software developed at the National Defense Research Institute (Åkersten & Gustafsson, 1978). Land-use mapping and an inventory of forest clearcutting were the first subjects of investigation (Orhaug, Wastenson & Åkersten, 1976 and 1977; Wastenson et al 1978 and 1981). In the LANDSAT experiments only data from the multispectral scanner system (MSS) were used. Using the same software airborne MSS data was included in the experiments during 1975, when a Swedish MSS campaign was carried out (Wastenson et al, 1978 and 1981).

To evaluate the classification results there was a need to rectify the scanner images (Larsson, 1980), digitize the ground truth data and carry out a pixelwise (pixel=picture element) evaluation (Arnberg, 1980). Programmes were developed and implemented on the Nord 10/S computer at the Laboratory - configuration is briefly described by Arnberg (1981 b) - together with programmes for training, classification and generalization of processed MSS data (Xu Guan-hua, 1981). Methods to select the most informative combination of wavelength bands and registration dates have also been developed (Arnberg, 1981 a).

Our first investigation of the possibilities of land-use mapping using LANDSAT and airborne MSS was performed in the Linköping region on the Image

Paper presented at a conference on Matching remote sensing technologies and their applications conveved in London by the Remote Sensing Society, 16-18 December 1981

Analysing System at the National Defense Research Institute. The results from this study were then compared with a multidate analysis of LANDSAT data.

Methods to monitor forest clearcuts using multi-temporal LANDSAT imageries are also presented. The methods were developed in co-operation with the Swedish Space Corporation on their image analysis system MDA (Mc Donald Dettwiler & Ass).

The last study in this paper deals with methods to integrate digitized map information in the LANDSAT multi-temporal classification and to select the most informative wavelength bands and recording dates. The first application of this method has been performed in an effort to classify mires from LANDSAT data.

VEGETATION AND LAND-USE MAPPING BY CLASSIFICATION OF MSS DATA FROM AIRCRAFT AND LANDSAT

Single-date analysis

A LANDSAT-2 scene from July 3, 1975 of Östergötland, Sweden, has been used for land-use mapping by supervised classification. Because severe disturbances were found in the digital data of band 6, this channel has been omitted. An area of 25 000 km^2 - about 7% of a LANDSAT scene - has been classified using 28 training areas and a classification system shown in Table 1.

Table 1: Classification system for land-use mapping using LANDSAT and airborne MSS data at Linköping, Östergötland, Sweden.

Level 1	Level 2	Number of training areas LANDSAT
1. Coniferous forest (including clear felled areas and sparse pine forest on bare rock)	1. Norwegian spruce forest	2
	2. Pine forest	1
	3. Mixed forest	1
	4. Clearcuts without scrub growth	1
	5. Clearcuts with scrub growth	2
2. Deciduous forest	6. Hardwood forest	2
	7. Other deciduous forest	1
3. Grassland	8. Grassland	1
4. Arable land	9. Arable land	9
5. Urban areas	10. Urban areas	4
-	11. Bare rock with sparse pine forest	1
6. Water	12. Water	3
7. Wetlands	13. Reedmashes (Phragmites)*	
	14. Other wetlands (mires)*	
8. Rejects	15. Rejects	

* not used in the LANDSAT classification

The classification results from LANDSAT have been compared with results from airborne MSS of a subregion of 700 km^2 recorded at an altitude of 6 km on July 6, 1975. A Daedalus scanner with 10 channels within the wavelength range 0.38-1.10 μ was used. In the analysis of airborne MSS data a total of 34 training areas and a slightly modified classification system were used. Different combinations of wavelength bands and number of training areas were tested.

The classified MSS images from aircraft and satellite have been rectified to fit the topographic map in the scale of 1:50 000 and the economic map reduced to 1:20 000 from its original scale 1:10 000. The result of the rectification is good for the LANDSAT data (standard deviation about 0.5 pixel), but in the airborne MSS data the maximum error is 5 pixels (75 m). Before the evaluation the classified data were generalized using a "clean up" programme.

The classification results were compared with the following "ground truth" maps:

1. Land-use map in 1:125 000 of the whole region based on topographic and geologic maps and air photo interpretation in panchromatic images in 1:30 000.

2. Vegetation map in 1:50 000 of a subregion (25 km^2) based on air photo interpretation in colour infrared photographs in 1:50 000 and on field checks.

3. Digitized vegetation map in 1:20 000 of a smaller test area (5 km^2). The map is based on air photo interpretation in colour infrared images in the scale of 1:20 000 and on very detailed field checks.

Map no 1 was only used for a calculation of the agreement in the areal distribution of the land-use categories between the LANDSAT map and the "ground truth" map. The comparison gives the following results (+ overestimation, - underestimation in LANDSAT):

coniferous forest	-0.9 %
deciduous forest	+1.6 %
arable land	-1.5 %
grassland	-3.0 %
urban land	+1.2 %
hydrography	-0.8 %

4.1 % of the LANDSAT pixels were rejects. Using these seven classes the LANDSAT classification gives an acceptable accuracy for land-use statistics.

The "ground truth" map no 2 was used for a manually performed evaluation of the LANDSAT classification presented in the scale of 1:50 000 and re-sampled to a pixel size of 40 x 40 m^2. The evaluation was done within 70 test sites of 4 x 4 pixels sampled by a window moved from the upper left corner. Only the first ten positions of each class not crossing boundaries in the ground truth map were accepted. Within the sampled test sites the number of correctly and incorrectly classified pixels were counted manually.

Map no 3 was used for a pixel-by-pixel computerized evaluation of a small part of the classified maps from LANDSAT and airborne MSS.

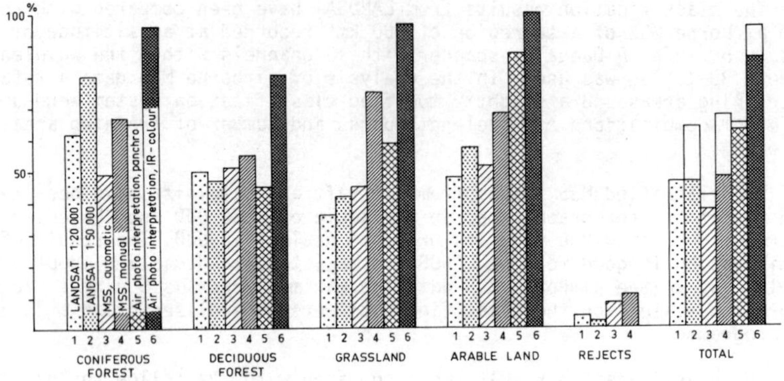

Figure 1 Comparison of the accuracy of land-use mapping achieved by different methods. Bars 1 and 2 from LANDSAT classification, 3 and 4 from airborne MSS classification, and 5 and 6 from air photo interpretation in panchromatic and infrared sensitive colour film. The overall classification "total" is given for level 1 (6 classes) and for level 2 (12 classes) in the classification system (Table 1). The open areas of the bars in the right diagram show the effect of reducing the number of classes from 12 to 6.

The results from the single-date analysis of MSS data from the two platforms are summarized in figure 1. Examples of ground truth and classified MSS maps are given in Plate I. The maps were plotted with an ink jet plotter directly as transparencies to the topographic and economic maps in the scales of 1:50 000 and 1:20 000, respectively.

In land-use mapping the accuracy of the single-date analysis of MSS data is not acceptable for applied mapping in Sweden. In figure 1 the results of air photo interpretation using panchromatic and colour infrared films (without any field checks) are compared with the results from the MSS methods. A comparison is made between different scales and evaluation methods for the two levels of the classification system.

Some notes are reguired for the diagram in figure 1. Water is not included in the evaluation. The accuracy for this class, however, is very good (90-100 %, depending on the size of lakes and rivers). The classification of water is made by an intensity level slicing in band 7. The class "wetlands" has not been represented by training areas in the LANDSAT tests. In the target area for airborne MSS only a few small mires occur.

The MSS classifications are all inferior to air photo interpretation in infrared colour images. In level 1, the results from the interpretation of panchromatic air photographs are as good as the LANDSAT classification at 1:50 000 and as good as airborne MSS map evaluated manually with regard to the remaining geometric distortions after rectification. The poor results in the computerized pixel-wise evaluation of airborne MSS data depend mainly on the geometric distortions. For this reason and due to the high costs, airborne MSS is not an alternative for land-use mapping in Sweden.

For operational use of LANDSAT data in Swedish terrain there is a need of greater accuracy. This can be achieved by multi-temporal analysis and/or a better geometric resolution in future satellites. The results now attained must be manually corrected by use of aerial photographs and maps, and in the future possibly integrated in the classification process. There is also a need for better methods of generalization with regard to the mapping scale.

Multi-date analysis

It is a well known fact to aerial photo interpreters that phenological changes in vegetation are of great importance in classification. This is also the case when using satellite images. To show the effect of different recording dates using the same classification procedure, four LANDSAT frames from the vegetation season 1975 were combined to a multi-temporal data set. Three of the frames (May 9, May 27 and July 21) were geometrically transformed to the fourth frame (July 3) using the MDA system at the Swedish Space Corporation.

There are great phenological differences between the four recording dates. On May 9 only the leafing of indifferent species and shrubs in the deciduous forest had started to come out in leaf. A part of arable land had been quite recently sown. Later, on May 27, even the leaves of the hardwood species (oak and ash) had come out and the crops had grown considerabley. In July the vegetation had reached its optimum.

In this study the same training set as that of the previously presented single-frame analysis has been used. 28 training areas were defined and these were later aggregated to 12 of the 13 classes shown in late I:D.

One important problem that immediately arises in multi-dimensional data sets is the redundancy of data or the correlation of variables. To speed up the classification of images it is urgent to reduce the dimensionality. Here the data has been put into the stepwise descriminant analysis of the Statistical Package for Social Sciences (SPSS) using the method "mahal". The variable (band) selection rule is to maximize the minimum squared Mahalanobis´ distance. The reason to select this method is based on the assumption underlying the use of maximum likelihood classification, namely a multivariate normal distribution of the data. Mahalanobis´distance is used in the calculation of the probability for a pixel to belong to a certain class in the maximum likelihood classification. The contours of constant probability Mahalanobis´distance to the actual mean value.

After establishing the discriminant functions, classification was made using all functions. Classification of the training set was performed using data from one, three and four dates in order to reveal the effects of the time dimension.

Table 2 shows a comparison of the classification results within the training areas using the single-frame data and the multi-temporal data. The overall correct classification result was 59.5 % for single-date analysis using 28 classes and three bands. When aggregating to 12 classes the value increased to 77.0 % due to erroneous classification within the main groups. The result of multi-temporal classification using 3 x 4 bands gives an overall correct classification of 77.6 % before and 91.2 % after aggregation from 28 to 12 classes. An impression of the separability of different classes is shown in Figures 2 and 3.

Table 2: A comparison of the classification results within the training areas using single-frame data (3 bands) and multi-temporal data (3 x 4 bands). All the discriminant functions have been used in the classification process. Figures in brackets represent single-frame data. The training areas are aggregated to 12 classes.

Class	Number of training areas	Number of pixels	Predicted group membership %											
			1	2	3	4	5	6	7	8	9	10	11	12
1. Spruce forest	2	52	79(52)		(8)		8(6)	10(14)	(2)	4	(14)		(6)	
2. Pine forest	1	28	4(32)	82(14)	(32)			7(7)					7(14)	
3. Mixed forest	1	40	(28)	5(28)	75(33)			3(3)					18(10)	
4. Clearcuts without scrub growth	1	40	3	3		83(90)		5(3)					8(8)	
5. Clearcuts with scrub growth	1	47	4			(4)	75(53)		(11)	11(14)	11(28)			
6. Hardwood forest	2	24	8(13)	4	4		4	75(42)			4(21)		(25)	
7. Other deciduous forest	1	23						13(17)	87(65)		(17)			
8. Grassland	1	26					4(4)		(8)	92(69)	4(19)			
9. Arable land	9	311	1(1)				3(5)	(1)	(2)	1(11)	94(87)	2(3)		
10. Urban area	5	264								4	4(17)	96(82)		
11. Bare rock with sparse pine forest	1	30	(7)	3(20)	20(20)			(13)					77(40)	
12. Water	3	167	(2)	1										99(98)
Total	28	1052	Overall correct classification 91(77)											

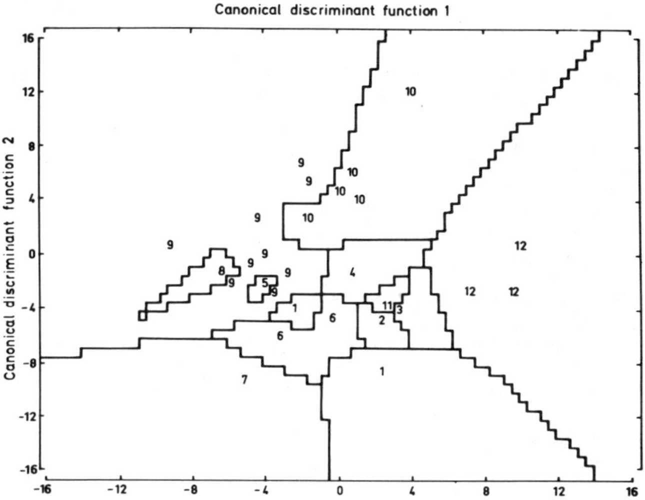

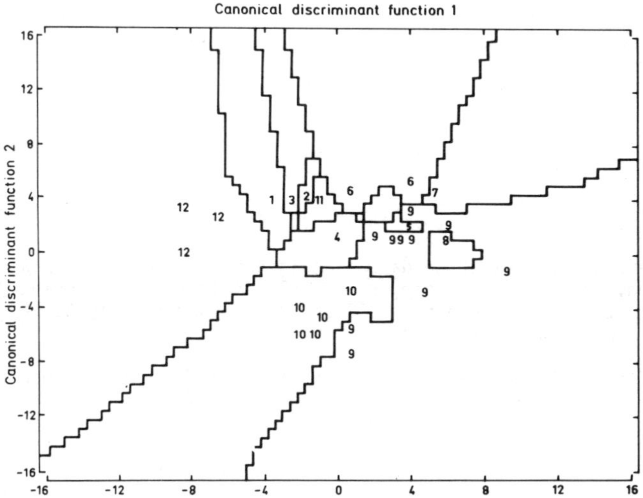

Figure 2 and 3: Territorial map assuming all functions but the first two are zero (Figure 2 = three bands from one date; Figure 3 = four bands from three dates). A number indicates a group centroid before aggregation according to the legend: 1 norwegian spruce forest, 2 pine forest, 3 mixed forest, 4 clearcuts without scrub growth, 5 clearcuts with scrub growth, 6 hardwood forest, 7 other deciduous forest, 8 grassland, 9 arable land, 10 urban areas, 11 pine forest on bare rock, 12 water.

Looking for eigenvalues greater than unity for the canonical discriminant functions, 6 functions are encountered explaining 97.2 % of variance.

Largest pooled within group correlation between, indecreasing order, canonical discriminant functions and LANDSAT bands are:

function 1:	band 6	3/7,	band 7	3/7		
function 2:	band 5	3/7,	band 4	3/7		
function 3:	band 5	27/5,	band 4	27/5,	band 6	27/5
function 4:	band 7	9/5,	band 6	9/5		
function 5:	band 5	9/5,	band 4	9/5		
function 6:	band 7	27/5				

The first 6 bands entered in the stepwise selection were band 6 27/5, band 7 9/5, band 6 3/7, band 5 3/7, band 4 3/7 and band 5 9/5. As band 6 from July 3 is severely affected by instrument errors it has been omitted. However, the discriminant analysis shows the importance of the red and infrared bands (5 and 7) according to the high correlation of these bands to the first six discriminant functions. In the following multi-temporal maximum likelihood classification, bands 5 and 7 from May 9, May 27 and July 3 have been used.

The fourth date (July 21) was, unfortunately, seriously affected by clouds. However, a four-date analysis has been performed using training areas free from cloud disturbances. The analysis is then restricted to spruce forest, clearcut areas with thicket, arable land, urban areas, pine forest on bare rock and water. The overall correct classification of the same areas increased from 77.4 % of three dates to 81.1 %. The eigenvalues of the first five discriminant functions explained 98.3 % of the variance in the three date set and 96.7 % in the four date set. The first four bands selected (band 7 3/7, band 4 3/7, band 5 3/7, band 5 9/5) in the stepwise procedure were the same in both cases and in the four date set band 5 of July 21 replaced band 4 of May 9 in the fifth position. Hence, the frame of July 21 contributed less than the frames recorded in May.

An example of the three-date classification is given in plate I:D for a subregion. In plate I:C a single-date classification is presented from the same area but using the bands 4, 5 and 7 recorded on July 3. The latter image is cleaned-up after evaluation.

In Table 3 the areal distribution of the different land-use classes is shown for that version of the single-date analysis that was not cleaned up. The statistics is based on an area of 512 x 512 pixels. The discrepancies from this classification are given for the two- and three-date classifications using bands 5 and 7. There is an obvious increase in the number of non-classified pixels (rejects) when more multi-dimensional data is used. The "signature" becomes more complicated for the training areas that may be less representative for the thematic class. However, the rejects can be decreased by choosing a higher value of sigma in the classification process.

A comparison between single-date and multi-temporal classification using sigma 2.0, shows an increase of unclassified pixels from 5,5 % to 22,2 %. The effect on water areas depends partly on the previously mentioned loss of one training area.

A positive effect of the multi-temporal classification is that the area of coniferous forest increases and the deciduous forests decreases. The latter has been over-represented in the single-date classification, especially

Table 3: Comparison between the areal distribution for single- and multi-date analysis for land-use classification of a 512 x 512 pixels scene.

Land-use class	Area (%) 1 date 3/7 level 1	level 2	Descrepancy (%-units) 2 dates (3/7, 9/5)		Descrepancy (%-units) 3 dates (3/7,9/5,27/5)	
Coniferous forest	16.8		+2.6		+4.4	
Spruce		9.4		-0.6		+0.5
Pine		2,8		+2.8		+4.0
Mixed		4.6		+0.4		-0.1
Clearcuts	10.7		+0.3		-1.1	
with scrub		8.7		-2.3		-4.0
without scrub		2.0		-2.3		+2.9
Deciduous forest	9.0		-0.4		-3.4	
Hardwood		6.9		+0.5		-2.1
Other		2.1		-0.9		-1.3
Grassland	4.0		-2.0		-3.1	
Arable land	27.7		-4.4		-7.4	
Urban land	4.9		-0.3		-1.1	
	10.5		-3.4		-4.9	
Water	8.9		-0.2		±0	
Rejects	5.5		+7.9		+16.7	

by scattered pixels within coniferous forests. These are now more correctly classified. Another positive effect of the multi-temporal classification is that the confusing of scattered pixels of crops and clear felled forest has disappeared.

The fidelity of the land-use classifications has been tested within the training areas (Table 4). There is an improvement of about 10 percent units when using multi-temporal classification into 28 classes. Using a classification system of 12 classes the overall correct classification is 86 % with the selected six wavelength bands. Further aggregation to 6 classes gives only marginal improvements.

The correct classification for the twelve classes using six bands is:

spruce forest	85 %
pine forest	50 %
mixed forest	65 %
clearcut without scrub growth	90 %
clearcut with scrub growth	28 %
hardwood	80 %
other deciduous forest	91 %
grassland	81 %
arable land	95 %
urban land	93 %
water	91 %

The multi-date study shows that the reduction of data from twelwe to the six best discriminating bands gives only a marginal loss in classification fidelity. It is obvious that the training areas for clearcuts with scrub growth and for pine forest are not representative for these thematic classes.

Table 4: The fidelity of land-use classification within the training areas for different combinations of wavelength bands and recording dates. The values in the two first columns are calculated from maximum likelihood classification, the right column from classification using discriminant functions.

	Overall correct classification		
	Single-date: July 3	Multi-date: May 9, May 27 and July 3	
	Bands 4,5,7	Bands 5,7	Bands 4.5.6.7
Original 28 classes	65 %	76 %	78 %
Aggregation to 12 classes (level 2, Table 1)	78 %	86 %	91 %
Aggregation to 6 classes (level 1, Table 1)	84 %	87 %	93 %

1) One of the training areas for water has been excluded

A problem with the multi-temporal classification results is the amount of unclassified pixels. In order to reduce the rejects a larger confidence interval has been tested within a subregion. Using 5 sigma the rejects increase from 0.1 % of the classified pixels in the single-date classification to 14.6 % in the multi-temporal classification. When using 10 sigma in the latter classification the rejects are reduced to 1.8 %.

Plate I:D shows an example of the multi-temporal classification using 10 sigma. The image is very generalized in comparison to the cleaned-up version of the single-date analysis with 2 sigma shown in Plate I:C.

A problem in maximum likelihood classification is the estimation of the parameters u and o for each class. In this very case the training areas have been selected within a window using a thresholding technique in one of the bands in order to mask out pixels belonging to other classes. A big risk with this method is a poor estimation of the population mean and variance. This risk will increase when the number of pixels is small. In addition, the fact that the proportion of boundary pixels to the total number of pixels of a training area will increase. This risk will even become bigger in a multi-date data set due to rectification errors. In this investigation this shows up as an improvement of the number of rejects within the training areas but the rejects increase in other parts of the image. Poor estimation of the population parameters is the reason why such a large confidence interval has been selected in the multi-temporal classification.

The accuracy in the multi-temporal classification outside the training areas is not evaluated. Only a visual comparison is carried out with a land-use map based on air photo interpretation. It is obvious that there are severe classification errors in the multi-temporal classification. Based on this study it is not possible to give recommendations about applied mapping. A more accurate choice of training areas and integration of already existing topographic maps in the classification process may give a better result.

MONITORING FOREST CLEARCUTS USING MULTI-TEMPORAL LANDSAT DATA

For forestry and military purposes there is a need for an annual recording of clearcut forest areas. Such a monitoring is also of great interest to environmental protection authorities.

Beginning in 1972 we tried to identify clearcut forest by visual interpretation of false colour LANDSAT images (Arnberg & Wastenson, 1973) but it was not possible to distinguish with and acceptable accuracy the clearcut areas from areas already reforested and from some types of mires. In 1974 we tried supervised classification of digital LANDSAT MSS data by comparing the results from two consecutive years (Orhaug, Wastenson & Åkersten, 1976). Due to insufficient resources for computerized co-registration of two LANDSAT scenes, visual comparison of the two frames was used. It was possible to identify the changes from the forested to the non-forested areas. However, identification of other changes in the landscape, such as succession of crops, reforestation or changes in distribution of scrub growth, were of too low accuracy for operational use.

In 1979 the MDA image analysis system became available at the Swedish Space Corporation for the first digital multi-temporal analysis of LANDSAT data for an inventory of clearcut areas in Sweden. We compared LANDSAT data from July 3, 1975; May 30, 1978 and May 31, 1979. An area of 625 km^2 at Hofors-Horndal in central Sweden was chosen as a test region.

When studying the changes in digital levels of some known areas of clearcut forests, it was established that the largest differences occurred in band 5 (the red part of the spectrum). The suggested method to map clearcuts is based only on differences between two years within this channel, and can be summarized in the following steps:

- The LANDSAT image from the last year is re-sampled to the geometry of the image from the first year.
- Within a subscene of 512 x 512 pixels from the first year the digital levels for known old forested areas are interactively determined on the terminal colour screen in band 5. Pixels with digital levels higher than a selected threshold value are masked and stored in a graphic plane as a binary mask.
- Threshold values for clouds are determined in the different LANDSAT scenes and the actual areas are stored in a second graphic plane.
- The threshold value for water is determined in band 7 and the hydrography is stored in a third graphic plane to be used as a reference to the topographic map.
- The differences between the digital levels in band 5 from two years are calculated pixel by pixel and threshold values are chosen for known clearcuttings carried out between the two dates. Pixel within these digital limits are classified as forest clearings.
- The stored binary masks are added to the classified image.
- The resultant image is generalized by a clean-up programme. Singular pixels get the dominating class of the surrounding pixels.
- The classified and cleaned image is re-sampled to fit the topographic map and an overlay is plotted using an ink jet plotter.

In plate II:A an example of an image processed according to the described method is shown. In Plate II:B the overlay has been printed on a topographic map. The ground truth data are then added as black lines marking the true boundaries of forest clearcuts carried out between the years 1975 and 1978.

The ground truth map is based on interpretations of air photographs and ortho photo maps from 1975 and 1978. The ground truth map was digitized as polygons and converted to a raster format with a pixel size of 16 x 16 m^2 according to methods described by Alm & Arnberg (1981). The LANDSAT image is re-sampled to the same pixel size and a pixel-wise evaluation is performed.

The evaluation is carried out only for the period of 1975-1978 within an area covered by 113 652 pixels. 98.6 % of the pixels are correctly classified in the two classes "clearcut" and "non-clearcut" forest areas. The total areal distribution of clearcut forest areas is underestimated by 7.2 % in the LANDSAT image. 80.8 % of the pixels classified as clearcut areas are verified by the ground truth data. 75.0 % of the existing clearcut area are indicated in the LANDSAT classification. In other words, the error of omission is 25 % and the error of commission is 19.2 %.

The classification results were evaluated without any agreement to the configuration. The number of clearcut areas indicated in the LANDSAT multitemporal image was counted. 66 of the 78 existing clearcut areas within the evaluated area were detected in the LANDSAT image. Of the remaining 12 areas all are smaller than 2.3 hectares nine are smaller than 1.3 hectares (2 pixels) and four are smaller than 1.0 hectares. By the used clean-up programme, areas of one or two pixels were removed. This was necessary due to co-registration errors of the two LANDSAT scenes (maximum error about 0.5 pixels) causing differences in digital levels of single pixels at boundaries between e.g. crop land and forest. These differences can be of the same order as between forest and clearcut areas. With a better resolution this kind of error can be reduced.

In conclusion, this investigation shows that the method used is applicable for detecting clearcut areas larger than 2.5 hectares (3 pixels). This is adequate for military and environmental monitoring purposes.

In order to get statistical information about the total amount of clearcut forest within a region the method gives an accuracy of about 93 %. For mapping purposes with a request of low position errors, only about 75 % of the existing clearcut area are correctly mapped. The error of commision is about 20 %.

CLASSIFICATION OF MIRES THROUGH INTEGRATION OF DIGITIZED DATA FROM A TOPOGRAPHIC MAP AND MULTI-TEMPORAL LANDSAT MSS DATA

In the summer of 1980 a study was initiated to investigate the possibility of classifying mires through integration of digitized data from the topographic map and multi-temporal data from LANDSAT MSS. Several classifications were made using different combinations of the four spectral bands from six recording dates.

The investigation area selected is situated some ten kilometres south of Gävle, Sweden, and includes a part of the Dalälven. In this area the river is composed of a system of bights which give rise to and influence several mires. The area includes mires of different types and sizes. From previous studies (Rafstedt & Andersson, 1981) a lot of field data was also available.

The classification system used was developed by Rafstedt and Andersson (1981 p. 24), and was originally adapted for interpretation of mires in colour infrared aerial photographs. The main groups in the system are bogs and fens which are subdivided into open and forested types, and further subdivided into different vegetation communities. In the actual study, however, the original classes were re-grouped to form two levels. The first level includes only three classes: forest covered mires, open bogs and open fens. Each of these main groups comparises three or four of the classes to the second level. Table 5 shows the classification system and the different classes used in the study. The number of training areas for each class, used in the classifications, is shown in brackets.

Table 5: System for classification for mires using digital LANDSAT data. The number of training areas used for the classes is given in paranthesis.

Level 1	Level 2
1. Forest covered mire	1. Bog with sparsely scattered pines (2)
	2. Fen with coniferous forest (2)
	3. Fen with deciduous forest (2)
2. Open bog	4. Bog, hummock vegetation, few dwarfed pines (2)
	5. Harvested peat bog
	6. Bog, tussocky or carpet-like vegetation (5)
3. Open fen	7. Fen with scrubs mainly willows (2)
	8. Fen, floodplain and similar (3)
	9. Fen, lawn-like vegetation (4)
	10. Fen, soft, carpet-like vegetation (1)

* This class has been added to the Rafstedt-Andersson system (1981)

Data sources and feature selection

MSS data from six different LANDSAT images were used in the study, distributed through the vegetation period as follows:

30/5 1978 LANDSAT 2 (received and processed by the ESRANGE station)
12/6 1979 LANDSAT 2 (received and processed by the ESRANGE station)
3/7 1975 LANDSAT 2 (processed by EROS)
21/7 1975 LANDSAT 2 (processed by EROS)
1/8 1978 LANDSAT 3 (received and processed by the ESRANGE station)
28/9 1979 LANDSAT 2 (received and processed by the ESRANGE station)

All the images were geometrically transformed to fit the image from 30/5-78. This means that all images may be superimposed. Radiometrically, however, no correction were made, so the values which were used are the digital levels of the raw radiance values.

In addition to MSS data, information from the topographic map was used in the classifications. For this purpose the printing plates of forested areas and mires from the topographic map were digitized, using the Optronics scanner at the National Defense Research Institute, and stored in raster format. The images were then re-sampled by the Swedish Space Corporation and geometrically transformed to the LANDSAT images.

Field data obtained by interpretation of colour infrared aerial photographs were digitized and transformed into raster format fitting the LANDSAT images. The computer programme used, MIDAS, is described by Alm and Arnberg (1981).

Based on the previous studies within the investigation area, 44 training areas were selected using the MDA system at the Swedish Space Corporation. For each training area, digital mean values, standard deviations, histograms, covariance matrices and correlation matrices, were obtained for the four MSS channels and the six recording dates. A first study of the statistical data gave an indication of the quality of the data and the variation within the classes. The 24 best training areas were selected for the classifications. In order to determine the minimum number of wavelength bands and recording dates required, the data were handled in two ways, firstly by means of computer programme, and secondly by judgement based on visual studies of the statistics.

The computer programme "ALLCOMB" (Arnberg, 1981 b) was used to calculate the minimum number ov wavelength bands and dates required for different combinations of training areas. Table 6 shows the ranking of the bands selected by this method.

Table 6 also shows the bands selected by the second method, which was based on visual studies of the statistics. Because of the large amount of data, this work was concentrated on finding the best channel combination for distinguishing the open bog classes from the open fen classes.

Furthermore, referring to mean value and standard deviation, two bands were found (band 5 from June 12, 1979 and band 5 from July 21, 1975), where the fen- and bog classes of the open mire area could be separated from each other by a digital level slicing. In Figure 4, the digital values for the training areas and the level for the slicing in band 5 from July 21 are shown.

Classification

The classifications were made on the MDA system at the Swedish Space Corporation, using the maximum-likelihood method. In addition to the MSS data, two masks with information from the topographic map concerning mire covered areas and forest covered areas were used. Several combinations of the selected bands, combined with the masks, were tested in different classifications. The main structure of the classifications can be summarized in the following points:

1. Supervised classifications of all mires on the topographic map using the mire mask.
2. The groups "open fens" and "open bogs" were separated from forest covered mires by combining the mire and the forest masks of the topographic map.

   2.1 Supervised classification of the forest covered area.
   2.2 Supervised classification of the open mire area.

3. The bogs and fens of the open mires were separated by a digital level slicing in band 5 from July 21, 1975.

   3.1 Supervised classification of the bogs.
   3.2 Supervised classification of the fens.

Table 6: Ranking of the bands required for separating different combinations of classes. In total 11 bands have been used in the combinations.

Dates	Visually selected bands				Bands selected by computer programme																				
	A The open bog and fen classes				The open bog and fen classes				All classes				The bog classes				The fen classes				The forest covered mire classes			
	band 4	5	6	7	band 4	5	6	7	band 4	5	6	7	band 4	5	6	7	band 4	5	6	7	band 4	5	6	7
30/5-78				4																4				
12/6-79		1		6				3				3				3			2				4	3
3/7-75				5								2												
21/7-75		3		2		1	2			1				1		2			1			2		1
1/8-78					6			5	6			5					5			3				
28/9-79								4				4												

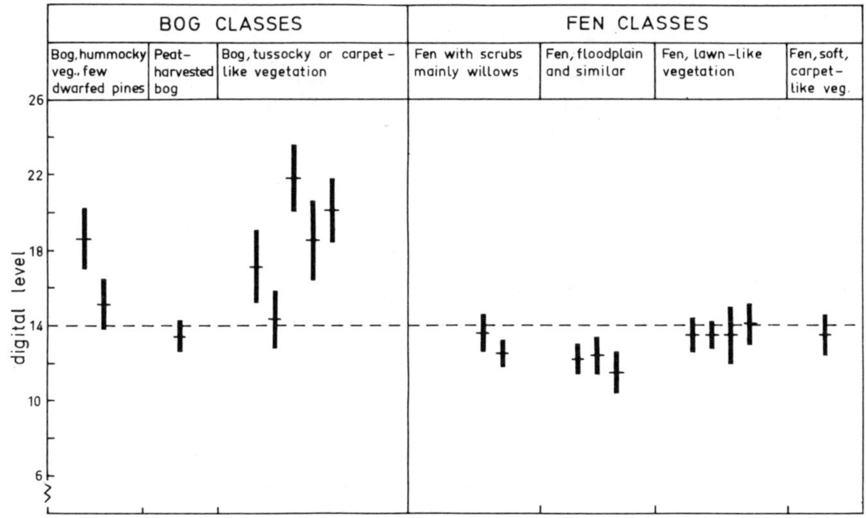

Figure 4: The bars in the diagram show the digital levels (mean value ± one standard deviation) in band 5 on July 21, 1975 for the training areas. The bog classes can be separated quite well from the fen classes by a slicing at level 14.

In alternatives 1 and 2.2 the following combinations of bands were tested:

- combination of bands selected by computer programme
- combination of bands selected by visual studies of the statistics
- combination of all existing bands in Table 5 (eleven bands in all)

For alternatives 2.1 and 3, bands selected by computer programme were used.

In all the classifications a threshold value of 5 sigma was used, but values of 3 and 4 sigma were also tested in some cases, causing an increase of the number of rejects.

Preliminary results

Since the evaluation of the classification results is not yet completed, only some preliminary results will be presented in this paper. Thus the evaluation presented here was performed for classifications where three of the combinations of bands shown in Table 6 (columns A, B and C) were used, and also for two classifications where a combination of all the eleven bands in Table 6 was used. The classifications were made at level two (Table 5), using at most 24 training areas distributed among ten classes and using a threshold value of 5 sigma. Before the evaluation an aggregation of the classes into three groups of level one was made.

A pixel-by-pixel comparison between the classifications and the digitized field data was then performed. The resolution of the field data is limited to 10 hectares, corresponding to about 20 pixels, implying that the field data are considerably more generalized than the LANDSAT data. There are some differences between the field data, based on interpretation of colour infrared aerial photographs and the distribution of mires on the topographic map. The LANDSAT classifications were made only in areas indicated as mires on the topographic map and, in consequence, the evaluation was possible only where accordance in areal distribution exists between the field data and the mires on the map. About 13 000 pixels in all, fairly equally distributed among the three classes, were examined.

Table 7 shows the agreement of the LANDSAT classification with the field data for three different classifications. The results are quite similar, with an overall classification agreement of 70 % for the 11-band alternative and 69 % for the 6-band alternatives. However, if all classes of level two are used in the evaluation, larger differences can be expected. Due to the high generalization level of the field data, such an evaluation will be made by using a different method, for example by checking control points selected at random.

In Table 8 the agreement of the LANDSAT classification with the field data for three other classifications are shown. Only the open mire areas were classified, using 18 training areas. This was accomplished by adding the mask of forested mires from the topographic map to the classification procedure. However, in comparison with the field data, 16 % of the classified "open mires" are actually forest covered mires due to differences between the field data and the distribution of forested mires on the topographic map, which has some errors.

The discrepancy can partly be explained by the higher generalization of the field data, but to vegetation interpreters, it is also a well-known fact that sharp boundaries between different vegetation communities seldom exist. The evaluation has been possible only where accordance in areal distribution exists between the open mires on the topographic map and in the field data. The number of pixels examined is about 6 800. The overall classification agreement with field data for the two open mire classes was improved to 90 % for the 11-band alternative and to 87 % respectively 85 % for the 6-band alternatives.

Finally classification by level slicing for the open mires from July 21 in band 5 gave an agreement with field data of 64 % for the two classes open bogs and open fens. 34 % of the disagreement are caused by one single bog area where peat has been harvested. In Figure 4 the training area for peat-harvested bogs is to be found at the same low digital level as the training areas for the fen classes.

A cartographic example of the results from a classification based on the six visually selected bands is given in Plate III:A. The original ten classes are aggregated into three. This colour image can be compared with the image of digitized field data shown in Plate III:B. Plate III:C shows an example of the results, from a classification using ten classes without aggregation. Field data for comparison are shown in Plate III:D.

At the present stage of evaluation the study shows that LANDSAT classification of mires into the three main groups: forest covered mires, open bogs and open fens, can be useful. The errors occur mainly between the open mire classes and the forest covered mires classes. This can, however, be expected considering the edge effects, especially when the mires are small.

Table 7: The agreement with the field data for three LANDSAT classifications, where different combinations of bands have been used. The original 10 classes have been aggregated into three classes: forest covered mires, open bogs and open fens. In the two first classifications, 1 % of the pixels was rejected for the forest covered mires. In all other cases the number of rejects can be ignored.

LANDSAT / Field Data	11 bands, a combination of all the bands in Table 5			6 bands visually selected (A in Table 5)			6 bands selected by computer programme (B in Table 5)		
	forest cov. mires	open bogs	open fens	fores cov. mires	open bogs	open fens	forest cov. mires	open bogs	open fens
forest cov. mires	<u>63</u>	28	8	<u>67</u>	25	7	<u>68</u>	26	6
open bogs	10	<u>84</u>	6	12	<u>80</u>	8	14	<u>80</u>	6
open fens	22	11	<u>67</u>	26	13	<u>61</u>	25	14	<u>61</u>
overall agreement		70			69			69	

Table 8: The agreement with the field data for three LANDSAT classifications, where different combinations of bands have been used. Only open mires have been classified. The original 7 classes have been aggregated into open bogs and open fens. The number of rejects can be neglected.

LANDSAT / Field Data	11 bands, a combination of all the bands in Table 5		6 bands visually selected (A in Table 5)		6 bands selected by computer programme (C in Table 5)	
	open bogs	open fens	open bogs	open fens	open bogs	open fens
Open bogs	94	6	92	8	91	9
Open fens	12	88	16	84	19	81
Overall agreement	90		87		85	

ACKNOWLEDGEMENTS

The investigations presented in this paper have been sponsored by the Swedish Delegation for Space Activity, the Swedish Natural Science Research Council and the Swedish Environmental Protection Board.

Prof L. Wastenson initiated the projects and was the project manager. The land-use mapping project at Linköping was performed by him and Dr M. Ihse in co-operation with the Swedish Research Institute of Defence. In the multi-date part of the project Dr W. Arnberg carried out the selection of the most informative combination of wavelength bands and registration dates. He also developed the methods for pixel-wise evaluation of classified data and was responsible for the computer cartography work in the project. The clearcut monitoring project was performed by L. Wastenson in co-operation with the Swedish Space Corporation. Miss L. Boresjö, B.Sc., assisted by Mr G. Alm, B.Sc., carried out the evaluation of the maps based on classified LANDSAT data. The mire inventory project was performed by L. Boresjö, using the MDA system at the Swedish Space Corporation and the Nord 10/S Computer System at the Department of Physical Geography, University of Stockholm.

REFERENCES

ALM, G., and ARNBERG, W., (1981): Map Information and Data Analysis System (MIDAS) at the Department of Physical Geography, University of Stockholm. (In preparation)

ARNBERG, W., (1980): Comparison of digitized thematic maps and computer processed MSS data from satellite and aircraft. - Proceedings of the first Scandinavian Conference on Image analysis. Studentlitteratur Lund, 219-225.

ARNBERG, W., (1981a): Multispectral reflectance measurements and signature analysis. - Research report 45, Department of Physical Geography, University of Stockholm, 39 pp.

ARNBERG, W., (1981b): Integration of maps and remote sensing data. Software framework at the Department of Physical Geography, University of Stockholm. - Geografiska Annaler, 63 A.

ARNBERG, W., and WASTENSON, L., (1973): ERTS-bilder över Sverige tolkade. Ny metod för hyggesinventering? - Forskning och Framsteg 1973:1, 15-19.

LARSSON, J., (1980): Rectification of digital images for remote sensing analysis. - Institute of Photogrammetry, Royal Technical Institute, Stockholm.

ORHAUG, T., WASTENSON, L., and ÅKERSTEN, S.I., (1976): Forest inventory using LANDSAT digital imagery. - FOA report A 30008-E1, October 1976, 58 pp.

RAFSTEDT, T., and ANDERSSON, L., (1981): Flygbildstolkning av myrvegetation. En metodstudie för översiktlig kartering. Summary: Air photo interpretation of mires. A methodological study. - SNV (Sw. Environmental Protection Board) PM 1433, 106 pp.

WASTENSON, L., ORHAUG, T., and ÅKERSTEN, S.I., (1978): Swedish experiences on forest inventory and land use mapping by automatic classification of digital MSS data from LANDSAT and aircraft. - ISP + IUFRO Symposium Freiburg in Br., July 1978. Int. Archives of Photogrammetry, Vol XXII-7, 683-694.

WASTENSON, L., IHSE, M., and ARNBERG, W., (1981): Vegetation and land use mapping by automatic classification of MSS data from aircraft and LANDSAT. Methodological studies in the Linköping area. - SNV (Sw. Environmental Protection Board) PM (in press).

XU GUAN-HUA, (1981): Digital Multiband Image Handling and Classification Package with Emphasis on LANDSAT Images. Programmes dedicated to China and Sweden (PDCS). - Research report 48, Department of Physical Geography, University of Stockholm (in press), 47 pp.

ÅKERSTEN, S.I., and GUSTAFSSON, L-E., (1978): Program library for handling and processing of remotely sensed multispectral data. Second revision. - FOA report G 30146-E1.

Part D

DATA HANDLING AND IMAGE PROCESSING

THE GROUND SEGMENT FOR A EUROPEAN OCEAN MONITORING SATELLITE ERS-1

M. Jones
ESA Computer Department
European Space Operations Centre
Robert-Bosch Str. 5
6100 Darmstadt
W. Germany

Abstract: At the present time the European Space Agency is making preparations for a European Remote Sensing Satellite Programme. Within this programme, the first satellite to be launched will be an ocean-monitoring satellite ERS-1, whose payload will principally comprise active microwave instruments, namely a Synthetic Aperture Radar (SAR), a scatterometer and an altimeter. To set up a proper basis for the discussion of the ground segment, the satellite and its payload are briefly introduced and the foreseen mission capabilities of the Programme are outlined. The paper then concentrates on that portion of the programme concerned with the ground facilities necessary for the control of the satellite on the one hand and payload data acquisition, processing, archiving and distribution on the other.

1. INTRODUCTION

The European Space Agency (ESA) is presently making preparations for a European Remote Sensing Satellite Programme. In particular, the thirteen states participating in this Preparatory Programme have agreed upon the nominal payload, thus permitting a preliminary definition of the Ground Segment to be made. This paper describes the Ground Segment so far defined, showing how the Mission Objectives, as they apply to the Ground Segment, will be fulfilled. The following important problems will be addressed:-

- the optimal control of satellite and payload;
- the distribution of large volumes of data;
- the need for relatively rapid delivery of certain thematic products for use in applications such as short-to-medium-term weather monitoring.

The plan of the paper is as follows: in §2, the basic assumptions of the ground segment design are discussed, covering satellite and payload characteristics and the intended mission capabilities. In §3, the mission requirements, which in part follow from these basic assumptions, are outlined. This is followed in §4 by a discussion of the chosen ground segment concept and in §5 the foreseen implementation of this

Paper presented at conference on Matching remote sensing technologies and their applications conveved in London by the Remote Sensing Society, 1981 concept is discussed, covering matters such as the actual location of the various facilities. The conclusion of the paper gives the present status and mentions other activities relevant to the ground segment definition.

2. BASIC ASSUMPTIONS

The main constraints on the design of the ground segment are the space segment and the mission objectives to be fulfilled by the whole system. These are described in some detail by Duchossois and Honvault (1981). However, both the Space Segment and the required Mission Capabilities will be briefly summarised here in order to permit a fuller understanding of the Ground Segment.

2.1 The Satellite and Its Payload

The Space Segment comprises the following:-

i) A multi-mission platform of the type used in the French National SPOT programme.

ii) A payload composed of:-
- C-band Active Microwave Instrumentation (AMI) with 3 operating modes (Imaging SAR, Wave Scatterometer and Wind Scatterometer);
- a Radar Altimeter (RA)operating in the Ku band;
- an additional scientific package, still to be decided.

iii) A recording system consisting of a tape recorder (possibly complemented by a magnetic bubble memory device) providing a storage capacity of the order of 5 Gigabits.

iv) An Instrument Data Handling and Transmission System (IDHTS) generating three logical data streams to be transmitted to the ground. These formats are as follows:-

Format A : Direct read-out low data rate instruments (overall data rate: 800 K-bits/s;

Format B : Raw SAR data (imaging mode of AMI - overall data rate about 100 M-bits/s (Mbps));

Format C : Playback of recorded data, consisting of the recorded output of low data rate instruments (overall data rate 6 Mbps).

2.2 Mission Capabilities

Generally, the ERS-1 mission should monitor the ocean surface and coastal regions with the aim of providing improved short-term or medium-term forecasts of both weather and ocean conditions. In particular, measurements of wind vectors, supplemented by measurements of wave spectra and wave height can be used in such forecasts, provided they are available within a few hours. All-weather high-resolution imagery (from the SAR) can be used to monitor ice coverage in polar or near-polar regions; supplementing this information with altimeter measurements can also give an indication of ice-sheet profiles. Details of the physical parameters to be extracted from ERS-1 payload data are given in Table 1, which includes all parameters currently envisaged.

Measurements of the sort described above can be used in various related fields of which physical oceanography, glaciology and climatology are the most important ones. Pollution monitoring is another possible area where remotely-sensed active microwave data could be

TABLE 1: FORESEEN ERS-1 SYSTEM CAPABILITY

Main Physical Parameters	Accuracy	Range	Horizontal Resolution	Coverage/ Repeat Cycle Delay	Main Contributing Instrument
Wind					
Speed	±2 m/s or ±10%	4 to 24 m/s	50x50 km^2 (Global)	Global/3 day	Wind Scatterometer
Direction	±20°	0 to 360°		3 to 6 hours	Altimeter (at nadir)
Wave Spectrum					
Direction	Angular resolution: 30°	0 to 180°	< 10 km^2 (Direction, Height, Wavelength)	Discrete measurement points. Grid size 900 km at equator in long, for 3 day repeat cycle.	Wave Scatterometer
Height	±20%				
Wavelength	Spectral resolution: 25%	100 to 1000 m		3 to 6 hours	
Sea State					
Sign. Wave Height	±0.5m or ±10%	1 to 20 m	< 10x10 km^2	Discrete measurement points Grid size 900 km at equator in long. for 3 day repeat cycle 3 to 6 hours	Altimeter
Currents					
Speed	±15 cm/s	15 cm/s		TBD	Altimeter
Direction	±20°	0 to 360°		TBD	
High Resolution Imagery					
All Weather	Grey level resolution 1 dB 2.5 dB		100x100m^2 30 x 30m^2	75 km swath width. 3 day repeat cycle over selected areas (important gaps in coverage) or complete coverage achievable in 35 days.	SAR (Imaging Mode)

useful, although research is at too early a stage to assess the full potential of such a satellite for pollution monitoring.

As concerns ground segment design, the most significant conclusion is that certain products will need to be generated within a few hours for use in operational weather forecasting or ocean monitoring systems. Outside this general area there are many long-term scientific activities where delay of delivery of data is less critical but where greater volumes of data or more sophisticated processing may be required (e.g. research into surface currents, internal waves and certain long-term research into ocean climatology models). A natural separation of the user community into "real-time" (response time 3-6 hours) and "non-real-time" users (response time < 24 hours) is thus apparent.

3. MISSION REQUIREMENTS

In order to fulfil the mission objectives, three types of services have to be provided, namely:-

i) Control and monitoring of the platform, payload and associated elements throughout all the mission phases. Control and monitoring must be carried out in such a way that both quality and quantity of the data produced by satellite and payload are maximised consistent with the performance capabilities of the satellite and its payload.

ii) Acquisition of data including verification of down-link integrity;

iii) Near real-time generation and distribution of selected products.

iv) Generation, archiving and distribution of fully-corrected or any non-urgent products in non-real-time.

The need to provide both near-real-time and non-real-time products follows from the discussion of §2.

4. FUNCTIONAL ELEMENTS OF GROUND SYSTEM REQUIRED TO SATISFY MISSION REQUIREMENTS

To fulfil the requirements of §3, three groups of tasks can be identified:-

- the overall management of the mission (mission management function);
- the technical operation of the satellite, including all the elements of the platform and the payload (mission control functions);
- the acquisition, processing, archiving and distribution of data, including the generation of thematic and other geophysical products (payload data management function).

The remainder of this section gives a general specification of these three tasks. This specification will be considered to describe the ERS-1 ground segment requirements. The three tasks are:-

Mission Management, including all aspects of planning and coordination of the mission. It will take into account user requirements and environmental, geographical and satellite- or payload-related constraints. The Mission Management function will also ensure quality control of products and services generated under the responsibility of the Programme.

Mission Control which is the main tool for providing the service (i). It includes:-

- operations and maintenance of the Control Centre and of the telemetry, tracking and commanding network;
- control of platform, its payload and its orbit, involving processing of telemetry and tracking data, payload data analysis (for a suitable proportion of the payload data) and generation of telecommands;
- conduct of day-to-day satellite systems scheduling, in particular that for payload operations, taking into account mission management planning and the statuses of both space and ground segments.

Payload Data Management which ensures the acquisition, processing, archiving and dissemination of data products to users. The tasks associated with both near real-time and delayed time activities are:-

- payload data acquisition;
- data preprocessing;
- generation of products from the payload data;
- quality control;
- archiving and retrieval;
- data distribution (both real-time and delayed).

The levels of product quality which the Payload Data Management function will provide are broadly defined in Table 2, where the delivery time requirements are also given.

The ground segment will therefore contain the following components:-

- a Mission Management function, responsible for overall mission coordination, combined in practice with a Mission Control Centre;
- a Mission Control System which includes a Mission Control Centre and appropriate ground stations for Telemetry Tracking and Control (TT and C) as well as a reference station for monitoring the payload data down-link;
- a Real-Time System, concerned with the handling of data played-back from the on-board recorder and requiring product generation and distribution within a few hours;
- a Non-Real-Time System concerned with data management activities having longer lead times such as generation, archiving and archive retrieval of high quality products.

4.1 Mission Management and Mission Control

Because of their close relationship the tasks of mission management and mission control are conveniently combined together within a Mission Management and Control Centre (MMCC). Within this entity there will be a mission management function which will principally be responsible for making payload utilisation plans, taking into account the ERS-1 mission profile and objectives, user requests, payload platform status and ground segment status. The MMCC will also be responsible for Mission Control, in which it will be assisted by appropriate ground stations. Specifically, the MMCC will be responsible for:-

- monitoring of satellite system/subsystems status by processing ERS-1 housekeeping telemetry data relayed from a telemetry station;
- analysis of a proportion of the payload data acquired from a reference payload acquisition facility (i.e. reference ground station). This analysis has the principal function of ensuring that the various control activities produce acceptable payload data and, if necessary, permits corrective action;
- commanding of the platform and payload systems via a telecommand station;
- reception of satellite tracking data from various tracking stations

TABLE 2: LEVELS OF PRODUCT QUALITY

LEVEL A :	RAW DATA Raw data as received from the satellite and recorded on HDDT* in satellite format.
LEVEL B : RT/DT	ANNOTATED RAW DATA/SYSTEM CORRECTED DATA Raw sensor data, time ordered, time tagged, earth located. Internal calibration tables are attached but not applied. Geometric auxiliary information will include earth location data, making use of the latest available orbits and attitude information, i.e. the corrections defined in the auxiliary products can be applied in order to generate system corrected data.
LEVEL C :	QUICK-LOOK DATA Data, in digital or photographic form, for payload monitoring and browse facility support. Depending on the sensor, the spatial resolution and radiometric quality could be lower than that of the final product.
LEVEL D RT/DT	REAL-TIME THEMATIC PRODUCTS Geophysical data, from low bit-rate sensors, after application of instrument transfer functions and removal of environmental effects. They will be extracted mainly from Level B data. They include, amongst others, wave spectra from SAR when this instrument is operated in its wave scatterometer mode.
LEVEL E : DT	PREPROCESSED PRODUCTS Preprocessed data, with principal instrumental, radiometric and geometric errors removed, taking advantage when possible of ground truth information and possibly using refined flight dynamics data based on reconstitution.
LEVEL F : DT	PRECISION THEMATIC PRODUCTS Geophysical parameters, extracted from Level E products, as appropriate for the instrument in question. Precision thematic products will be earth-located.

RT = Real-time (normally hours)
DT = Delayed time (from one day up to several weeks)
*HDDT stands for High Density Digital Tape

and generation of predicated and reconstituted (i.e. a posteriori) satellite position data and distribution of these data to payload data acquisition and processing facilities. In order to fulfil these functions, the MMCC will be complemented by:-

i) The TT and C Ground Stations for telemetry, tracking and command of the satellite.

ii) The Data Acquisition Facility (DAF) which will act as a reference station, as well as acting as a data acquisition station (regarding this latter function, see §4.2.1). This means it will be built to carefully controlled specifications and will be supplied with suitable equipment for monitoring the X-band down-link from the satellite. Because of the decentralised nature of the ground segment (see §5), which could include a number of stations not under ESA responsibility, the ability to distinguish between down-link faults and faults in user ground stations is important.

4.2 Payload Data Management

4.2.1 The Real-Time System

For obvious practical reasons, acquisition of playback data will be concentrated at one station, this in turn facilitates location of its processing at one place. The real-time system is composed of:-

i) The playback data acquisition facility (PDAF) which serves as the prime facility for receiving the data stored on the on-board tape recorder. The data will be recorded after reception. It is assumed that all low data-rate instruments will store all their data on the recorder, i.e. the PDAF would be the only station concerned with this category of data.

ii) The real-time processing facility (RTPF) which processes the data acquired in the Playback Data Acquisition Facility (PDAF) generates data products (mainly thematic products) within several hours. Data exchange between the PDAF and the RTPF needs to be at real-time rate.

iii) The real-time data-acquisition facility (RDAF). Since imaging SAR data cannot be recorded on-board, it is essential to ensure the desired coverage by means of suitably distributed real-time data acquisition stations. They will receive real-time payload data, record them using high density data tape recorders and produce quick-look products in near real-time. The location of the RDAFs is primarily determined by the desired sensor coverage since the output of the imaging SAR cannot be recorded on-board. In case of on-board tape recorder failure, the RDAFs would have to acquire the data stream coming from the low data-rate instruments and retransmit it to the RTPF.

iv) The Data Distribution System for Real-Time Users
The availability in near real-time of products up to level D (Table 1) is an essential prerequisite for the success of the ERS-1 mission since the value of certain of these data can decrease very rapidly with time. A two-level transport system from the real-time processing facilities (RTPF & DAF) to the end users is proposed as illustrated in Figure 1. This system involves:-
- prime user centres which will, in general, further process and annotate ERS-1 products with non-spaceborne information before sending them to end-users using local facilities such as facsimile, telex lines, radio links, etc.;

- end-users who, particularly in the case of open ocean applications, are not readily connected to a general purpose communication network, although they are often linked by a specialised network to prime centres.

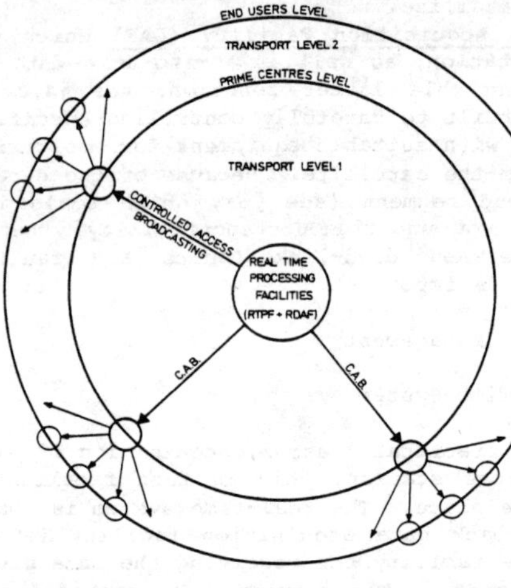

Figure 1: ERS-1 Distribution System Concept

4.2.2 The Non Real-Time System

Finally, the following activities will be necessary, although they will not be supported by ERS-1-dedicated facilities under ESA's responsibility:-

- data archiving and retrieval of data from the archive;
- the generation of products for users whose applications do not demand near real-time access to the data as well as the production of high-quality products which cannot be generated in real-time (because e.g. the correction factors are only known a postiori or the processing itself involves inherent delays, as in ground control point (GCP) extractions for geometric image correction).

These non real-time activities can be handled by one or several Preprocessing and Archiving Facilities (PAFs) which would receive data from the stations for further processing such as:-

- cataloguing, archiving and retrieval;
- SAR precision processing;
- dissemination of high precision products on media such as high density digital tape (HDDT), CCT, photographs or paper (e.g. graphic hardcopies, computer printouts);
- generation of thematic products for users not requiring a real-time response;
- quality control covering monitoring of both quality of the products and of the service itself (as measured by available quantities of the various products and end-to-end delay times).

5. SYSTEM IMPLEMENTATION

The ERS-1 ground segment is illustrated in Figure 2. It can be seen that it is structurally simpler than the conceptual discussion of §4 would suggest. This simplification is brought about because certain of the facilities can be conveniently co-located or combined and certain facilities will not necessarily be under the responsibility of the ERS-1 programme.

As concerns funding by the ERS-1 Programme, the proposed ground segment will be limited to Northern European facilities comprising:-

- A combined mission management and mission control facilities located at the European Space Operations Centre (ESOC), Darmstadt, West Germany and known as the MMCC (Mission Management and Control Centre); the prime TT and C station will be co-located with the DAF at Kiruna, North Sweden;
- One station located, for coverage purposes, at Kiruna, North Sweden; this station will combine the functions of TT and C, PDAF and RDAF. This station will also act as a reference station. With the foreseen orbit, a high proportion of the orbits include a pass over Kiruna (10 out of the total of 14 orbits per day);
- A Real-Time Processing Facility (RTPF), whose location is as yet undecided. However, it should be noted that because of its real-time role, it must either be co-located with the DAF or connected to it by a wide-band line with speed of ~ 1 Mbps;
- For the Data Distribution System two possibilities exist, namely:-
  - transmission via land-lines to regional links, as currently used by meteorological services;
  - transmission via satellites.

  The former solution is likely to suffer from severe data-rate limitations. The latter solution is still relatively new for wide-band digital data transmission; thus the use of ESS (European Specialised Service on the European Communication Satellites (ECS)) is still subject to satisfactory negotiations with EUTELSAT, the relevant controlling body. In this context, the technical success of SPINE (Space Informatics Network Experiment) is worth mentioning. This has demonstrated the feasibility of highspeed (1-2 M-bps) digital data communications via satellite links (on ESA's Operational Test Satellite (OTS)) using terminals with small (3m) antennae.

It is clear that the DAF at Kiruna can only provide limited (i.e. Northern European) coverage for the SAR, although the other instruments (all low bit-rate) have full orbit coverage because of the on-board recorder. It is thus envisaged that other ground stations (RDAFs) may participate; these ground stations may be those of ESA's Earthnet, or may be national ones, although the costs would have to be borne by the Earthnet Programme or national budgets respectively. Processing and Archiving Facilities (PAFs) could be provided on a similar, regional, basis; no PAF is presently foreseen within the budget of the ERS-1 Programme.

6. CONCLUSIONS

During the ERS-1 Phase B activities, the full system definition of the ground segment will be developed to the stage necessary for proper implementation. The Phase B studies are expected to commence early in 1982.

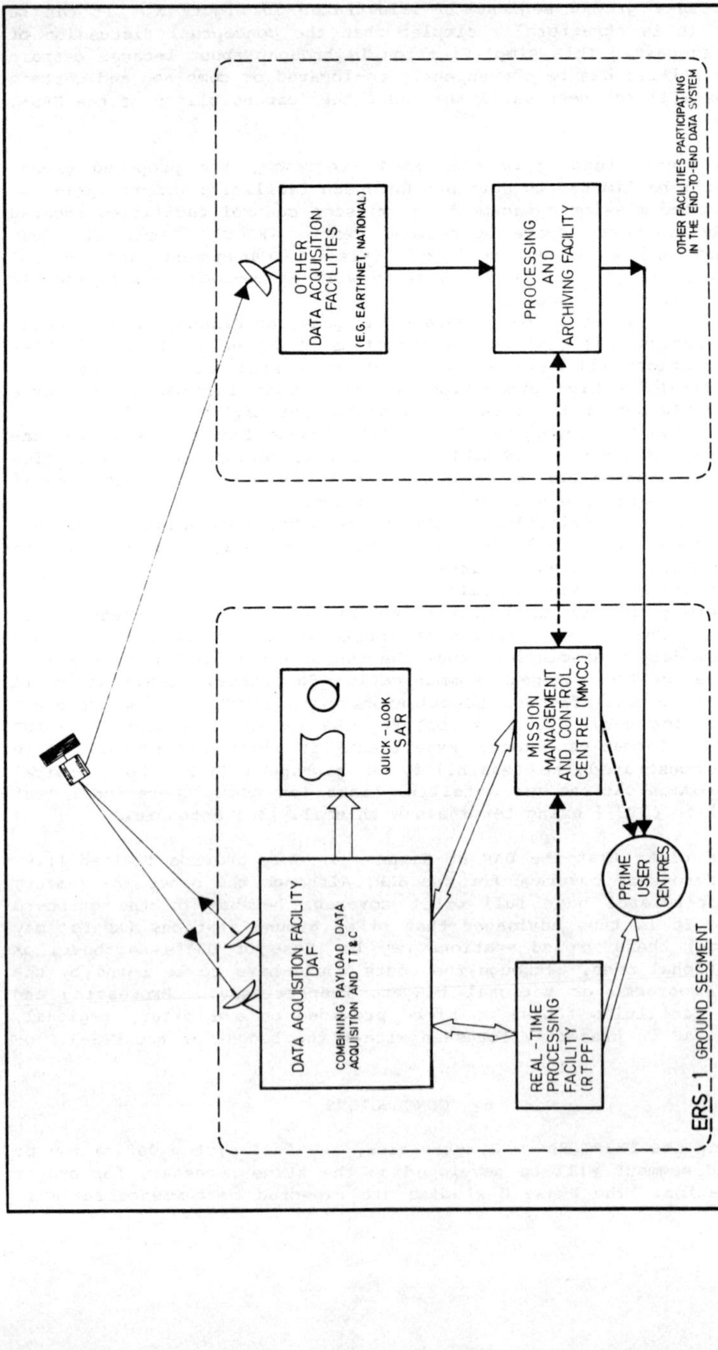

Fig.2 SCHEMATIC DIAGRAM OF ERS-1 GROUND FACILITIES

It will be apparent from the discussions in the foregoing sections that the ERS-1 ground segment will, like the space segment, present a considerable technological challenge over a wide field ranging through highspeed data acquisition and recording, the technology of large-scale data-distribution and high-throughput data processing, to name but a few key areas. ESA is accordingly carrying out research and development in the relevant areas, for example:-

- the SPINE experiment mentioned above will continue to give further experience in data transport technology;
- various approaches are being investigated to the problem of high-throughput SAR processing. The indications are that use of ERS-1 SAR data for ice-monitoring in northern regions (e.g. along the Scandinavian coastline) will require production of scenes (corresponding to around 40 km x 40 km) within 5-10 minutes. Apart from being a problem which will have to be solved if the full potential of the SAR data produced by ERS-1 is to be realised, this demanding area can be regarded as a paradigm of highspeed data processing in the remote sensing area;
- technologies for the initial recording and subsequent archiving of large volumes of data are being studied, where a key aspect is that of reliability.

7. REFERENCES

DUCHOSSOIS, G. and HONVAULT, G., (1981), The First ESA Remote Sensing Satellite ERS-1, Proc. 15th Int. Symposium on Remote Sensing of the Environment, Ann Arbor, Michigan.

PREPROCESSING OF AIRBORNE REMOTE SENSING DATA.

PART 1: THE PRESENT SITUATION

Peter Hoogeboom,
Physics Laboratory TNO,
The Hague, The Netherlands

Abstract: Studies on microwave surface scattering in The Netherlands have indicated the need of accurate radar systems for applications in remote sensing. A SLAR system with digital recording was developed and is now being used for several programs. This system was designed with special attention for speckle reduction and system accuracy. The digital data are processed in a computer with algorithms for geometric and radiometric corrections. In the future also aircraft position and attitude measurements will be used in these correction algorithms (see ref. 3). Examples of the results are shown.

1. INTRODUCTION

In the Netherlands, microwave remote sensing related programs were carried out during a number of years. These programs emphasized the various aspects of interaction between sensor and object. Groundbased measurements were carried out on agricultural crops and over the seasurface. The goal of these measurements was to get a better understanding of the nature of the reflected microwave signals. We so gained insight in the possibilities of radar for classification, oil detection, etc. (e.g. ref. 1, 2). But we also found that the use of radar has its own specific problems and limitations. For instance the speckle in the images combined with the relatively small differences in reflection coefficients for various crops, demonstrated to us the need for extremely accurate systems that must be well calibrated, and do not show much speckle.

In early years we used a SLAR (= Side Looking Airborne Radar) system to study the possibilities of these instruments for remote sensing. However, the requirements that followed from our groundbased measurement programs could not be met by this system. In fact it was a simple imaging radar system, which was not suited for the type of accurate measurements we wanted to perform. Therefore it was decided to build a new SLAR system with digital data recording and accurate absolute signal handling. By using a short pulselength and adequate recording and processing of the received signal, the speckle in the images can be greatly reduced, and the accuracy increased.

Paper presented at a conference on Matching remote sensing technologies and their applications convened in London by the Remote Sensing Society, 16-18 December, 1981.

In this paper the processing of the data will be treated and some of the results will be shown. In a second paper (ref. 3) the future plans with respect to further correction of the images with use of aircraft attitude and position information will be treated.

2. System configuration and properties

The heart of the system consists of a conventional transceiver as used in a shipsradar. This transceiver was slightly modified to improve linearity and to achieve an absolute output signal. The original IF amplifier was replaced by a better module with a dc coupled, low drift detection circuit (Regular imaging system often use cheap ac coupled circuits, which are suited for imaging, but not for measuring). The specifications of the system are given in table 1. The PRF of the system is low in comparison with conventional imaging radarsystems. The value has been optimized from microwave scattering theories taking into account the speed of the carrier and the resolution of the antenna. The higher value of the PRF in a conventional system is often necessary to produce a reasonable intensity on a PPI.

Another remarkable feature is the short pulselength (50 nanosec.) used. Although a higher value is available (250 nanosec.) in the case of imaging at a long range and/or low reflective targets, the use of the short pulselength is preferred. The reason for this is not the higher geometric resolution that can be obtained, but the reduction of the speckle in the image. An area, that is instantaneously covered by a 250 nanosec. pulse, can be divided into 5 areas, each of which can be covered independently by a 50 nanosec. pulse. In this way we obtain 4-5 looks, reducing the speckle by a factor of approx. 2. The longitudinal movement of the aircraft produces about 15 looks (independent samples) per pixel. This yields the most important gain in speckle reduction of a SLAR system in comparison to SAR systems, which often produce only 1 to 4 looks per pixel. Our system gives about 70 independent samples per pixel of 15 x 15 m, thus reducing the speckle by a factor of 8 - 9.

The output signal of the system is digitized by an 8-bit analog to digital converter. Samples are taken at 50 nanosec. interval time (20 MHz), corresponding with 7.5 m slantrange. The high speed digital datastream is recorded on a high density digital tape-recorder. The maximum recording time is about 30 minutes per tape, corresponding to an image with dimensions of 15 x 160 km.

frequency	: 9.4 GHz
transmitted power	: 25 kW
pulselength	: 50/250 ns
polarisation	: HH
antenna beamwidth	: .9 degr.
PRF	: 100 Hz
recording	: 2048 x 8-bit samples
sample frequency	: 20 MHz (7.5 m)

Table 1: Specification of the system parameters

Apart from the radar signal, several aircraft attitude and position parameters are recorded. Both signals are recorded on the same tape with different repetition rates. To enable a combination of these separate datastreams in a later stage, a timeframe (IRIG - B time format) is added to every datablock. The aircraft attitude measurements will be used in future to compensate for unwanted aircraft motions (e.g. roll, yaw, drift), whereas the aircraft position measurements can be used to adapt the along track scale of the image to a map. These corrections on the image will be done during processing afterwards in a digital computer system.

3. Sensor oriented corrections

After the digital information is converted from the high density tape to a standard computer compatible tape, the dataprocessing can take place. Up to now only a part of the desired corrections could be carried out, because the previously mentioned aircraft attitude and position information is not yet available in a format, which is convenient for further processing. However, another class of corrections could be applied, namely the sensor oriented corrections. Two types of corrections can be distinguished, namely geometric and radiometric corrections.

3.1 Geometric corrections

For a number of corrections we need to know the exact height of the aircraft above the ground. Since the recording of the output signal of the system starts just before the transmission of the pulse, it is possible to use the system as an altimeter. The height above the ground can be determined accurately, by taking a moving average of the timelapse between the transmission of the pulse and the reception of the first groundreturn.

The signal is sampled equidistant. This means that the distance to the aircraft increases linearly with the pixelnumber. The resulting image will show a distortion, especially in the region near the aircraft. By resampling the signal, using the distance of the pixel to the groundtrack of the aircraft as a basis, the distortion can be eliminated. The height of the aircraft above the ground is needed in the resampling algorithm. Figure 1 gives an example.

The scale of the image in range direction is determined by the resampling algorithm. In general this scale is related to the original sample interval. In azimuth direction the scale is dependent on the true ground speed of the aircraft, the PRF of the system and the number of successive lines that are averaged. Since the PRF is much higher than a 100 percent coverage of the area would require, this averaging of lines is necessary to produce square pixels. At the present time the true ground speed of the aircraft is only available as a pilots reading of the instrument. Therefore the number of lines to be averaged, is selected manually in the computerprogram. As a result the scalefactor in azimuth direction is constant. Correction for eventual speed variations can not yet be applied. Also the correction for the drift angle and its variation (yaw) has to wait until the aircraft attitude measurements are made available.

3.2 Radiometric corrections

The radiometric corrections, that are applied to the received signal, are all originating from the radar formula. In general this formula can be written as:

Figure 1: Example of a slantrange image (a) and a groudrange image (b), (no radiometric corrections applied).

Figure 2: Example of a fully corrected image.

$$P_r = \text{constant} \cdot \frac{P_t \cdot G^2(\theta) \cdot \gamma \cdot \sin\theta}{R^3 \cdot \cos\theta}$$

With
P_r = received power
P_t = transmitted power
G = antenna gain
R = distance to reflecting target
γ = reflection coefficient of reflecting area
θ = incidence angle of radarbeam with respect to horizontal

By rewriting the formula gamma can be expressed in the other parameters. In this form the formula can be used as a correction algorithm. The antenna gain is not a constant as one might expect, but a function of the incidence angle. This function has to be determined accurately, which is very difficult, because the antenna pattern may change due to close obstacles when the antenna is mounted under the aircraft. Also one has to take into account that the antenna in fact integrates over the two dimensional antennadiagram. So a diagram measured only in the main direction is not useful. In our case the two dimensional antennadiagram was measured first. This diagram was integrated, to produce the necessary correction function. The result was compared with a function, produced from an image. If, in the previously described correction algorithm, the antennagain is left out, then the resulting image will show the square of the antennafunction multiplied by the reflection coefficient. In some cases, the reflection coefficient is nearly constant for a wide range of incidence angles. For instance, this is true for sugarbeets, which is an easy recognizable crop, because of its very high reflection (see ref. 2). So, in this case we can determine the antennafunction by measuring the reflection of sugarbeets at various incidence angles. The comparison of this function with the previously measured and integrated diagram, showed only minor differences.

The radiometric accuracy of the resulting pixels is highly dependent on the attitude of the airplane and the measurement of the height. Rolling of the aircraft causes light and dark stripes on the image, because the antenna correction function is used with the wrong (shifted) incidence angle. In the future we hope to overcome this problem by applying the measured roll angle in the correction algorithm. The measurement of the height as described in 3.1 is easy to do, especially in our perfectly flat country.

4. Some results

Application of the described correction algorithms leads to fine results, if unwanted aircraft movements are small. Figure 2 gives an example of a fully corrected image. This image was taken over the Flevopolder, an agricultural area in The Netherlands. The greytone varies only slightly with range in this image. Because of the high radiometric accuracy of this image, it can be used for classification experiments, as mentioned earlier in this paper. Figure 3 gives an example. The image on the left is in fact a crop type map of the test area on the same scale as the radarimage. The image on the right is produced from a part of figure 2. From this part the average radar reflection per field is calculated. The difference between the highest and the lowest reflection of these crops is only about 8 dB. By density slicing, the crop types can be distinguished. Sugarbeets and potatoes are 100 percent correctly classified, with no missed

Figure 3: Example of radar classification of crops (grazing angle from 7 to 15 degrees).

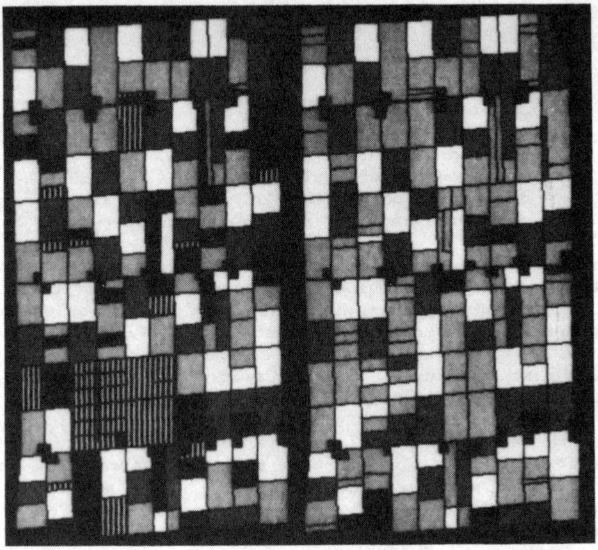

Crop type map Classified radar image

Striped = unknown (in crop type map)
Black = not classified or obstacle (e.g. farm)

Dark grey = potatoes
Light grey = winterwheat
White = sugarbeets

Figure 4: SLAR image taken in turbulent weather.

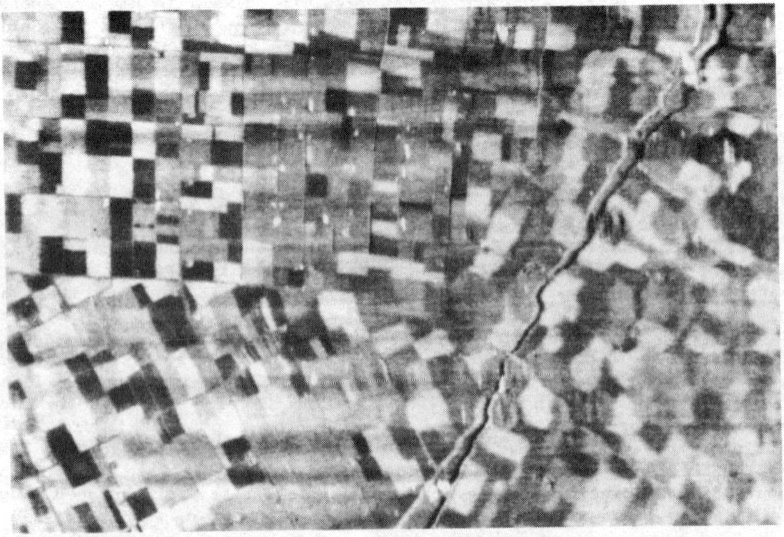

or added fields. The various types of grain cannot be separated in this image. If the levels of a class are set for full classification of winter-wheat, the dominant type of this group, then also other grains and crop types, like onions and peas are classified as winterwheat. Therefore the succes percentage for wheat is lower and depends on the crop type distribution. It is our expectation, that the additional use of other incidence angles and/or imagery at different moments in the growing season will improve the classification results further.

The need for further corrections of the images is clearly demonstrated in figure 4. Our main objective for the near future will be to incorporate aircraft attitude and position measurements in the correction algorithms. This will improve the quality of the imagery further and bring us one step closer to the operational use of microwave remote sensing.

5. References

De Loor, G.P. and P. Hoogeboom (1982): Radar backscatter measurements from platform Noordwijk in the Northsea, IEEE J. Oceanic Engineering OE-7, No.1.

De Loor, G.P., P. Hoogeboom and E.P.W. Attema (1982): The Dutch ROVE Program, IEEE Trans. Geoscience Electronics GE-19, No.1.

Binnenkade, P., Preprocessing of airborne remote sensing data. Part 2: future developements, these proceedings.

PREPROCESSING OF AIRBORNE REMOTE SENSING DATA
PART 2: NEW DEVELOPMENTS

P. Binnenkade
National Aerospace Laboratory NLR
Amsterdam, The Netherlands

Abstract: The Netherlands remote sensing community has at its disposal a number of sensor systems operated from airborne platforms. Thusfar it was not possible to correct the acquired imagery for platform motion and position. Presently NLR within a multidisciplinary team is effecting the combination of aircraft data with SLAR imagery to obtain geometrically and radiometrically corrected data for further user-defined processing and interpretation (PARES). (See also ref. 1)

1. Introduction

The National Aerospace Laboratory NLR participates in the Netherlands remote sensing developments and is primarily involved in airborne operations and data processing. Together with the remote sensing sensor systems, the NLR aircraft are equipped with instrumentation to determine accurately aircraft position and attitude with INS, multi-DME, etc. (fig.1). These data, as well as SLAR or scanner data, are digitally recorded on HBR-tape on board and form the basis of the present development of preprocessing algorithms.

2. Aircraft motion

During most of the runs of a measurement programme any aircraft deviates from an ideal track. Furthermore the longitudinal axis of the aircraft will not always coincide with the direction of forward motion and the thus introduced driftangle will distort the acquired imagery. Also, turbulence and variations in wind will result in yaw-, pitch- and roll-movements and cause errors. The result of these three sources of error is clearly visible and has already been demonstrated in figure 4 of Part 1 of this presentation. However, even in stable flight conditions these variations are intertwined and may not always be negligible (fig. 2).

3. Functional solution

The flow diagram in figure 3 shows the sequence of checks and activities that have already been or are now being developed.

The indicated blocks 1 through 8, 10, 12 and 15 have, to some extent, been discussed (ref.1). Most of them have already been implemented.

A sequence of activities similar to those in the blocks 1 through 8 has been identified for the recorded data on aircraft position and attitude. Computer-compatible tapes (CCT) of both sets of data have been related to the same time code from an internal, on-board clock.

Thus, a new dataset can be generated with both radar data and suitably interpolated position and attitude data. NLR has developed standards from other flight programmes to derive this data set.

Paper presented at a conference on Matching remote sensing technologies and their applications convened in London by the Remote Sensing Society, 16-18 December, 1981.

Figure 1 List of available aircraft parameters

NR	PARAMETER	ABR	TRANSDUCER	UNITS	RANGE
1	FLIGHTNUMBER	FLT	COUNTER		0/999
2	EVENTMARKER	EM	SWITCH		0/1
10	HEADING	HDG	ARC COMPASS	DEG	360
11	PRESSURE ALTITUDE 1	PA1	ROSEMOUNT	M	-600/80K
14	INDICATED AIRSPEED1	IAS1	ROSEMOUNT	M/SEC	
17	ANGLE OF PITCH	AP	VG 14	DEG	+/-90
18	ANGLE OF ROLL	AR	CG 14	DEG	360
19	ACCELERATION X AXIS	AX	DONNER	M/S-2	+/-10
20	ACCELERATION Y AXIS	AY	DONNER	M/S-2	+/-10
21	ACCELERATION Z AXIS	AZ	DONNER	M/S-2	+/-100
27	RADAR ALTITUDE	RA10	PLESSEY	M	0/300
30	STAGNATION TEMP.	TAT	ROSEMOUNT	DEG	+/-60
31	CABIN TEMPERATURE	TC	STEM	DEG	+/-60
300	FRAME COUNT	FRC	DIDAU		0/16384
301	TRUE HEADING COURSE	HDGTC	LTN 58	DEG	0/360
302	TRUE HEADING FINE	HDGTF	LTN 58	DEG	
303	N/S VELOCITY C	NSC	LTN 58	M/SEC	0/168.5
304	N/S VELOCITY F	NSF	LTN 58	M/SEC	
305	E/W VELOCITY C	EWC	LTN 58	M/SEC	0/168.5
306	E/W VELOCITY C	EWF	LTN 58	M/SEC	
307	LATITUDE C	LATC	LTN 58	RAD.	+/-PI
308	LATITUDE F	LATF	LTN 58	RAD.	
309	LONGITUDE C	LONC	LTN 58	RAD.	+/-PI
310	LONGITUDE F	LONF	LTN 58	RAD.	
311	GROUND SPEED	VG	LTN 58	M/SEC	0/500
320	DME DISTANCE C	DMEC	KDM7000	NM	0/200
321	DME DISTANCE F	DMEF	KDM7000	NM	
322	DME STATION	DMES	ADAU		
400	TCG TIME OF DAY	TD	TCG	H	0/24
401	MEASUREMENT NUMBER	MIN	TCG		0/999
500	INS ARINC DATA	INS	LTN 58		
600	SLAR	SLAR	DECCA RADAR		

Source: NLR

Figure 2 Typical roll-behaviour

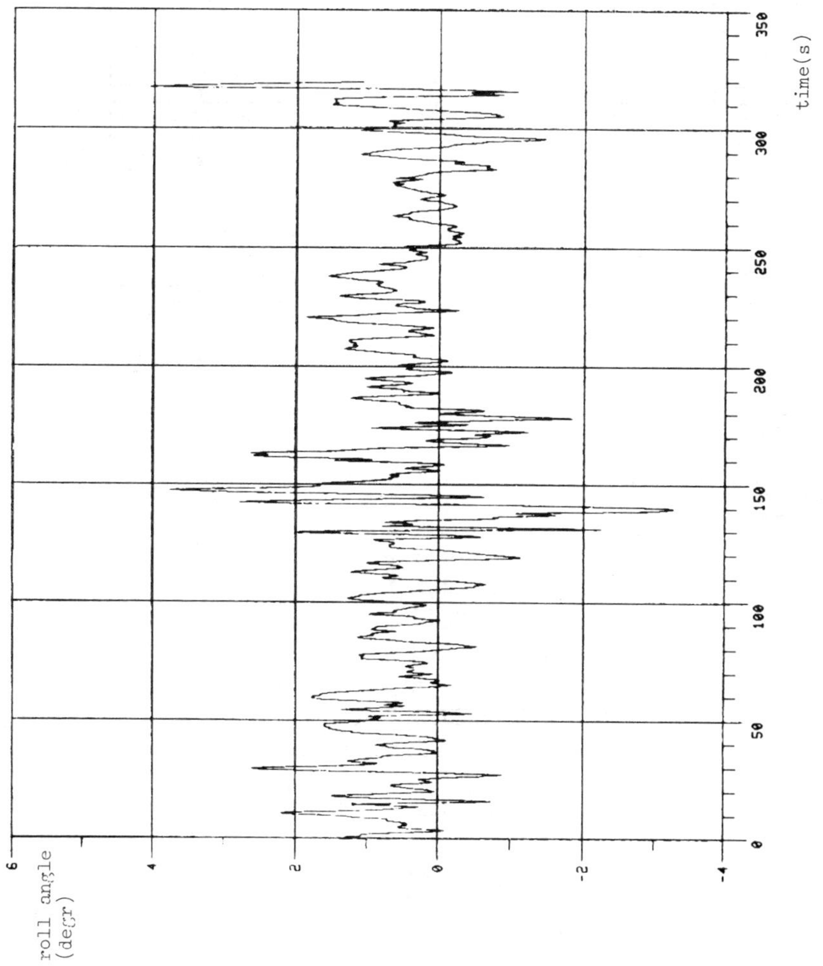

Source: NLR

Figure 3 : Functional diagram of PARES-activities

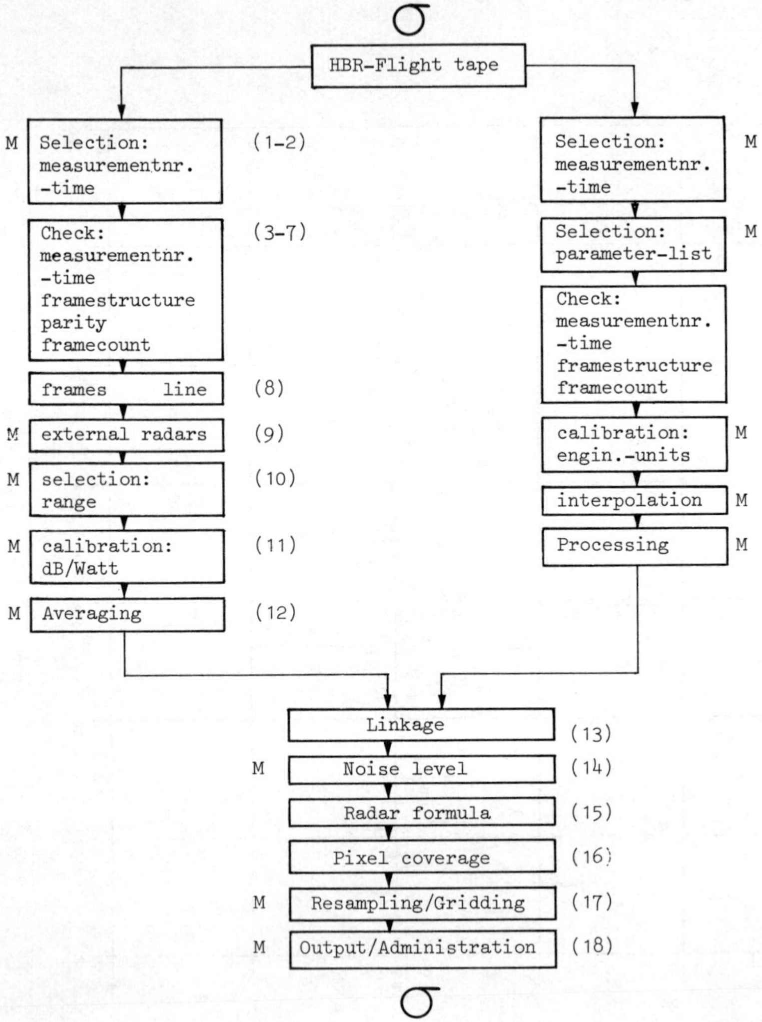

Source: Author

Because most blocks are designed to be "menu-susceptible", they can be executed, not-executed or partly executed, depending on the user-defined parameters which are inserted into the programmes.

An interesting feature is the interference from other,external radar sources (fig.4). Their presence has been noted for quite some time, whereas the necessity of removal is not always stipulated. However, especially in geographic areas with dense treffic such as ships, aircraft and airports, these signals can become a source of inaccuracy when further processing is required.

In a typical radar line, signals from external radars can be characterized both by their excessive signal strength and by their once-only occurence in relation to previous and subsequent radar lines. These signals also show little variation within themselves. These characteristics are used to remove them from the data set. Their values may be replaced by interpolated signal data.

Block 16 and 17 contain the most time-consuming and delicate processing. In order to determine the exact pixel location on the ground, a number of transformations have to be executed. Hence a foursome of coordinate systems is defined:

1. Aircraft coordinate system, originating in the center of gravity of the aircraft and parallel to its main axes.
2. Radar antenna coordinate system, originating in the front end of the slotted wave-guide.
3. Inertial coordinate system, linked to the inertial navigation system LTN 58.
4. Terrestrial coordinate system, based on a suitable geodetic grid.

A series of transformations is thus derived which determines the location of a pixel in the terrestrial system as a function of the received return signal at a given aircraft position and attitude (ref.2).

Subsequently a suitable resampling algorithm is chosen to achieve homogeneous pixel coverage and registration thereof on CCT for further user-defined processing and information extraction.

4. Generalization

The development of preprocessing algorithms for SLAR data will provide an insight which is necessary to ensure the future generalization of these techniques toward airborne thermal infrared linescanning, as operated at NLR.

Furthermore, the design and construction of a prototype multispectral CCD-scanner, which has recently been initiated in The Netherlands, will lead to one further application of preprocessing algorithms in the future.

5. Further processing

Computation of blocks 13 through 18 is intended to be implemented on the new interactive data processing system for remote sensing data (RESEDA , fig.5), which will be installed in March 1982 at NLR.

Thus, it is expected that the development of preprocessing techniques (PARES) and the installation of the RESEDA-system at NLR will provide a new stimulus to the operational use of remote sensing in The Netherlands.

Figure 4 Influence of external radars (raw data)

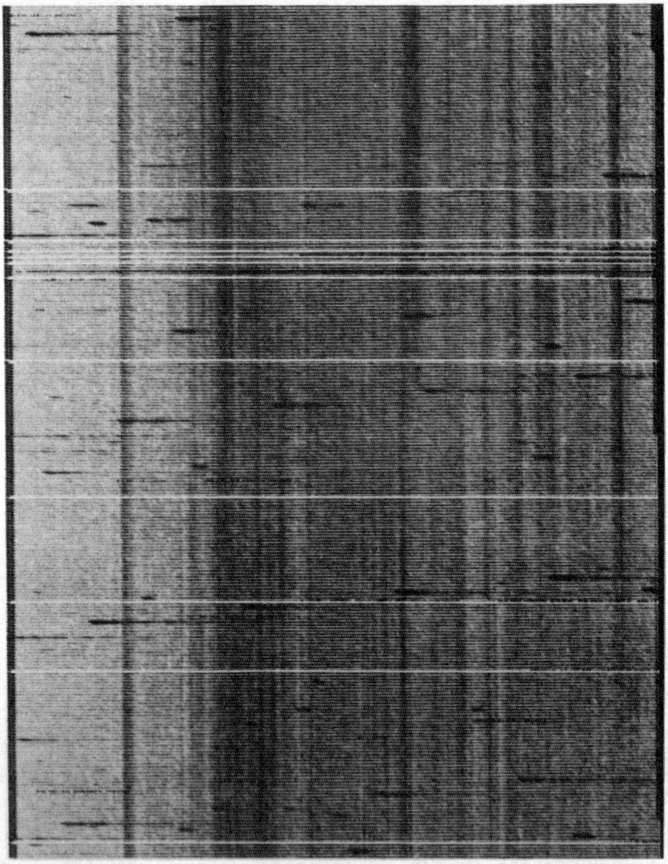

Source: NLR

Figure 5 The RESEDA data-processing system

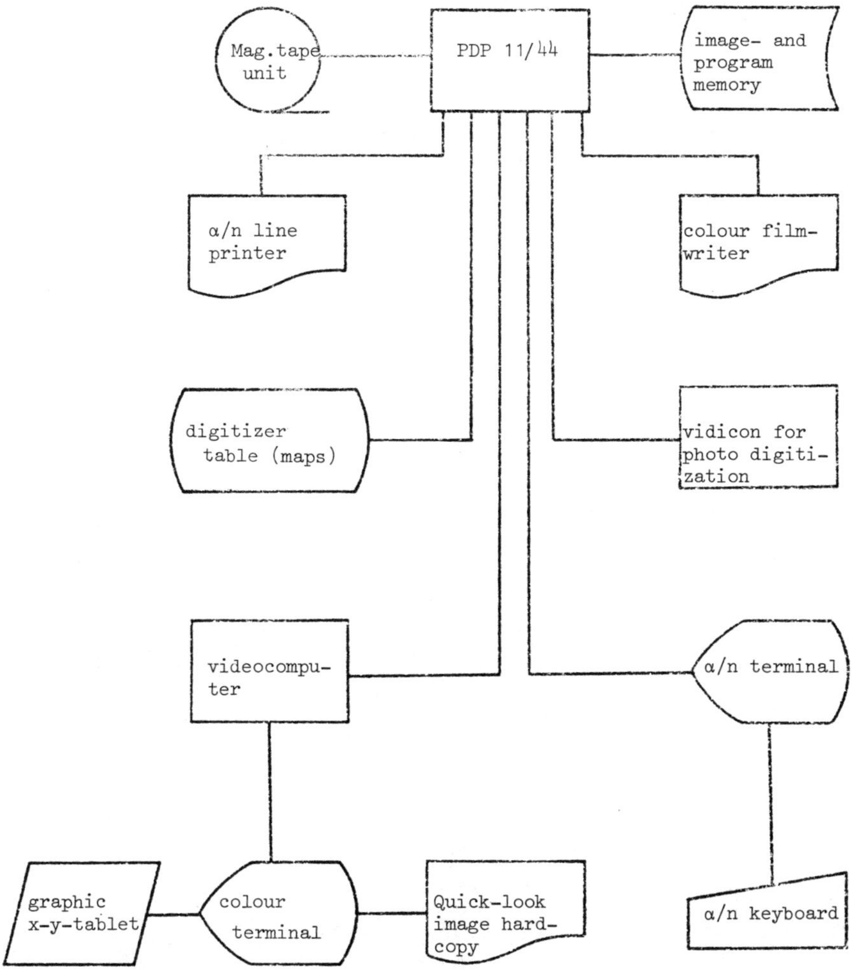

<u>Source</u>: NLR/DIPIX

References

1. Hoogeboom, P.: Preprocessing of airborne remote sensing data. Part 1: The present situation, these proceedings.

2. Moek, G., Kroon, R., Veugen, T.: Geometric correction of SLAR-imagery, October 1981, internal PARES-communication.

A MICROCOMPUTER SYSTEM FOR PROCESSING AND ANALYSIS OF REMOTELY SENSED DATA

M J Adams
Department of Spectrochemistry
The Macaulay Institute for Soil Research
Craigiebuckler
Aberdeen
AB9 2QJ

Abstract: Much of the information obtained from remote sensing techniques is recorded in analogue format, and, for most applications involving the digital analysis of such data, a mainframe computer or dedicated minicomputer system is used to control the processing and analysis. This is often necessary because of the large amounts of digital information required to reproduce accurately the resolution of the original analogue data. The increasing computing power and availability of microcomputers, however, can provide a useful and inexpensive alternative. This paper describes such a microcomputer based system employed for the processing and analysis of remotely recorded data. The system is used for the study and display of airborne-recorded infrared data and the analysis of crop and soil reflectance spectra (ca. 350 nm to 1000 nm) obtained with the aid of a portable spectrometer. It is demonstrated that with efficient data packing within the available RAM of the system and by digitising information as required, a microcomputer can provide an efficient and rapid tool for remote sensing studies.

1. INTRODUCTION

Instrumental methods of analysis involve a sequence of signal data-domain conversions, (Malmstadt et al., 1974). The Majority of transducers produce an analogue signal output which is recorded in this mode, although direct digitization of the signal in the measuring instrument is becoming more common. Whilst digital recording can assist in the elimination of recorder noise, amplifier drift and dynamic distortion in the signal processing electronics, the domain conversion from analogue to digital imposes its own constraints. The storage density of signals from digital recordings is commonly inferior to analogue techniques and digitization entails an increase in complexity in the signal chain electronics,e.g. it is necessary to sample the analogue signal very rapidly to achieve a digital copy of the data at its original resolution. If the data is presented to the user in digital, computer compatible format then the choice of the system processing unit is dictated, in part, by the selected format of the data. The analysis of Landsat tapes is an example of this

Paper presented at a conference on Matching remote sensing technologies and their applications convened in London by the Remote Sensing Society, 1981 situation. When the data recording is controlled by the user, however, a wide variety of processing systems is available. At the Macaulay Institute for Soil Research, a study is currently being undertaken of the use of a microcomputer-controlled unit for the processing and analysis of remotely sensed data recorded in analogue mode on magnetic tape. Two examples are discussed; they are the study and display of airborne recorded thermal emission data and the analysis of crop and soil reflectance spectra, obtained with the aid of a portable spectrometer.

2. DATA-PROCESSING INSTRUMENTATION

The recording and data processing system has been described in detail elsewhere, (Adams and Ritchie, 1981). The analytical data is recorded in analogue format, at the highest resolution required, on ¼" magnetic tape using a portable reel-to-reel instrumentation recorder. Digitization and data processing are conducted subsequently in the laboratory with the aid of a medium speed (70 μs), 8-bit analogue-to-digital, A-D converter controlled by an Apple II microcomputer. A twin floppy-disk unit is employed for data and program storage, and output is via a 9" monitor or, for colour display, a Tektronix video display unit, (model 4027). Hard-copy output of isothermal data and reflectance spectra is obtained from a thermal printer and a flat-bed plotter.

For all applications, the control of the A-D unit, data formatting, processing and analysis are undertaken by machine-coded routines for the 6502 microprocessor within the computer.

3.1. Case I - THERMAL INFRARED IMAGERY

During the evening of the 11th December 1980 a thermographic survey of the City of Aberdeen was conducted by Clyde Surveys (Scotland) Ltd., using an airborne, quantitative Daedalus line-scanner. Line-scanning of emitted thermal, (10-14 μm), radiation was undertaken in the cross-track direction by means of a rotating single-sided mirror. A radiant temperature sensitivity of ± 0.2°C could be achieved with the system, and the signal from the infrared detector was recorded, in analogue format, on magnetic tape.

To achieve high-density storage of digital data within the microcomputer, 16-level (4-bit) slicing of the analogue signal is employed with double-density serial packing of the data within the available RAM of the microcomputer. The slicing levels employed depend on the scene under investigation and the required thermal resolution, e.g. a 16-level linear density slice of the original data provides a radiant temperature resolution between adjacent levels of 0.625°C. The use of double-density packing techniques allows 64,000 data points to be retained in less than 32K of RAM. With a spatial resolution of approximately 2.5 m x 2.5 m per data point, a scene of an 800 m x 500 m area can be available within the computer for analysis and display. Contrast manipulation of the digitized data is rapidly achieved and, if required, the analogue scene can be resampled. The floppy-disk units allow up to six scenes to be on-line, in packed format, for further studies. Typical analysis routines include a temperature distribution output for the scene examined, and for isothermal plots, a nearest-neighbour smoothing program.

3.2. Case II - SPECTRAL REFLECTANCE DATA

Two general uses of spectral reflectance data are being investigated. Firstly, considerable work has been undertaken at the Macaulay Institute in the analysis of Landsat data for land-cover mapping, (Stove and Bagot, 1980). In this work ground-truth measurements are of considerable importance in complementing the satellite imagery. Secondly, interest is being expressed in portable, hand-held reflectance measuring instruments, employing optical bandpass filters, for crop and soil analysis. The possible applications of any such instrument are dependent on the spectral characteristics of the optical filters used. Because of the wide variety of applications for such reflectance studies, a portable spectrometer has been assembled at the Macaulay Institute capable of recording spectra in the wavelength range 350 - 1000 nm. The instrument is based upon a commercially available, miniature, grating monochromator with a high-speed photodiode detector and amplifier unit. Wavelength scanning is achieved by continuous rotation, of the grating, at 10 Hz, providing a spectral recording period of approximately 7 ms. The use of a fast-scanning monochromator allows spectra to be recorded free from problems due to relatively slow changes in ambient light levels. The collecting optics of the spectrometer enables a 1 m^2 target to be examined with the instrument mounted about 2 m above the sample. A sample spectrum is recorded 300-400 times on magnetic tape. To increase the signal-to-noise ratio of the recorded data, the spectra are summed and averaged on digitization. The digital spectra are stored on floppy disk for further processing. Analytical techniques employed include quadratic smoothing and spectral correlation, this latter technique involving ratioing spectra of interest to determine wavelength regions of similarity and difference. Computed spectral profiles of suitable optical filters can then be fitted to the spectra to select the optimum filters for an application.

4. SUMMARY

A microcomputer-based system has been assembled at the Macaulay Institute for the processing and analysis of analogue recorded, remotely sensed data. In the case of an airborne thermographic survey the data is retained in analogue format on magnetic tape. A scene is digitized as required, to the necessary specification, thus minimising the computer memory storage requirements. For the analysis of reflectance spectra, the analogue recording technique is employed as a temporary storage process the signal-averaged spectra being stored in digital format on floppy disk.

The analogue recorder - digital computer system is a flexible instrument and allows a wide variety of data to be recorded and made available for computer analysis.

5. REFERENCES

ADAMS, M.J., and RITCHIE, P.F.S., "Processing and Analysis of Thermographic Data with the aid of a Microcomputer". Computer Appl., 1981 (In Press).

MALMSTADT, H.V., ENKE, C.G., CROUCH, S.R., and HORLICK, G., "Electronic Measurements for Scientists", W.A. Benjamin Inc., (1974).

STOVE, G.C., and BAGOT, K.E., "Current Resource Mapping and Environmental Monitoring Applications using Landsat MSS Data in Scotland". Proc. 6th Can. Symp. Rem. Sens., 551 (1980).

SEP IMAGE ANALYSIS SYSTEM : VIPS 16/65

L. Salter ; G. Maincent ; D. Sarrat
Space Division
SOCIETE EUROPEENNE DE PROPULSION
27200 VERNON - FRANCE

Abstract : The VIPS 16/65 System (VIZIR IMAGE PROCESSING SYSTEM) has been especially designed for users, who plan to have their own ground receiving station and/or processing center of images transmitted from Earth Observation Satellites, such as LANDSAT D or D' and SPOT, and therefore require a computer assisted DIGITAL ANALYSIS SYSTEM adapted to such a station.

It is a digital processing system of image data from mechanical scanner (LANDSAT Satellite) and push-broom scanner (SPOT Satellite).

The VIPS 16/65 mainly includes :

- one powerful data processing configuration based on a mini-computer, with its operating system and peripherals necessary for the development of application software,
- one or several colour interactive video consoles,
- one laser beam recorder (VIZIR (Black and White) or VIZIRCOLOR (colour film)),
- one image digitizer,
- powerful peripherals : array processor, 300 Mbytes magnetic disk, 1600/6250 bpi magnetic tape recorder,
- software for applications organised around a monitor with main capabilities,
- multi-users, multitemporal works, geometrical and radiometrical corrections, image manipulation, supervised and unsupervised classifications, image to map or image to image corrections, ground truth acquisition, offset output...

This new system has a full compatibility with the image data from every type of optical cameras, now in use or planned to be used in the NASA/LANDSAT and CNES/SPOT programmes.

Paper presented at a conference on Matching remote sensing technologies and their applications convened in London by the REMOTE SENSING SOCIETY, 16-18 December 1981.

1. FOREWORD

In the coming years Remote Sensing will be in expansion following the launch of U.S. LANDSAT D and D', French SPOT series, and European Spatial Agency satellites.

These new resources will provide data of an improved spectral and spatial resolution. As a matter of fact, a new field of possibilities will be open in thematic as well as cartographic applications.

SEP European Society of Propulsion, France, develops a ground receiving and pre-processing station for LANDSAT D and SPOT data.

As a complement for this station, the VIPS 16/65 (VIZIR IMAGE PROCESSING SYSTEM 16/65) set of systems allows the digital processing of the images produced by the station, in order to meet remote sensing user's requirements, according to the present state of technology.

2. FUNCTIONAL CHARACTERISTICS

2.1. The main definition specifications

The definition of the VIPS 16/65 system has taken into account the following specifications :

. VIPS 16/65 must be a user-oriented image processing system for THEMATIC OR CARTOGRAPHIC APPLICATIONS OF REMOTE SENSING.

. VIPS 16/65 must be able to operate as a stand-alone system or as an extension of a SEP SPOT/LANDSAT D Ground Station. In the latter case, it is considered as including at least the GCP (Ground Control Points) acquisition subsystem of the station.

. VIPS 16/65 has not to be only one system, but rather a set of systems fully compatible with each other, all of them built from a common hardware and software kernel.

. VIPS 16/65 must be easy to use :
- it must accept data and produce results under a conventional form :
  1) input : photographs, topographic and thematic maps, ground truth,
  2) output : thematic maps (with offset printing capability), photographic restitutions, statistical results,
- the manipulation of the system must need neither computer experience nor special training : a high level procedural language must be implemented to mask the machine operating system, to command the execution of the catalogued procedures and to develop new processing procedures (written in command language).

. VIPS 16/65 must be capable of expansion to suit the users needs and must provide the user with all the software tools allowing the development of specific processing primitives (written in FORTRAN).

. VIPS 16/65 must look to on the future and be designed to-day to be able to process the data from the future satellites (SPOT, LANDSAT TM...).

2.2. The design guidelines

The functional catalogue has been elaborated by analyzing the main steps in the processing of a given image.

1) Step 1 : input.

The user has chosen the data he wishes to process (sensor, date, channels, area) ; first, he must introduce them into the system.

Three possible sources :
- magnetic tape,
- a photograph,
- a map.

One destination only :
- magnetic disk.

2) Step 2 : preprocessing.

Sometimes, data must be radiometrically or geometrically corrected before being analyzed by the interpreter or the computer :
. radiometrical corrections related to the sensor (calibration, MTF corrections...),
. radiometrical corrections related to external effects (haze removal, sun angle, ...),
. geometrical corrections related to the sensor and its platform : ephemeris and attitude errors, scanning defects...,
. geometrical corrections related to external effects : earth rotation, earth curvature, cartographic projection, relief...

3) Step 3 : segmentation.

The segmentation consists in dividing the image into connected parts by means of one of the four following criteria :
- spectral thresholds in one or several channels : for instance, threshold between water and earth modes in MSS-7 channel,
- previous classification : the user is allowed to analyze or subdivide one or several classes of a previous classification,
- textural criteria such as granularity, directivity, edgeness, frequential contents,
- criteria independent of the radiometrical contents of the image : ground-truth polygons, thematic zoning (administrative, geological...).

4) Step 4 : photo-interpretation.

The aim of a photo-interpretation function is to improve the presentation of the image in order to facilitate its visual or automatic interpretation.
In particular, photo-interpretation functions allow the user to determine the best spectral signature, in relation to the subject of interest.
The results of photo-interpretation processing are obtained by combination of the initial data :
- either within a channel, by means of spatial filtering,
- or between two or several channels.

5) Step 5 : classification.

Faced with the amount of data to interpret, the user is sometimes discouraged... and may wish some schematisation of the information : here is the goal of the classification, which always works in two steps.

a) training, which consists in determining the number of classes and their statistical parameters, on a training area.
The training can be supervised or not supervised ; if it is supervised, it needs ground truth acquisition.
After the training, it is often necessary to analyze the quality of the training :
- spectral homogeneity of each class (histograms),
- spectral separability between classes (divergence matrices),
- comparison with the ground truth (confusion matrices).

b) classification, which assigns each pixel of the whole image to the classes determined by the training.

6) Step 6 : output of

a) images : raw, preprocessed or processed data.
The image can be edited on :
- magnetic tapes,
- VIZIR restitutions (with offset printing capability),
- electrostatic printer.

b) statistical results : histograms, mean spectra, covariance matrices, surfaces covered by classes within administrative regions, ...

2.3. List of the functions (to-day)

2.3.1 I/O functions

INPUT	NAME	FUNCTION	OUTPUT
MSS SCENE ON CCT (NASA, TSP, ...)	LOAMSS	MSS IMAGE LOADING	MSS SUBSCENE ON DISK (SUBSCENE FORMAT)
SPOT SIMULATION SCENE ON CCT (CNES)	LOASIM	SPOT SIMULATION IMAGE LOADING	SPOT SIMULATION SUBSCENE ON DISK
VIPS FORMAT SCENE ON CCT	LOAVPS	VIPS FORMAT IMAGE LOADING	ANY SUBSCENE ON DISK
BW OR COLOUR PHOTOGRAPH ON SCANDIG	SCDDSK	SCANDIG TO DISK DIGITIZATION	SUBSCENE ON DISK
BW OR COLOUR PHOTOGRAPH ON SCANDIG	SCDTAP	SCANDIG TO TAPE DIGITIZATION	VIPS FORMAT SCENE ON CCT
ANY SUBSCENE ON DISK	SAVIPS	ANY IMAGE UNLOADING	VIPS FORMAT SCENE ON CCT
SUBSCENE ON DISK	SAVSSC	CATALOGUING A SUBSCENE ON TAPE	SUBSCENE ON CCT
SUBSCENE ON TAPE	LOASSC	RE-LOADING A SUBSCENE ON DISK	SUBSCENE ON DISK
SUBSCENE ON DISK	DSKVIZ	DIRECT RESTITUTION ON VIZIR LBR	400 x 400 MM2 PHOTOGRAPHY
SUBSCENE ON DISK	PREVIZ TAPVIZ	OFF-LINE RESTITUTION ON VIZIR LBR (DISK → TAPE → VIZIR)	400 x 400 MM2 PHOTOGRAPHY

INPUT	NAME	FUNCTION	OUTPUT
CLASSIFIED IMAGE ON DISK	PROVIZ OFFVIZ	OFF-LINE RESTITUTION ON VIZIR LBR (DISK → TAPE → VIZIR)	N OFFSET FILMS FOR PRINTING
ANY SCENE ON CCT	QUICTP	GREY LEVEL QUICK LOOK OF A CCT (ON ELECTROSTATIC PRINTER)	BW SCALED RESTITUTION ON EPR
ANY SUBSCENE ON DISK	GREYPR	GREY LEVEL RESTITUTION OF A SUBSCENE (ON ELECTROSTATIC PRINTER)	BW SCALED RESTITUTION ON EPR
CLASSIFIED SUBSCENE ON DISK	SYMBPR	GRAPHIC SYMBOLS RESTITUTION OF A CLASSIFIED SUBSCENE (ON ELECTROSTATIC PRINTER)	BW SCALED RESTITUTION ON EPR
MAP ON DIGITIZER	DIGITE	MAP DIGITIZATION	SUBSCENE ON DISK

2.3.2 Preprocessing functions

INPUT	NAME	FUNCTION	OUTPUT
RAW MSS IMAGE ON CCT OR DISK	BLKMSS	BULK GEOMETRICAL CORRECTIONS OF A RAW MSS IMAGE (ASPECT RATIO AND EARTH ROTATION)	BULK CORRECTED MSS SUBSCENE ON DISK
STRIPED MSS IMAGE ON CCT OR DISK	CALMSS	STATISTICAL CALIBRATION OF A STRIPED MSS IMAGE	CALIBRATED MSS SUBSCENE ON DISK
SEGMENTED SUBSCENE ON DISK (AND TRIADE DISPLAY) AND MAPS ON DIGITIZER	MAPGCP	ACQUISITION OF IMAGE/MAP GROUND CONTROL POINTS AND COMPUTATION OF A POLYNOMIAL DEFORMATION MODEL	DEFORMATION MODEL ON DISK
SEGMENTED RAW AND REFERENCE SUBSCENES ON DISK (AND TRIADE DISPLAY)	IMARCP	ACQUISITION OF IMAGE/IMAGE REFERENCE CONTROL POINTS AND COMPUTATION OF A POLYNOMIAL DEFORMATION MODEL	DEFORMATION MODEL ON DISK
SEGMENTED RAW SUBSCENE AND DEFORMATION MODEL ON DISK	RECTIF	GEOMETRICAL RECTIFICATION OF A SUBSCENE BY POLYNOMIAL DEFORMATION MODEL (NEAREST NEIGHBOUR OR BICUBIC LAW)	RECTIFIED SUBSCENE ON DISK

2.3.3 Segmentation functions

INPUT	NAME	FUNCTION	OUTPUT
ANY SUBSCENE ON DISK	THRESH	BUILDING OF THE MASK BY MULTISPECTRAL THRESHOLDING	ONE CHANNEL SUBSCENE ON DISK
CLASSIFIED SUBSCENE ON DISK	CLASEG	USING THE CLASSIFIED SUBSCENE AS A MASK	IDEM INPUT
ANY SUBSCENE ON DISK	POLYTR	DIRECT ACQUISITION OF GROUND-TRUTH POLYGONS ON COLOUR DISPLAY	ONE CHANNEL SUBSCENE ON DISK
ANY SUBSCENE ON DISK AND REFERENCE DOCUMENT ON DIGITIZER	POLYNU	ACQUISITION OF GROUND-TRUTH POLYGONS BY CORRELATION BETWEEN THE REFERENCE DOCUMENT (MAP, PHOTOGRAPHY, ...) AND THE IMAGE	ONE CHANNEL SUBSCENE ON DISK
ANY SUBSCENE ON DISK	TEXTUR	SEGMENTATION BY TEXTURE CRITERIA	ONE CHANNEL SUBSCENE ON DISK

2.3.4 Batch photo-interpretation functions

INPUT	NAME	FUNCTION	OUTPUT
. XS OR XT SUBSCENE ON DISK . OPTION : MASK IN SUBSCENE FORMAT	ADD	$A_1.X_1(i,j) + A_2.X_2(i,j)$	SUBSCENE ON DISK
IDEM ADD	MULT	$A_1.X_1(i,j).A_2.X_2(i,j)$	IDEM ADD
IDEM ADD	RATIO	$A_1.X_1(i,j)/A_2.X_2(i,j)$	IDEM ADD
IDEM ADD	ATARAT	$\text{ARC TG}\left[A_1.X_1(i,j)/A_2.X_2(i,j)\right]$	IDEM ADD
IDEM ADD	NORDIF	$\frac{A_1.X_1(i,j) - A_2.X_2(i,j)}{A_1.X_1(i,j) + A_2.X_2(i,j)}$	IDEM ADD
IDEM ADD	LCOMBI	$\sum_K A_K.X_K(i,j)$	IDEM ADD
IDEM ADD	KLOEVE	KARHUNEN - LOEVE TRANSFORMATION (NORMALIZED OR NOT)	PRINCIPAL COMPONENTS SUBSCENE ON DISK
IDEM ADD	CORANA	CORRESPONDENCE ANALYSIS (CHI SQUARED METRICS) OF SPECTRA/CHANNELS ARRAY	MULTIFACTORIAL IMAGE SUBSCENE ON DISK
. XS OR XT SUBSCENE ON DISK . CLASSIFIED SUBSCENE ON DISK	DISANA	DISCRIMINANT ANALYSIS OF SPECTRA CLASSES/CHANNELS ARRAY	MULTIFACTORIAL IMAGE SUBSCENE ON DISK
IDEM ADD	STDRD	NORMALIZATION OF THE DYNAMIC RANGE ($\overline{M}$ = 128, σ = 128/3)	IDEM ADD

INPUT	NAME	FUNCTION	OUTPUT
ANY SUBSCENE ON DISK	FILTER	FILTERING BY 3 x 3 KERNEL DEFINED BY USER	SUBSCENE ON DISK
IDEM FILTER	AVRAGE	SMOOTHING BY MOVING AVERAGE ON 3 x 3 NEIGHBOUR	IDEM FILTER
IDEM FILTER	MEDIAN	SMOOTHING BY MOVING MEDIAN ON 3 x 3 NEIGHBOUR	IDEM FILTER
IDEM FILTER	SHARP	EDGE ENHANCEMENT (GRADIENT)	IDEM FILTER
IDEM FILTER	EDGES	EDGE EXTRACTION (LAPLACIAN)	IDEM FILTER
IDEM FILTER	LOCOPT	LOCAL OPTIMIZATION OF THE DYNAMIC RANGE	IDEM FILTER
IDEM FILTER	ZOOM	ZOOM WITH INTERPOLATION	IDEM FILTER
IDEM FILTER	SAMPLE	SAMPLING WITH AVERAGE	IDEM FILTER

2.3.5 Interactive photo-interpretation functions

INPUT	NAME	FUNCTION	OUTPUT
. IMAGE ON DISK OR TRIADE 80 MEMORY . OPTION : MASK ON DISK OR TRIADE 80 MEMORY	BAW	BW DISPLAY OF AN IMAGE, WITH INTERACTIVE CONTRAST ENHANCEMENT (STRET)	TRIADE 80 DISPLAY
IMAGE IN TRIADE MEMORY	STRET	INTERACTIVE CONTRAST ENHANCEMENT, WITH THE OPTIONS : LIN, LOG, EXP, ISO, PIECE, TAB	. TRANSFER FUNCTION IN TRIADE RAM . TRIADE 80 DISPLAY
IDEM BAW	PSCOL	PSEUDO-COLOUR DISPLAY OF AN IMAGE, WITH "STRET" AND INTERACTIVE BUILDING OF THE COLOUR TABLE (CTBUIL)	IDEM BAW
IMAGE IN TRIADE MEMORY	CTBUIL	INTERACTIVE BUILDING OF THE COLOUR TABLE, WITH THE OPTIONS : INICOL, PALET, PSKRO	. COLOUR TABLE IN TRIADE RAM . TRIADE 80 DISPLAY
IDEM BAW, BUT WITH 3 CHANNELS AT LEAST	FACOL	FALSE (OR TRUE) COLOR DISPLAY OF A MULTIPLE IMAGE WITH "STRET"	IDEM BAW
IDEM BAW, BUT WITH 2 CHANNELS AT LEAST	KROCOL	COLOUR DISPLAY OF A TWO CHANNELS IMAGE WITH THE "KROSSCOLOR" TABLE	IDEM BAW
. 2 CHANNELS IN TRIADE MEMORIES . KROSSCOLOR TABLE IN TRIADE RAM	KROMAN	SIMULATION OF ARITHMETICAL OR LOGICAL OPERATIONS BETWEEN TWO CHANNELS BY MANIPULATION OF THE KROSSCOLOR TABLE	IDEM BAW

INPUT	NAME	FUNCTION	OUTPUT
IMAGE IN TRIADE MEMORY	ZOOM	VIDEO ZOOM OF THE DISPLAYED IMAGE	IDEM BAW
IMAGES AND TABLES IN TRIADE	COM80	INTERPRETED LANGUAGE FOR IMAGE MANIPULATION (CONTROL STATEMENTS, PROCESSING STATEMENTS	IMAGES AND TABLES IN TRIADE

2.3.6 <u>Batch classification functions</u>

INPUT	NAME	FUNCTION	OUTPUT
. XS OR XT IMAGE ON DISK (SUBSCENE FORMAT) . GROUND-TRUTH MASK ON DISK (SUBSCENE FORMAT)	PARAM	COMPUTATION OF STATISTICAL PARAMETERS OF GROUND-TRUTH CATEGORIES : - HISTOGRAMS (OPTION) - MEAN SPECTRA - COVARIANCE MATRICES	LIST OF MEAN SPECTRA AND CO-VARIANCE MATRICES (WITH CLASSES POPULATIONS), ON DISK
LIST OF MEAN SPECTRA AND COVARIANCE MATRICES OF INPUT CATEGORIES OR CLASSES	SEPAR	SEPARABILITY ANALYSIS OF THE GROUND TRUTH CATEGORIES OR STATISTICAL CLASSES : - PROJECTION OF THE ELLIPSES ONTO THE PLANES OF THE CHANNELS - COMPUTATION OF DIVERGENCE MATRICES AND HIERARCHICAL CLASSIFICATION OF CATEGORIES OR CLASSES - MERGING OR ELIMINATION OF CATEGORIES OR CLASSES - SELECTION OF THE MORE DISCRIMINANT CHANNELS	LIST OF MEAN SPECTRA AND COVARIANCE MATRICES OF OUTPUT CATEGORIES OR CLASSES, ON DISK
. XS OR XT IMAGE ON DISK . LIST OF MEAN SPECTRA AND COVARIANCE MATRICES OF CATEGORIES OR GROUPS OF CATEGORIES ON DISK . OPTION : MASK (SUBSCENE FORMAT) ON DISK	LIKHOD	MAXIMUM LIKEHOOD CLASSIFICATION	CLASSIFIED IMAGE ON DISK (SUBSCENE FORMAT)
. XS OR XT IMAGE ON DISK . OPTION : TRAINING MASK, ON DISK	DATCOD	DETERMINATION OF CODING LEVELS OF SPECTRA (PREPARATION OF AN UNSUPERVISED TRAINING : SEE 'CODANA - HAC - AGREG')	LIST OF CODING BOUNDARIES (ON DISK)

INPUT	NAME	FUNCTION	OUTPUT
. XS OR XT IMAGE ON DISK . LIST OF CODING BOUNDARIES (ON DISK) . OPTION : TRAINING MASK (ON DISK)	CODANA	CORRESPONDENCE ANALYSIS OF THE BINARY ARRAY : PIXELS/ CODED CHANNELS	LIST OF EIGEN-VALUES AND FACTORIAL COMPONENTS OF THE CODED CHANNELS (ON DISK)
LIST OF EIGENVALUES AND FACTORIAL COMPONENTS OF CODED CHANNELS (ON DISK)	HAC	HIERARCHICAL ASCENDING CLASSIFICATION OF CODED CHANNELS AND MANUAL DETERMINATION OF CODED CHANNELS GROUP	LIST OF CODED CHANNEL GROUPS (ON DISK)
. LIST OF CODED CHANNELS GROUPS (ON DISK) . XS OR XT IMAGE . OPTION : TRAINING MASK (ON DISK)	AGREG	ITERATIVE AGGREGATION OF SPECTRA ROUND THE CLASS CENTRES INITIALIZED BY THE 'HAC' FUNCTION (ON THE TRAINING AREA)	. LIST OF MEAN SPECTRA AND COVARIANCE MATRICES OF FINAL CLASSES . OPTION : CLASSIFIED TRAINING AREA ON DISK
. XS OR XT IMAGE ON DISK . LIST OF MEAN SPECTRA AND COVARIANCE MATRICES, ON DISK . OPTION : CLASSIFICATION MASK ON DISK	LUPTAB	LOOK UP TABLE CLASSIFICATION OF THE WHOLE IMAGE	CLASSIFIED IMAGE ON DISK
. XS OR XT IMAGE ON DISK . CLASSIFIED IMAGE ON DISK . OPTION : ADMINISTRATIVE MASK ON DISK	CLSTAT	SPECTRAL AND SPATIAL STATISTICS OF THE CLASSES : - MEAN SPECTRA - COVARIANCE MATRICES - POPULATION AND COVERED SURFACE - HISTOGRAMS (OPTION)	LIST OF SPECTRAL AND SPATIAL PARAMETERS OF THE CLASSES
. CLASSIFIED IMAGE ON DISK . GROUND TRUTH MASK ON DISK	CONFUS	COMPUTATION OF CONFUSION MATRICES ON TEST POLYGONS	CONFUSION MATRICES (ON LISTING)
. CLASSIFIED IMAGE ON DISK . LIST OF MEAN SPECTRA AND COVARIANCE MATRICES OF THE CLASSES ON DISK	REGULA	CONDITIONAL SMOOTHING OF THE CLASSIFICATION	. SMOOTHED CLASSIFIED IMAGE ON DISK

2.3.7 Interactive classification functions

INPUT	NAME	FUNCTION	OUTPUT
. TWO CHANNELS IMAGE ON DISK AND ON DISPLAY (WITH KROSS-COLOR TABLE) . GROUND TRUTH MASK ON DISK AND ON DISPLAY . CODING BOUNDARIES OF THE CHANNELS	CONST	ELABORATION OF THE CLASSIFICATION TABLE AND VISUAL ANALYSIS OF THE TRAINING POLYGONS	. NORMALIZED CROSS HISTOGRAMS OF THE CATEGORIES (ON DISK)
. TWO CHANNELS IMAGE IN TRIADE MEMORIES . NORMALIZED CROSS HISTOGRAMS OF THE CATEGORIES	BAYES	SIMULATION OF A BAYES CLASSIFICATION, BY CHANGING THE COLOUR TABLES TO MATCH THE CLASSIFICATION TABLE OBTAINED FROM NORMALIZED CROSS HISTOGRAMS	. COLOUR TABLES IN TRIADE 80 . CLASSIFIED IMAGE ON DISPLAY (PSEUDO COLOUR PRESENTATION) . OPTION : CLASSIFIED IMAGE ON DISK (SUBSCENE FORMAT)

2.3.8 Service functions

INPUT	NAME	FUNCTION	OUTPUT
	LOGIN	BEGINNING OF A USER SESSION	
	LOGOUT	END OF A USER SESSION	
	SAVE	SAFE KEEPING OF THE WORK CONTEXT OF A USER	
	RESTOR	RESTORATION OF THE WORK CONTEXT OF A USER	
SUBSCENE OF DISK	HISTO	COMPUTATION OF THE HISTOGRAM OF A SUBSCENE	SUBSCENE ON DISK
	AREA	SPECIFICATION OF THE WORK ZONE AND OF THE WORK CHANNELS IN THE SUBSCENE	
SUBSCENE ON DISK	SSCHAN	EXTRACTION OF N CHANNELS FROM A SUBSCENE	SUBSCENE ON DISK
SUBSCENE ON DISK	SSCZON	EXTRACTION OF A ZONE FROM A SUBSCENE	SUBSCENE ON DISK
SUBSCENE ON DISK	SSCSEG	SEGMENTATION OR DESEGMENTA-TION OF A SUBSCENE	SUBSCENE ON DISK
TWO SUBSCENES ON DISK	SSCMRG	SPECTRAL MERGING OF SUBSCENES (SAME DIMENSIONS)	ONE SUBSCENE ON DISK
TWO SUBSCENES ON DISK	SSCSTI	SPATIAL STICKING OF SUBSCENES (SAME CHANNELS) : - EITHER IN DIRECTION OF LINES - EITHER IN DIRECTION OF COLUMNS	ONE SUBSCENE ON DISK

INPUT	NAME	FUNCTION	OUTPUT
SCENES DIRECTORY (ON DISK)	SCLIS	LIST OF THE SCENES ARCHIVED ON TAPES	ALPHANUMERIC DISPLAY OR LINE PRINTER
SUBSCENES DIRECTORY (ON DISK)	SSCLIS	LIST OF THE SUBSCENES STORED ON DISKS OR TAPES	ALPHANUMERIC DISPLAY OR LINE PRINTER
PARAMETER LISTS DIRECTORY (ON DISK)	PARLIS	LIST OF THE PARAMETER LISTS STORED ON DISKS	ALPHANUMERIC DISPLAY OR LINE PRINTER
SCENES DIRECTORY (ON DISK)	SCCAT	CATALOGUING A SCENE TAPE RECEIVED FROM OUTSIDE	SCENES DIRECTORY (ON DISK)
. SUBSCENES DIRECTORY . EVENTUALLY THE SUBSCENE ON DISK	SSCCRE SSCCLO SSCOPE SSCDEL SSCREN	TO CREATE/CLOSE/OPEN/DELETE/ RENAME A SUBSCENE	. SUBSCENES DIRECTORY . EVENTUALLY THE SUBSCENE ON DISK
SUBSCENE ON DISK	SSCDUM	DUMP OF THE HEADER FILE OF A SUBSCENE	ALPHANUMERIC DISPLAY OR LINE PRINTER
. PARAMETER LISTS DIRECTORY . EVENTUALLY, THE LIST ON DISK	PARCRE PARCLO PAROPE PARDEL PARREN	TO CREATE/CLOSE/OPEN/DELETE RENAME A LIST OF PARAMETERS	. PARAMETER LISTS DIRECTORY . EVENTUALLY, THE LIST ON DISK
LIST OF PARAMETERS ON DISK	PARDUM	DUMP OF A LIST OF PARAMETERS	ALPHANUMERIC DISPLAY OR LINE PRINTER

3. VIPS ARCHITECTURE

The architecture of VIZIR Processing System shown in the following block diagram is mainly composed of a minicomputer and several specific peripherals. The proposed system is in fact very modular so that it is possible to upgrade it as a function of the capacity to store the informations (mass memory), as a function of the computation capacity and as a function of interactive capacity.

3.1. Minicomputer

The selection of a minicomputer depends on the used criteria and the weight put on these criteria.

For this VIZIR Processing System the main criteria that have been considered are :

- input/output capability : DMA input/output data flow speed and interrupt system efficiency,
- mass memory capability : possibility of connecting large and high speed mass memory like 300 Mbytes disks, 6250 BPI 125 IPS CCT drive and digital videodisc,
- computation capability : transfer and elementary operation instruction

efficiency, wired or firmware double accuracy floating operations, cache memory etc...
- software operating system efficiency and specially the real time capability (interruption system efficiency) and large mass memory supporting capability,
- origin of the product : exportations and importations of minicomputers are very much controlled by governments and it is very important to take this criteriums into account for international trade market.

For VIPS 16/65 SEP has selected SOLAR 16/65 or 16/75 minicomputers. It is a very powerfull 16 bits double bus minicomputer with the following main features :

- CPU with possibility of FAO (Floating point, double accuracy wired operator) and cache memory,
- 64 Kwords high speed sc memory (420 ns),
- mean speed sc memory (700 ns),
- high speed memory bus (16 Mbytes/s),
- I/O high speed processor : 900 Kwords/s,
- I/O mean speed processor : 600 Kwords/s via I/O specific bus,
- 10 to 300 Mbytes dual access disc,
- 1600 or 6250/1600 BPI, 125 IPS CCT drive,
- multi-task and multi-user operating system (MPES) with input/output monitor (IOCS) and File manager (FMS).

An other VIPS using a 32 bits computer is actually on study at SEP an will be proposed very soon as a new product in VIPS Range.

3.2. Mass Memory

The mass memory proposed by SEP in the base configuration of VIPS 16/65 is composed of :

- 3 discs of 20 Mbytes on the same controller. One of them (DM1) is used as software system support and the two others (DM2 and DM3) as data storage memory,
- 1 or 2 1600 BPI, 75 IPS CCT drive.

3.3. Operator system interface

The operator system interface is composed of :

- one system console for the operator dialog with the software system. It is a VDT 10 comprising a video display and an alphanumerical type written,
- one user console for operator dialog with the whole application software except the software related to TRIADE and digitizer devices. It enables the operator to select the wanted function and to give it the good parameters. It is also a VDT 10 comprising a video display and an alphanumerical type written.
- one user console for operator dialog with the whole application software except the software related to VIZIR and SCANDIG devices. It consists in a alphanumerical display and a keyboard connected both to the host computer and the local processor of the TRIADE,
- one line printer for making hard copy of dialog and programs.

3.4. Image digitizer

The image digitizer proposed in VIPS 16/65 system is a device that enables to analyse an image on a photographic film and to transform it into a digitized image that can be acquired and stored by the computer.

The photographic film is fixed on a rotating drum. A photometric head gives an electrical signal as a function of the transparency of an area limited by an adjustable aperture. The scanning along one axis is obtained by the rotation of the drum and along the other axis by the step by step motion of the head.

The specific features of this digitizer are the following :

- aperture scanning diameter : 25 - 50 - 100 - 200 microns,
- maximum rotation speed : 12 rotation/sec,
- maximum image format : 350 x 400 mm,
- digitizing accuracy : 8 bits,
- optical density range : 0 to 3 D,
- transmission or reflexion analysis mode (photographic film or paper),
- black and white or colour analysis mode,
- automatic computer control excepted sample mounting and unmounting.

3.5. Electrostatic printer

The electrostatic printer is used in VIPS 16/65 system for printing 3 types of informations :

- graphics and alphanumeric data for giving the result of computation (histogram for instance),
- alphanumeric data,
- image data for giving a quick look hardcopy of the image before or after processing.

The main features of this printer are the following :

- printer mode : 1 000 lines/minut,
  132 characters/line,
- plotter mode : . resolution : 200 nibs/inch,
  . plot speed : 2 inches/s,
  . paper width : 11 inches,
- restituter mode : 512 pixels/line.

3.6. VIZIR laser beam recorder

This laser beam recorder is used to produce image on photographic film. It makes the transformation of the processed digital image into a photograph with a very good accuracy in radiometry and geometry.

The photographic film is fixed on a drum rotating on magnetic bearings that give very regular and very accurate scanning.

The film is printed by a circular spot obtained by a modulated laser beam through an optical system. The rotation of the drum gives the line scanning and the step by step moving head the second axis scanning.

This equipment enables, with colour option consisting mainly in adjonction of two more independant modulation chains (blue and green), to produce a colour image on photographic film by printing simultaneously the three spectral bands (blue, green and red in one pass).

This device produces images of high quality in geometry and radiometry in black and white and specially in colour where the superimposition of the three spectral bands is perfect compared to the other methods.

The main features are listed in the following table :

VIZIR PERFORMANCES

CARACTERISTIQUES CARACTERISTICS	NOIR ET BLANC BLACK - WHITE	COULEUR COLOR
Format image utile Size of images	405 x 405 mm2 510 x 510 mm2 (4 images of 240x240 mm2) 510 x 750 mm2 (6 images of 240x240 mm2)	
Support image	Film : 2400 SO115	Film : EKTACHROME 200 ou : KODAK 2483
Laser	Hélium Néon 6328A R	Hélium Néon 6328A R Argon {Blue {Green
Modulateur avec boucle d'asservissement photométrique Servo-Looped Modulator	Type Electro-optique ou Acousto-optique	Type Acousto-optique Acoustico-optical
Machine de développement film automatique associée Automatic film processor	VERSAMAT	COLENTA KREONITE
Grandissement sens ligne Enlargement along the line	1 à 15 1 to 15	1 à 15 1 to 15
Correction du Gamma Gamma correction	EPROM	EPROM
Adaptation de la caractéristique (filtre, isodensités, contraste) Contrast enhancement	RAM (4 pages)	RAM
Nombre de spots différents Number of spot diameter	3	2
Positif négatif Positif or negatif	Oui Yes	Oui Yes
Information binaire codée sur Numerical data	8 bits	8 bits
Interface mini-calculateur Minicomputer interface	Oui Yes	Oui Yes
Nombre de points/ligne maximum Number points/lignes	32 400 Option 38 400 Option 60 740	16 200 Option 19 200 Option 30 370

CARACTERISTIQUES CARACTERISTICS	NOIR ET BLANC BLACK - WHITE	COULEUR COLOR
Nombre de lignes/film Maximum Number of lines/film	32 400 Option 38 400	16 200 Option 19 200
Vitesse du tambour Speed of the drum	3-20 tr/s round/sec	3-20 tr/s round/sec
Vitesse d'inscription en lignes/mn Writting speed l/mn	180 à 1 200 l/mn 360 à 2 400 l/mn	180 à 1 200 l/mn 360 à 2 400 l/mn
Durée d'inscription de l'image SPOT de 40,5 microns Writting duration of 1 image with spot 40 m	8'	8'
Temps de développement avec séchage Processing duration	15 mn	50 mn
Vitesse de transfert MAX Flowrate kilo pixel/second	Electro-optique 450 Kp/s Acousto-optique 200 Kp/s	Acousto-optique Acousto-optique 200 Kp/s s/chaque voie
Diamètre de spot possibles Spot diameters	*27 m-54 m-81 m	27 m - 54 m
Dynamique des Ø de Spot Ratios Ø Max/Ø Min	3	2
Nombre de niveaux de densité Number of grey levels	64	Sur chaque voie $(64)^3$ color shades $2.6\ 10^5$
Valeur de la non linéarité en densité sur film Density non-linearity	D = $\pm$ 0,03	D = $\pm$ 0,03
Niveau de densité optique Range of density	0,2 to 2,2	Mini 0,3 Max 2,8
Précision géométrique Geometrical accuracy on (240 x 240 mm)	$8\ 10^{-5}$ (0.008 %)	idem
Valeur du Jitter	+ 5 microns	idem

3.7. Coordinate digitizer

This peripheral is used in VIPS 16/65 for entering coordinates of specific points from a map or for digitizing maps.

The main features of this device are :

- format : 84 cm x 120 cm,
- digitizing accuracy : $\pm$ 0,1 mm,
- resolution : $\pm$ 0,02 mm.

3.8. Triade 80 colour display

Triade 80 colour display is a new interactive console with a very flexible architecture. It can be considered as an intelligent peripheral capable of working in "stand alone".

It is composed of 512 x 512 x 10 bits memories (up to 12) for image data and auxilary corresponding data, graphical memories (512 x 512 x 4 bit) and real time video processors that enable to make transfer and processing of data for visualisation on a colour monitor.

The data are processed in real time under control of the operator in interactive mode or under control of computer.

Data managing in manual and computer controlled modes is controlled by a microprocessor.

It is an interactive image visualisation equipment controlled by computer but with a large range of internal processing capabilities that makes it very flexible to use with computer and very powerfull.

4. SOFTWARE DESCRIPTION

The following figure describes the software structure of VIPS 16/65.

The software consists of five main layers.

4.1. Supervisor software

The highest level layer constitutes the interface between the user and VIPS : this interface is assumed by the SILEX language which allows the user to have access to all the system ressources through :

- commands, by means of which the user may manage his own objects (subscenes, parameter lists, ...),
- procedures, by means of which he may program a sequence of commands. The programming language of the procedures is SILEX.

With SILEX, the user may work in two different modes :

- execution mode : the user asks for the execution of commands or procedures,
- edit mode : the user develops his own procedures.

SILEX is present on all the systems of the VIPS series.

Its main characteristics are :

- procedural language derived from PASCAL language with simple syntax,
- block structured language,
- simple or structured data (arrays or records) manipulated either directly or with the help of pointers,
- manipulation of character strings,
- extended set of control statements : IF THEN ELSE, SELECT, WHILE, REPEAT, FOR, EXIT,
- set of statements for syntactical analysis allowing SILEX to be a metacompilation language,
- call for primitives written in another language (FORTRAN, Assembler...),
- transmission of procedure parameters by position or key-word and management of implicit parameters.

TEST SOFTWARE						
SUPERVISOR SOFTWARE	COMMAND LANGUAGE : SILEX					
APPLICATION SOFTWARE	INPUT/ OUTPUT	PREPROCESSING	SEGMENTATION	PHOTO-INTERPRETA.	CLASSIFICATION	SERVICE
	LOAMSS LOASIM LOAVPS LOASSC SAVIPS SAVSSC SCDDSK SCDTAP DSKVIZ PREVIZ TAPVIZ PROVIZ OFFVIZ QUICTP GREYPR SYMBPR DIGITE	BLKMSS CALMSS MAPGCP IMARCP RECTIF	THRESH CLASEG POLYTR POLYNU TEXTUR	ADD MULT RATIO ATARAT NORDIF LCOMBI KLOEVE CORANA DISANA STDRD FILTER AVRAGE MEDIAN SHARP EDGES LOCOPT ZOOM SAMPLE — BAW STRET PSCOL CTBUIL FACOL KROCOL KROMAN COM80	PARAM SEPAR LIKHOD DATCOD CODANA HAC AGREG LUPTAB CLSTAT CONFUS REGULA CONST BAYES	LOGIN LOGOUT SAVE RESTOR HISTO AREA SSCHAN SSCZON SSCSEG SSCMRG SSCSTI SCLIS SSCLIS PARLIS SCCAT SSCCRE SSCCLO SSCOPE SSCDEL SSCREN SSCDUM PARCRE PARCLO PAROPE PARDEL PARREN PARDUM
	COMMON MODULES					
I/O UTILITY SOFTWARE	DISKS	TRIADE 80	IMAGE TAPES	ELECTROSTATIC PRINTER	VIZIR	SCANDIG
	IMAGE OBJECT MANAGEMENT MONITOR : MGO	SUBROUTIRES LIBRARY MGT 80	READ/WRITE UTILITIES	GREY LEVELS OR SYMBOLS RESTITUTION UTILITIES	GREY LEVELS OR SYMBOLS RESTITUTION UTILITIES	DIGITIZATION UTILITIES
BASE SYSTEM SOFTWARE	FILE MANAGER SYSTEM	PERIPHERAL DRIVERS				
	OPERATING SYSTEM					
HARDWARE						

SILEX is a interpretated language, the compilation and execution unit of which is the procedure ; the interpretor is re-entrant resulting in a multi-user system.

The command syntax is the following :

```
FUNCTION FROM < input objects > TO < output objects >
PARAM 1 =
PARAM 2 =

PARAM n =
```

Example :

```
BAW  FROM  SSCMSS  TO  ARRAY 1
BAND = 4
ZONE = 1, 512, 1, 1001, 1512, 1.
```

4.2. Application software

The size of this software depends on the importance of the system.

The processing programs are written in FORTRAN language and constitute the primitives called by the SILEX procedures.

Primitives contain neither dialogue sequence nor OPEN/CLOSE commands of the image objects.

4.3. I/O utility software

This software is the interface between the application software and the peripheral drivers and easies the task of I/O programming for the primitives.

MGO is a monitor type routine written in assembler ; it is re-entrant and is implemented on all the systems of the VIP series.

The remaining software consists of subroutine libraries generally written in FORTRAN.

4.4. Base system software

This is the software sold by the computer manufacturer (SEMS for SOLAR minicomputers).

4.5. Test software

This software allows all levels of the system to be tested from the hardware level to the supervisor level.

5. SPECIAL FEATURES OF VIPS 16/65

Among the various features of VIPS 16/65 system we can select four particularties that makes it very interesting for the user.

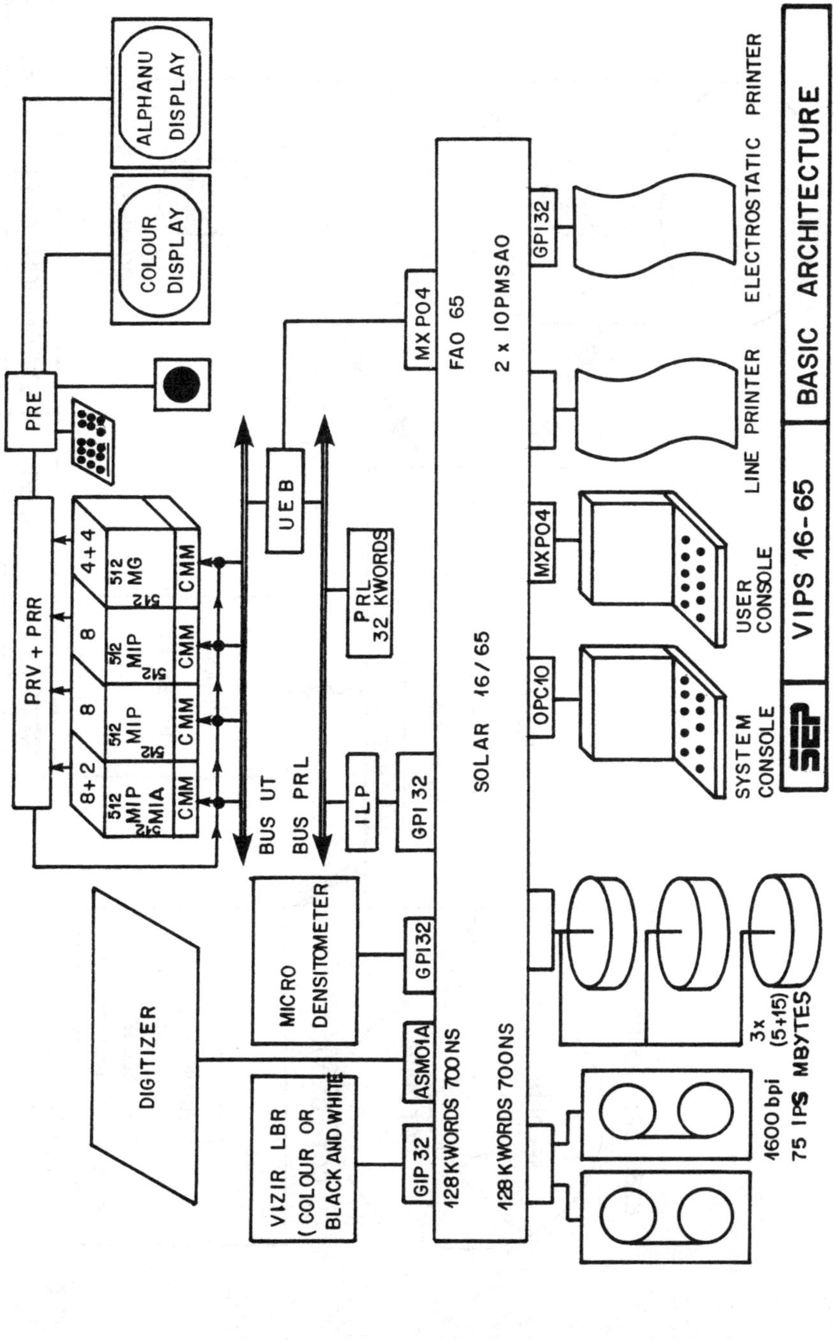

ALPHANU DISPLAY
COLOUR DISPLAY
PRE
PRV + PRR
8+2
512 MIP MIA
8
512 MIP
8
512 MIP
4+4
512 MG
CMM
CMM
CMM
CMM
UEB
PRL 32 KWORDS
BUS UT
BUS PRL
ILP
GPI 32
MX PO4
FAO 65
2 x 10PMSAO
SOLAR 16/65
128 KWORDS 700 NS
128 KWORDS 700 NS
GPI32
MXPO4
OPC10
DIGITIZER
MICRO DENSITOMETER
GPI32
ASMO1A
VIZIR LBR (COLOUR OR BLACK AND WHITE
GIP 32
ELECTROSTATIC PRINTER
LINE PRINTER
USER CONSOLE
SYSTEM CONSOLE
3x (5+15) MBYTES
1600 bpi 75 IPS
VIPS 16-65
BASIC ARCHITECTURE

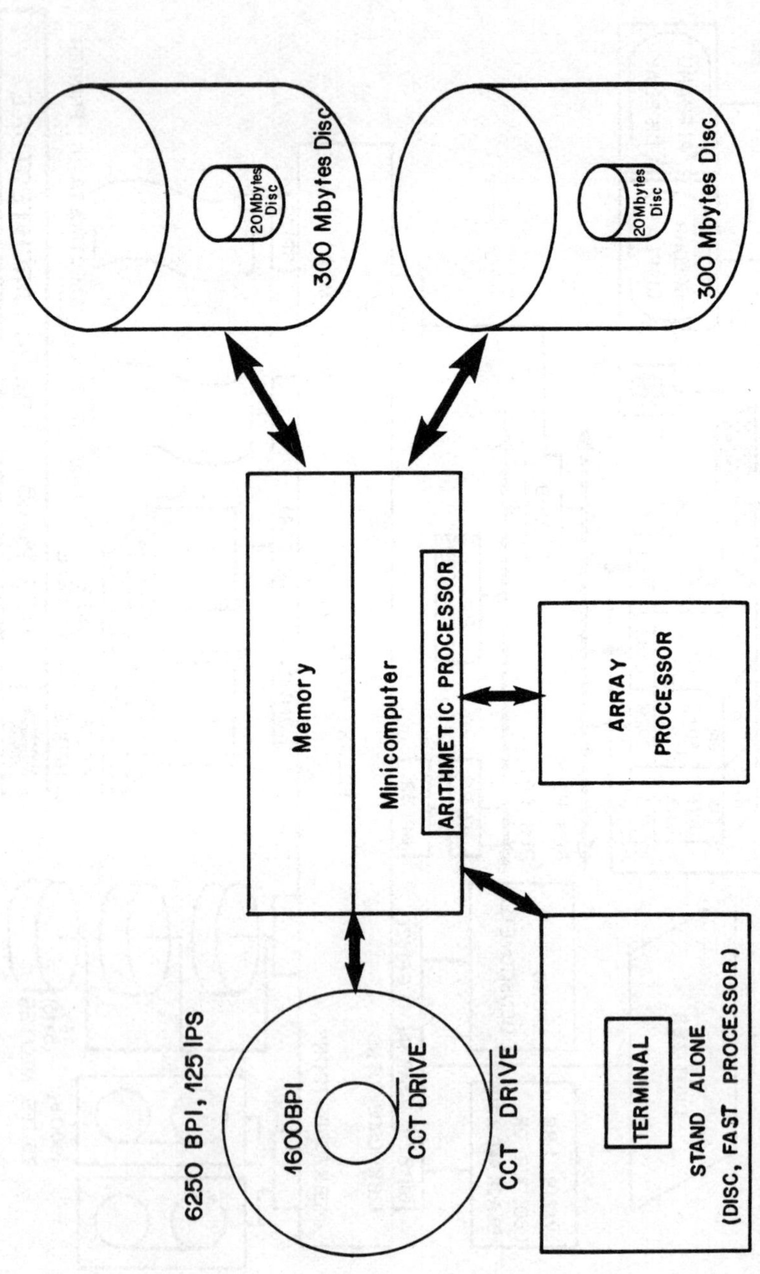

VIPS 16/65 EXTENSION CAPABILITIES

5.1. Very interactive, accurate and easy dialog language

This operator dialog gives to the system very good qualities of operationality and makes it very easy to upgrade.

5.2. Large possibility of extension

The VIPS 16/65 system can be in upgraded in three main ways :

* mass memory can be widely inlarged by connecting one or several 300 Mbytes discs and 6250 BPI 125 IPS CCT drives,
* capability of computation by adding to the system a very powerful array processor. We propose AP120B(FPS)processor that is very reliable and has a very efficient operating system and a very large mathematical library,
* the stand alone feature of the TRIADE console can be improved by adding local mass memory (up to 50 Mbytes) and fast processors.

5.3. Very high level peripherals

VIPS system is using for imagery applications very new and very accurate specific peripherals (TRIADE 80 interactive console, color Vizir, etc...).

5.4. Range of products.

SEP will propose very soon a complete range of systems either on French Solar Computer (SEMS) or on US computers (VAX, SEL, ...), with possibility for the user to upgrade his configuration by adding peripherals or memory.

INSIGHT : AN IMAGE PROCESSING SOFTWARE PACKAGE

J Brumfitt and P Redstone
Space Systems and Image Processing Division
Logica Ltd
Cobham Park
Downside Road
Cobham
Surrey KT11 3LX

Abstract : There are many types of image processing hardware systems which are used in remote sensing, incorporating conventional and array processors, image displays etc. Some of these come with built-in software or firmware but this is inevitably specific to the particular hardware. This paper describes the requirements,design philosophy and implementation of INSIGHT, a portable image processing software package which has been developed by Logica. INSIGHT provides the framework within which image processes can be developed. It has the following features :

- a high level command language, recognising images as data types
- an optimised image data accessing technique
- a portable display interface
- multi-tasking architecture

An example of INSIGHT in a remote sensing application is also given.

1. INTRODUCTION

With the recent growth in the number of image processing applications there has been a corresponding increase in the range of image processing hardware offered by manufacturers, including displays, video digitizers, array processors etc. Some of these come as peripherals to general purpose computers and are generally supplied with simple driver software to control the device without supporting image processing itself. Some manufacturers supply complete turnkey systems including fairly comprehensive software packages but these are invariably non-portable and tailored to the capabilities of the particular hardware, perhaps without being extendable to provide new image processes.

Paper Presented at a conference on Matching remote sensing technologies and their applications convened in London by the Remote Sensing Society, 16-18 December 1981

This paper describes the design philosophy and implementation of INSIGHT, an image processing software package designed to run on a general purpose computer linked to special peripherals such as image displays or array processors. A simplified example of the use of INSIGHT in a remote sensing application is also given.

2. DESIGN REQUIREMENTS

The following design requirements are essential to any general purpose image processing system :

- user friendly interface(s), providing ease of use but allowing full use of the capabilities of the hardware
- large library of standard image processing functions, easily extended by users to include special fuctions
- high degree of portability between both different types of host computer and different special purpose peripherals
- interactive and batch modes using similar user interfaces

Any interactive system should be designed to be user friendly, particularly in image processing applications where users without computer experience may wish to use the system. Examples of such applications are medical image enhancement and remote sensing image interpretation. As well as this user interface there is also a need for a lower level interface suitable for the development of special processes, perhaps using special purpose hardware. In practice this requires a programming environment very similar to the general purpose program development environment provided on the host computer but supported by libraries containing utility functions to simplify the generation of new processes.

The system should possess a large library of commonly used image processes. No matter how comprehensive the standard library there will always be new functions to implement, especially where special peripherals are involved. Therefore it is important that new processes written by the user can be added to the system without, for example, rebuilding the whole system.

The system should be designed to be portable so that users changing or upgrading from one hardware configuration to another can have the same user environment. Of course for a commercial system this also increases the number of possible installations.

On general purpose computers with typical size images some processing operations on complete images can take several hours. Ideally such processes will have been developed using small parts of the images for testing.

It is therefore essential that the system supports both interactive and batch processing with very similar user interfaces to simplify the running of processes developed in interactive mode in a batch environment.

3. INSIGHT ARCHITECTURE

Having listed the main requirements of an image processing system, the following paragraphs discuss the features of INSIGHT which allow it to meet these requirements.

INSIGHT was developed under RSX11M running on a PDP 11/34 and is mostly written in ANSI FORTRAN with small amounts of machine dependent FORTRAN and Assembler. However, it is easily tailored to run under other multitasking operating systems on many types of host computers with many types of peripherals. The description below uses RSX11M terms but applies to all similar operating systems.

The main architectural feature of INSIGHT is that it runs as a number of different tasks as shown in figure 1. Ignoring the two branches of the task structure, there are three types of task. The terminal task handles all communications via the terminal with the user. It also produces a log of all dialogue which is a useful record of the sequence of processes which have been carried out. It is also the master task which, when started, automatically starts up the rest of the system. When it is run it installs and runs the first control task.

The control task implements the command language user interface. It sends messages requesting a command line to the terminal task which sends back a command line which has been prompted from the user. The high level language is discussed below and is similar to BASIC but incorporates images as data types. The control task itself performs no image processing. Individual low level image processes (such as adding two images together or performing a linear filter operation) are implemented in separate tasks called primitives. When a command line is typed which requires an image process to be performed then the specified primitive is installed and run by the control task. Any parameters which are required by the primitive are passed to it from the control task. Parameters can also be returned. A large number of primitive tasks exist and new ones are easily added. They are developed in the host computer program development environment which is thus the second user interface. Primitives are therefore different to the control task since only one version exists of the latter. They can also communicate directly with the terminal via the terminal task, which is useful for interactive primitives and debugging purposes.

Figure 1 INSIGHT task structure

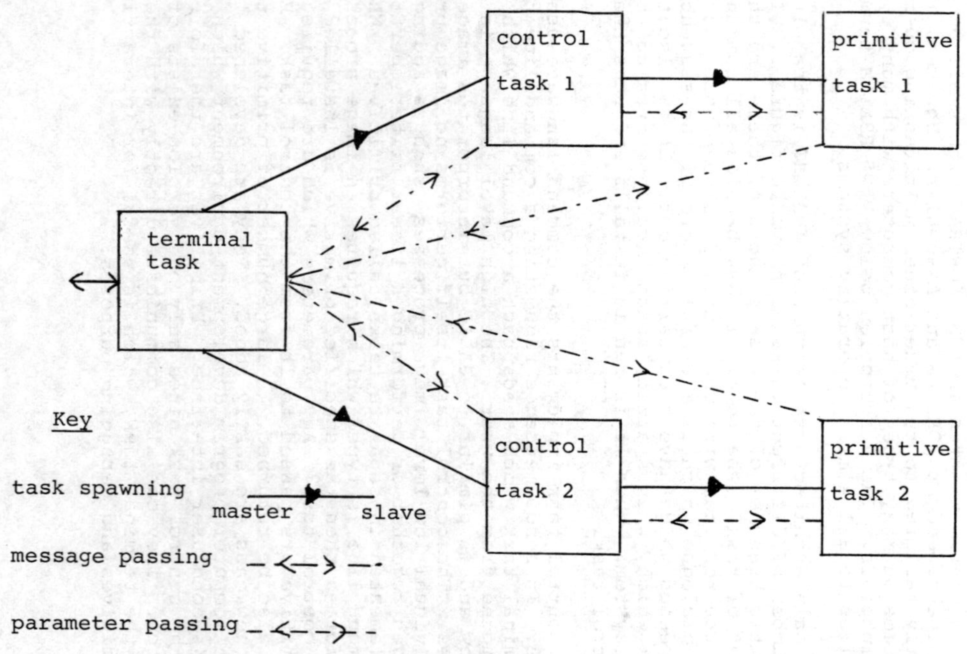

4. INSIGHT COMMAND LANGUAGE

This is similar to BASIC, but incorporates images as data types. In this short paper it is impossible to show all its features, but a short example is given below.

```
SET TOT = 0
FOR I = 1,9; ASK 'MASK VALUE', MASK(I); SET TOT = TOT+MASK(I)
TYPE 'MASK TOTAL IS', TOT
ASK 'NORMALISE TO ONE (YES OR NO)' NORM
IF(NORM);SET SCALE=FLOAT(1/TOT); GOTO LABEL
ASK 'SCALE FACTOR', SCALE
LABEL: COM THIS LINE DOES NOTHING
LOADPRIM LINFIL
LINFIL < INIM > OUTIM  ( < MASK,SCALE > OVERFLOW)
TYPE 'NUMBER OF OVERFLOWS', OVERFLOW
```

This sequence shows the execution of a 3 x 3 linear filter primitive (LINFIL) with all parameters being requested interactively. The SET keyword allows assignment to variables, FOR supports looping and TYPE and ASK perform terminal input and output. The IF statement allows conditional testing and branching using the GOTO keyword. The keyword LOADPRIM loads a primitive which is then executed by the next command line. This takes an input image with name INIM and performs the process LINFIL on it to give an output image OUTIM. The variable MASK contains the mask weights (it is an array), SCALE has the scale factor and the number of overflows is returned in the variable OVERFLOW.

Groups of statements can be contained within procedures which are called in exactly the same way as primitives. For example if a group of statements above were bracketed by the lines

```
PROC  PROCESS1 < INIM > OUTIM
   :
   :
RETURN
```

then the complete sequence could be invoked by the lines

```
LOADPROC     PROCESS1
PROCESS1    < IM1 >IM2
```

This also shows how image names can be passed between procedures : IM1, IM2 are the actual image names whereas INIM, OUTIM are dummy image names. Procedures can be nested which makes the command language very powerful. Procedures can be used to simulate new primitives and can also be used to schedule batch sequences of processes.

There is one other feature of the command language which is noteworthy. Image names can be concatenated with a window specification, for example :

DISPLAY <IM1 [100, 110, 40, 50 : 4, 5, 7]

The DISPLAY primitive writes an image to the display. The window specification in this case indicates that a region of 100 x 110 pixels starting at offset (40,50) in IM1 is to be displayed, taking spectral bands 4,5 and 7. Window specifications can, of course, be totally or partially omitted when sensible defaults are assumed. The concept of windowing is extremely powerful since it allows the development of new processes on small parts of an image, giving fast interactive response, prior to trying the process on a complete image.

5. PRIMITIVES

As well as programming the system using the command language, new primitives can be developed using the host computer FORTRAN development environment. This is simplified by the use of various utility libraries including :

- intertask communication
- parameter fetching/returning
- image data accessing
- error handling

The intertask communication routines support the running of the task under command from the control task, as well as dialogue with the user. The parameter utilities allow fetching and return of parameters which were given in the command line. The image data accessing routines implement highly efficient storage of images in disk files. Images can have bit precisions of 1,8,16 or 32 bits and have up to 16 spectral bands. The accessing routines implement the image windowing in a very simple manner such that the user is almost unaware that an image is windowed - all primitives can therefore process windowed images as well as complete images. The accessing routines also implement an extremely efficient input/output optimising algorithm which ensures a maximum processing speed. Finally the error handling routines support a comprehensive error reporting system which allows recovery from all errors.

If special peripherals are incorporated in the system then new utilities to handle these can be written. In many cases manufacturers supply such utilities to, for example, drive array processors or displays.

The most important point concerning primitives is that the user is free to do what he wants. Since primitives are small, separate tasks they are quickly generated and debugged. Most primitives contain about 100 lines of code but much of this is standard between different primitives and the actual process generally takes only 20 to 30 lines of code.

6. EXAMPLE

A simple example process for Landsat MSS images is given below.

```
FOR I = 4,7,; NORM < RAW[ :I] > NORMIM[ :I]
KARLOEVE<NORMIM [ 4,5,6,7] > COMPON [ :1,2]
```

This is a two stage process to first normalise the individual bands of a Landsat image and then find their principle components, taking the two most significant bands. The process as expressed in the high level language is very simple and would probably be expressed in a procedure to perform the complete sequence. It will be noted that image acquisition and display of the result is not shown, but this is performed using simple primitives.

7. SUMMARY

A powerful image processing software package has been described and a simple example of its use given. INSIGHT was developed for in-house use in Logica's Space Systems and Image Processing Division where it is being used in consultancy projects in space, civil, defence and industrial application areas. It is also being supplied in individually tailored forms to clients, either as a software package or as part of a complete turnkey system including both hardware and software.

8. ACKNOWLEDGEMENTS

The authors are grateful to Logica for supporting the development of INSIGHT as part of Logica's strong commitment to image processing. INSIGHT would not have been completed were it not for the valiant work by the project team including Vartkes Goetcherian, Jonathan Hughes, Neil McEvoy, Nigel Tucker and Gordon Vince.

METEOROLGOICAL SATELLITE DATA ANALYSIS SYSTEMS

E Seeley, MacDonald Dettwiler & Assoc, Richmond BC, Canada

MacDonald Dettwiler's Meteorological Satellite Data Analysis System allow simultaneous reception, analysis and output of satellite and conventional meteorological data. The systems receive data from polar-orbiting (TIROS-N/NOAA, AVHRR AND TIP) and geostationary (ie GOES VISSR, METEOSAT VISSR) satellites. The system contains two components: the satellite data reception (WIPS) subsystem and the meteorological data analysis (METDAS) subsystem.

The reception subsystem includes the antenna, bit synchronizer, several disks, image recorder and a MDA built mini computer with high - throughpu capabilities. Multiple satellite data can be received in parallel. The image database is stored on a dual ported disk, which provides rapid access to full scene satellite imagery by both the WIPS and METDAS subsystem.

The meteorological data analysis (METDAS) subsystem allows the meteorologist to interactively manipulate and merge data from satellite and conventional sources. The subsystem is built around a VAX, with several disks and the meteorological work station. The meteorological work station includes a CRT and a Video Display Processor. Here the meteorologist can display and analyse single and multiband imagery and animation sequences of several contigous satellite scences. Data links provide input of standard meteorological data, ie temperature, pressure, dew-point and wind vectors. These meteorological fields are displayed as contours or streamlines on top of satellite imagery. In addition, the meteorologist can interactively extract meteorological data from the satellite bands itself, ie cloud coverage and sea surface temperature.

The system is designed in a modular fashion in order to allow future extensions and provide the meteorologists with an integrated forecast tool.

AN ADAPTIVE ALGORITHM FOR ON-BOARD COMPRESSION OF REMOTELY SENSED DATA

N.A. McEvoy
Space Systems and Image Processing Division
Logica Ltd
Cobham Park
Downside Road
Cobham
Surrey KT11 3LX

Abstract: An image data compression algorithm designed for use with the Ocean Colour Monitor (OCM) to be flown on a European Space Agency Earth Resources Satellite is described. The algorithm is known as "Rice Machine 2" and was originally developed at the Jet Propulsion Laboratory, Pasadena, California. It consists of the application of cascaded Hadamard transformations to the image data, followed by entropy coding of the non-d.c Hadamard coefficients.

The algorithm was simulated in software and applied to Nimbus-7 Coastal Zone Colour Scanner (CZCS) images, principally of the Southern North Sea. The results of these extensive simulations were used to examine the trade-off between compression factor and reconstructed image quality.

1. INTRODUCTION

In a single rotation of the OCM mirror, 4 lines of 2120 pixels are scanned in 10 spectral bands. Since the bit precision for each pixel in all bands is 12, this represents approximately 1.0 megabits of image data for each scan. The period of rotation of the OCM mirror is 294 milliseconds, and therefore the OCM generates image data at an approximate rate of 3.6 megabits per second.

Paper presented at a conference on Matching remote sensing technologies and their applications convened in London by the Remote Sensing Society, 16-18 December, 1981

An Earth Resources Satellite flying an OCM will necessarily be out of reach of ground stations for a large proportion of the time in which it is collecting image data. It is therefore necessary to store this data using an on-board tape recorder. In this situation data compression is clearly desirable, since given a fixed size recorder and a compression factor of (say) four then four times greater ground coverage is possible. A data compression unit may also be of use in reducing the band width required for real time transmissions of image data.

2. RICE MACHINE 2

An adapted version of the Rice Machine 2 (RM2) algorithm, developed at the Jet Propulsion Laboratory (Rice, 1975), was simulated in software by Logica, in order to assess the feasibility of using the algorithm as part of an OCM data reduction unit. A brief description of the algorithm, together with a discussion of its properties is given below. A complete block diagram is presented in Figure 1.

The RM2 algorithm is applied to square blocks of 64 x 64 pixels. The first stage is to represent each 2 x 2 portion of the block in terms of coefficients of the four orthonormal Hadamard basis vectors :

$$H_0 = \frac{1}{4}\begin{pmatrix} 1 & 1 \\ 1 & 1 \end{pmatrix} \qquad H_1 = \frac{1}{4}\begin{pmatrix} 1 & -1 \\ 1 & -1 \end{pmatrix}$$

$$H_2 = \frac{1}{4}\begin{pmatrix} 1 & -1 \\ -1 & 1 \end{pmatrix} \qquad H_3 = \frac{1}{4}\begin{pmatrix} 1 & 1 \\ -1 & -1 \end{pmatrix}$$

As a result of applying this transformation, four 32 x 32 arrays of coefficients (C_0, C_1, C_2, C_3), corresponding to the four basis vectors, are generated. The first set of coefficients (c_0) represent the original image, but at reduced resolution. The vectors H_1, H_2, and H_3 are edge detectors, and for natural images the distributions of the C_1, C_2, and C_3 sets of coefficients are unimodal about zero. Such distributions are conductive to efficient variable length entropy coding, which is the means by which data compression is achieved using the RM2 algorithm.

Figure 1. RM2 Block Diagram

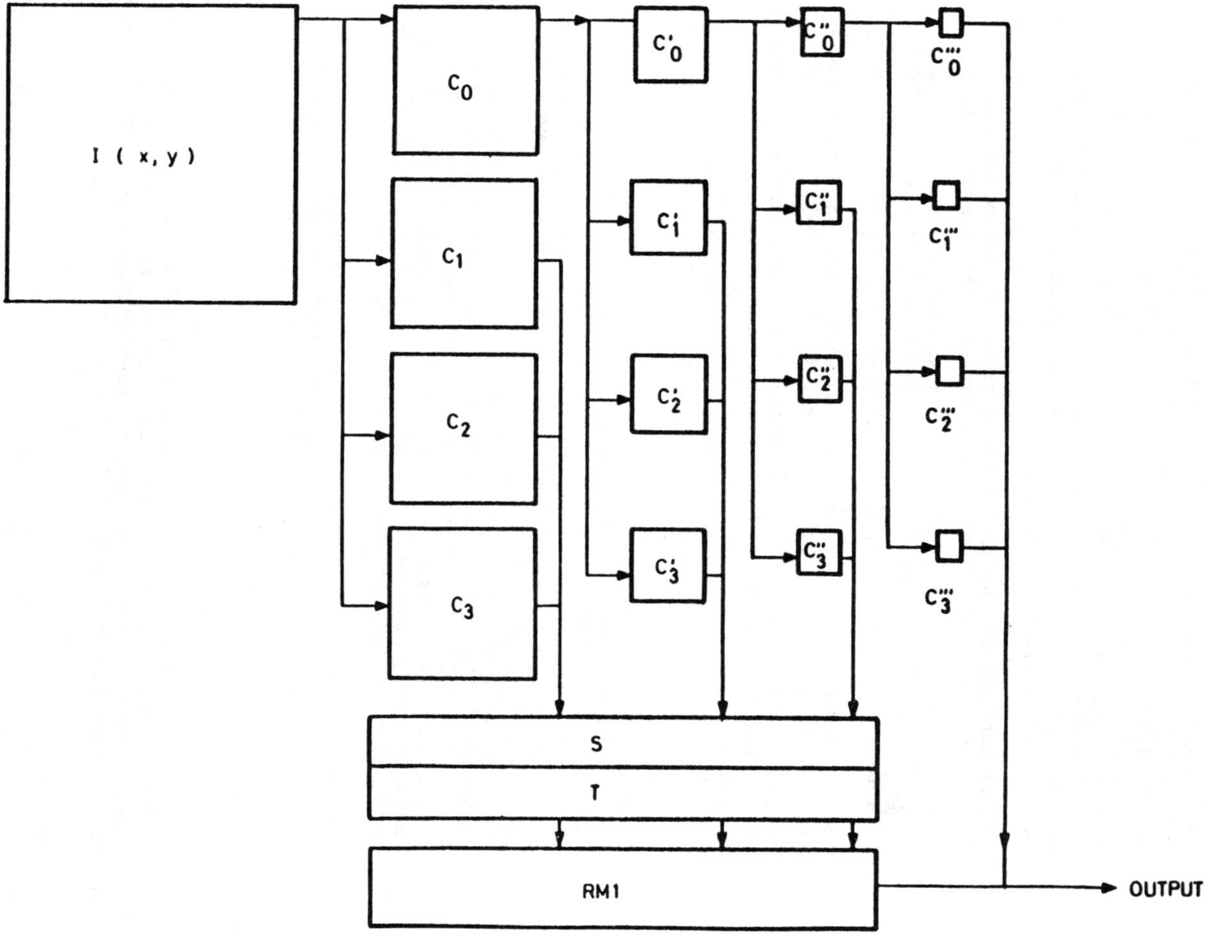

The method used to perform the entropy coding is based upon the Rice Machine 1 (RM1) algorithm (Rice and Plaunt, 1971),which was originally developed to compress DPCM data in a fully reversible manner. Its performance is such that the number of bits used to code any unimodal distribution is only slightly greater than the entropy of the distribution, which is the theoretical lower bound for any predictive coding technique (Shannon's theorem).

Having applied the Hadamard transformation and coded the C_1, C_2 and C_3 coefficient distributions, the process may be repeated by applying the Hadamard transform to the 32 x 32 array of C_0 coefficients. This process will yield four 16 x 16 arrays of Hadamard coefficients, C_0, C_1, C_2 and C_3. The C_0 coefficients represent the original image averaged over 4 x 4 pixel blocks, whilst the C_1, C_2 and C_3 distributions again contain information of edges within the image (but this time at a lower frequency) and are again unimodal about zero. They are therefore also coded using the RM1 algorithm. In all, the process of applying the Hadamard transformation and coding the resulting C_1, C_2 and C_3 distributions is applied four times.

As decsribed so far, the RM2 algorithm is completely reversible, that is to say the original 64 x 64 pixel image can be exactly reconstructed from the coded output. (Small errors due to arithmetic noise may occur if the Hadamard coefficients are constrained to be of a fixed bit precision) However, for typical CZCS images RM2 gives a compression factor of approximately two when operated in this mode, compared to a requirement for the OCM data reduction unit of approximately four. In order to achieve greater compression and to allow flexibility in choosing a particular compression factor, RM2 includes two approximation techniques which are applied to the coefficient distributions prior to RM1 coding.

The first approximation is to reduce the degree of linear quantisation of the distributions by applying the following shift and round operation to each coefficient :

$$c(s) = \frac{c}{|c|} \left\lceil 2^{-s} \; |c| + 0.5 \right\rceil$$

Where $\lceil n \rceil$ denotes the largest integer less than or equal to n. A large value for the shift parameter (s) implies a coarse quantisation of the given coefficient distribution. The process is illustrated in Figure 2 for s = 1.

Figure 2. Reduction in Quantisation

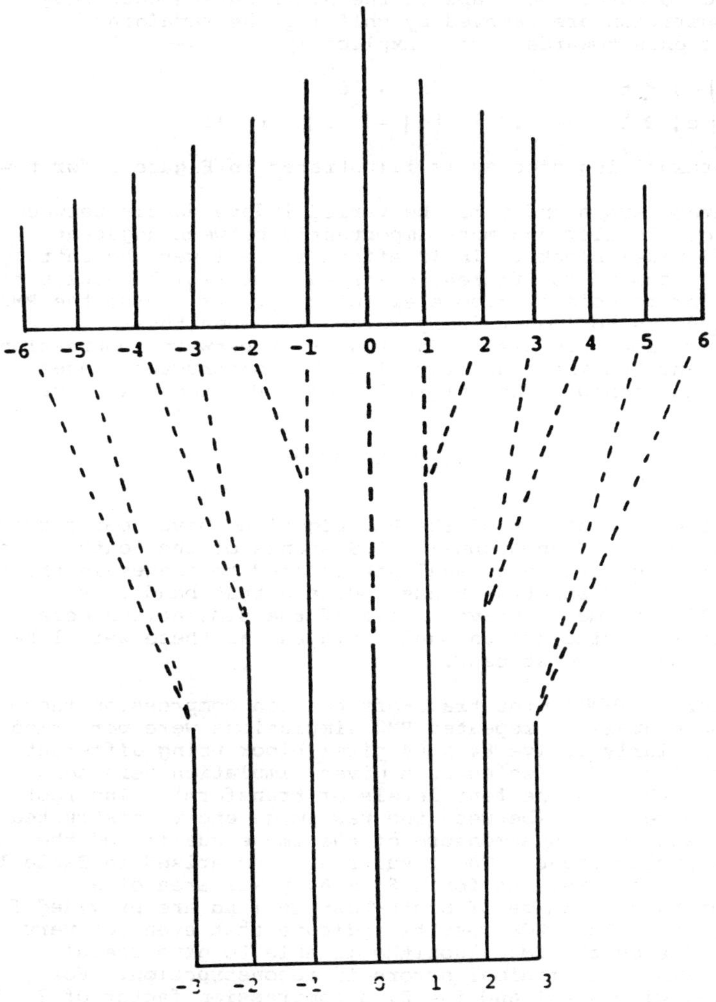

The second approximation is to threshold the distributions after the shift and round operation. All coefficients whose modulus is less than the threshold parameter (t) are equated to zero. The gaps in the distribution caused by this operation are removed by shifting the remaining coefficients towards zero. Explicitly,

For $|c| < t$, $c = 0$

For $|c| \geqslant t$, $|c| = |c| - t + 1$

The thresholding process is illustrated in Figure 3 for t = 2

The parameters s and t may be varied independently between the transform levels, and more importantly between adjacent 64 x 64 pixel blocks. Their effect is to lower the entropy of the Hadamard coefficient distributions, which enables the RM1 coder to code them more efficiently. Of course the RM2 algorithm is no longer fully reversible when these approximations are used. The trade-off between compression factor and the image quality of the reconstructed images (as a function of s and t) is considered in the next section.

3. SIMULATIONS OF RM2

Extensive simulations of the RM2 algorithm have been carried out, particularly on Nimbus-7 CZCS scenes of the Southern North Sea. Fourier frequency analyses applied to images in the six different CZCS spectral bands indicate that band 1 is generally the most active. Most of the simulations were therefore carried out on band 1 images, as these were likely to represent a worst case.

In order to assess the trade-off between compression factor and image quality, repeated RM2 simulations were performed on a particularly active 64 x 64 pixel block using different values of s and t (which in a given simulation were held constant through the four levels of transform). The root mean square error (between the raw image and reconstructed image) was used as a measure of the image quality of the reconstructed image. The results are summarised in Table 1. The equivalent results for a 64 x 64 pixel area of a Landsat band 4 image of South-East England are provided for comparison. The CZCS results indicate that even for very active scenes the RM2 algorithm is able to give useful compression with minimal errors in reconstruction. For example, with s = 1 and t = 2, a compression factor of 3.27 is achieved with a root mean square error of less than one on a 256 - level grey scale.

Figure 3. Thresholding

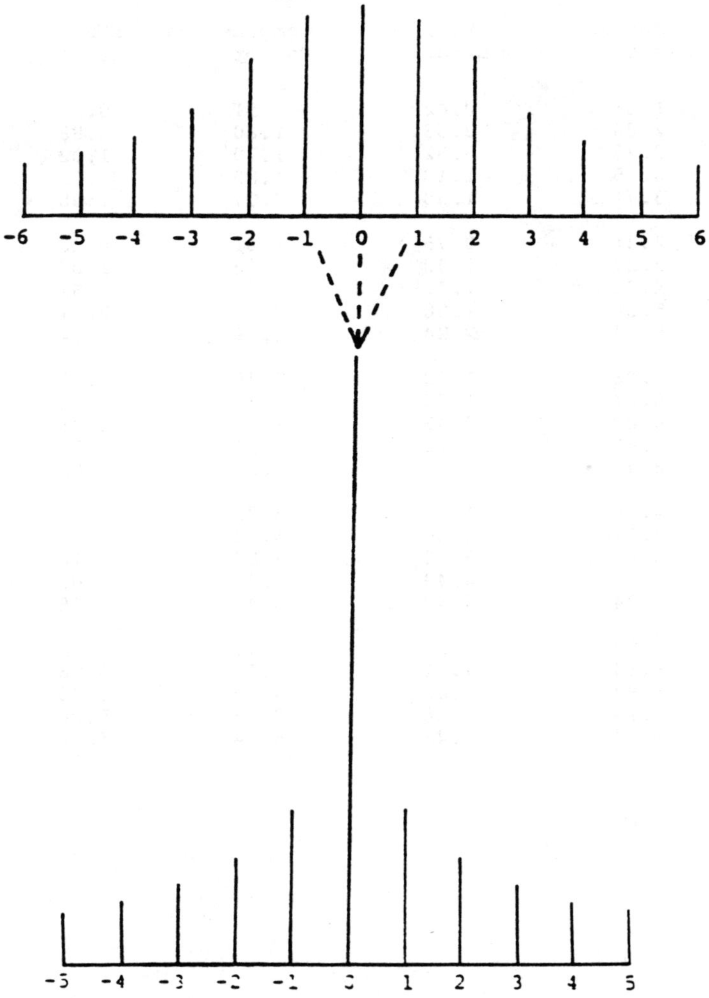

Table 1. Compression Factor/RMS Error Trade-off

		Nimbus-7 CZCS		Landsat	
s	t	Compression Factor	RMS Error	Compression Factor	RMS Error
0	1	1.86	0.42	1.51	0.83
0	2	2.26	0.58	1.60	0.88
0	3	2.75	0.82	1.70	1.02
0	4	3.25	1.13	1.85	1.21
0	5	3.71	1.38	2.07	1.58
1	1	2.21	0.75	1.81	0.88
1	2	3.27	0.92	2.08	1.07
1	3	4.33	1.41	2.52	1.61
1	4	5.24	1.88	3.19	2.41
1	5	6.04	2.24	3.79	2.94
2	1	3.08	1.02	2.40	1.15
2	2	5.41	1.72	3.46	2.15
2	3	6.89	2.45	4.76	3.26
2	4	7.86	3.02	6.11	4.01
2	5	8.41	3.47	6.92	4.62
3	1	4.76	1.66	3.72	1.88
3	2	7.77	2.80	6.38	3.79
3	3	8.66	3.65	7.86	4.76
3	4	8.90	4.49	8.26	5.05
3	5	8.94	4.49	8.45	5.28
4	1	6.87	2.56	4.92	3.29
4	2	8.84	4.01	8.58	4.96
4	3	8.94	4.48	8.89	5.33
4	4	8.94	4.48	8.93	5.38
4	5	8.94	4.48	8.94	5.40

The RM2 algorithm was applied to a complete 512 x 512 band 1 CZCS scene of the North Sea. Values of s = 1, t = 2 were used for all the 64 blocks within this area. Overall, a compression factor of 5.09 was achieved with a root mean square reconstruction error of 0.87. An image was formed of the difference between the raw and reconstructed image, and was contrast expanded. This image showed that such errors which did occur in the compression/decompression process were distributed randomly over the image, rather than occurring systematically. The detailed results of this simulation are given in Table 2, which shows the number of bits required to code each 64 x 64 pixel block, together with the associated compression factor and root mean square reconstruction error. From these figures it is clear that the code lengths vary considerably between inactive and active blocks in the image (from 3913 to 15984 bits).

4. FURTHER WORK

The variability of the output data rate noted above is a disadvantage of the RM2 algorithm in that on-board tape recorders and telemetry systems require fixed data rates. Therefore a mechanism has to be found by which the RM2 output data rate can be smoothed. Such a mechanism will calculate new values of s and t to give a desired output rate, based upon the performance of the coder over previous blocks. Extensive simulations using large swaths of CZCS images will be required to define a procedure which will reliably perform this function.

In the simulations performed to date, statistical measures of image quality, together with the visual properties of reconstrcuted images, have been used to assess the acceptability of RM2 compressed imagery. Whilst the use of these criteria provides encouraging results, it is obviously desirable to apply automatic user interpretation algorithms to RM2 compressed imagery before a final decision on the acceptability of RM2 data compression is made.

It is hoped that work to resolve both these outstanding problems will commence shortly at Logica.

5. SUMMARY

The studies of the RM2 algorithm carried out at Logica have demonstrated its feasibility for implementation in an OCM data reduction unit. Further work will enable the specification of a complete data reduction unit based on RM2 to be integrated within the Earth Resources Satellite data handling system.

Table 2. Variability of Output Data Rate

BITS	4117	3951	4080	4188	4240	4148	3961	4245
COMP	7.959	8.295	8.031	7.824	7.728	7.900	8.273	7.719
RMS	0.765	0.768	0.841	0.839	0.830	0.839	0.794	0.793
BITS	3913	4161	4108	4250	4165	4076	5907	6202
COMP	8.374	8.875	7.977	7.710	7.867	8.039	5.547	5.283
RMS	0.735	0.803	0.809	0.801	0.798	0.750	0.777	0.825
BITS	4056	3931	4060	4063	4143	4205	5117	7788
COMP	8.709	8.336	8.071	8.065	7.909	7.793	6.404	4.207
RMS	0.788	0.832	0.831	0.829	0.802	0.833	0.924	0.869
BITS	3984	5184	4389	4231	4109	5467	12150	15984
COMP	8.225	6.321	7.466	7.745	7.975	5.994	2.697	2.050
RMS	0.781	0.865	0.870	0.848	0.819	0.823	0.910	0.870
BITS	4922	5231	4993	4453	4231	8715	10563	11584
COMP	6.657	6.264	6.563	7.359	7.745	3.760	3.102	2.829
RMS	0.891	0.927	0.923	0.829	0.882	0.919	0.917	0.916
BITS	4673	5919	5550	5561	5672	6150	10177	13228
COMP	7.012	5.536	5,904	5.892	5.777	5.328	3.220	2.477
RMS	0.895	0.925	0.953	0.852	0.854	0.894	0.918	0.924
BITS	6438	5977	5810	6295	9414	9781	8815	7903
COMP	5.090	5.482	5.640	5.205	3.481	3.350	3.717	4.146
RMS	0.890	0.927	0.954	0.918	0.907	0.886	0.941	0.923
BITS	6425	9719	8795	7575	10101	9006	13258	12877
COMP	5.100	3.372	3.726	4.326	3.244	3.638	2.472	2.545
RMS	0.946	0.866	0.928	0.926	0.926	0.883	0.901	0.867

6. ACKNOWLEDGEMENTS

The work described in this paper was carried out for British Aerospace under European Space Agency contract 4415/80/NL/PP(SC).

The author is grateful to Paul Redstone of Logica for his help and encouragement in the study of data compression algorithms.

7. REFERENCES

RICE, R.F., (1975) : An Advanced Imaging Communication System for Planetary Exploration SPIE Seminar Proceedings on Efficient Transmission of Pictorial Information,. August 21-22, 1975

RICE, R.F, and PLAUNT, J.R., (1971) : Adaptive Variable-length Coding for Efficient Compression of Spacecraft Television Data, IEEE Trans, Comm. Tech. Vol COM-19, No 6, December 1971

SPATIAL RESOLUTION OF REMOTELY SENSED DATA AND ITS EFFECT ON CLASSIFICATION ACCURACY

A R Harrison, Department of Geography, University of Reading

The efficiency of classification algorithms applied to remotely sensed imagery is largely dependent on the statistical attributes of the data which, in turn, are related to the spatial resolution of the data; via the Modulation Transfer Function (MTF) of the imaging system, and the inherent spatial frequencies of the terrain. Through the analysis of the statistical properties of the digital data, the relationship between spatial resolution and the efficiency of classification algorithms for the extraction of earth resources data can be established.

The paper describes a study based upon an empirical simulation where high resolution multispectral scanner data was progressively degraded to assess the effects of spatial resolution on the ability to classify land cover types using per-pixel digital classification techniques. The high resolution data was degraded by applying a two-dimensional digital filter to the image followed by the addition of a specified level of noise. Four simulated data sets were produced in this way at resolutions of 5m, 10m, 20m and 40m. The statistical properties of the data at the varying spatial resolutions are being investigated and related to the performance of the classifer.

The results are set in the context of improvements in the spatial resolution of imagery from earth resources satellites in the 1980s and, provide a basis for understanding the performance of classification algorithms as a function of spatial resolution.

THE UTILITY OF SMALLER LANDSAT GROUND RESOLUTION LEVELS (PIXELTOWN, U.S.A.)

by Richard F. Hyde, Nicholas J. Vesper,
Irvin A. Goldblatt, *et al*.

Holcomb Research Institute
Butler University
Indianapolis, Indiana 46208

ABSTRACT: A progress report is presented describing a continuing project at the Holcomb Research Institute (HRI) to construct a database that will synthesize scenes observed by future space satellites. This is accomplished through a digital database of solar reflectance terrain, fashioned after a typical Midwestern city. The database has been constructed with a one square meter resolution which can be aggregated to synthesize SPOT, ERS-1, TM, and any other ground resolution level (pixel size or spatial resolution) one meter or larger. Likewise, any spectral bands can be related to this system, permitting complete control of spatial and spectral characteristics of future space scanners. The initial purpose of the project is to identify the optimum spatial resolution necessary to provide the information required by major data users. Rather than launching new generations of satellites, then painfully learning the useful applications, it is preferable to proceed from an analysis of the Pixeltown database to suggestions for the design specifications of multispectral scanner systems.

I. INTRODUCTION

The concept of Pixeltown was first introduced at the ERIM Conference in the spring of 1981[1]. The primary purpose of the project was to construct a database that would allow synthesizing map information to be acquired by future space satellites. This database permits creation of any pixel size (ground or spatial resolution) of one meter or larger. A digital terrain of solar reflectance values was seen as a reasonable means of synthesizing the scenes that future satellites would "see" as they passed over Pixeltown, U.S.A. Of paramount interest was the production of synthesized Landsat scenes with resolution unit sizes of 10, 20, 30, 40, and 80 meters.

Paper presented at a conference on Matching remote sensing technologies and their applications convened in London by the Remote Sensing Society, 16-18 December, 1981.

The Holcomb Research Institute has constructed this database, fashioned after a typical Midwestern city in Indiana. The model city contains every imaginable land-use/land-cover feature, including sidewalks, bushes, trees, swimming pools, golf course, and, of course, Pixeltown Downs, a suburban racetrack.

The geographic area of the database is 64 square miles (over 100 sq. km), and was assembled like a mosaic, one square meter at a time. The matrix dimensions of 12,960 x 12,960 meters sums to 167,960,000 cells or pixels. Assignment of four typical spectral reflectance levels matching Landsat was performed for every cell, yielding a sum of 671,184,000 reflectance values. Any number of bands can be created with any spectral qualities, allowing for complete spectral variability. Just as important as controlling the spectral characteristics of the database is the ability to adjust the spatial resolution, or the ground resolution capability.

Progress toward completing the MSS-like CCT's has been steady, albeit slow because of the sheer volume of data and the numerous calculations and manipulations. Since generation of the 20-meter aggregation required many continuous cpu hours, verification of every parameter was required prior to committing our computer to this purpose. Progress to date in reaching our goals has brought us to a very important milestone. The 20-meter data have been produced and all software is performing properly. Maps produced of the 20-meter data look extremely useful, but at this time no comparisons can be made with other of pixel sizes until they are generated.

II. METHODOLOGY

The Pixeltown data set can be compared to a framework from which a wide variety of data bases can be derived. To each of its 167,960,000 cells an identification number (tag value) is assigned to each cell from the 69 land-cover tag values that are possible. Since each land-cover is uniquely defined by this tag it is possible to associate with any cell an n-tuple of numbers which could represent n-bands of a remote-sensing device, or, conversely, to associate to each tag a new land-cover interpretation. In this latter case, a polygon that might be interpreted as a corn field could be revalued as a bare soil area.

A. Data Base Construction

The initial development of Pixeltown consists of a largely pastoral region surrounding a suburban-like town (Pixeltown, of course). There are three parts to its construction: design, encoding, and computer generation. A particular data base implementation involves associating n-tuples to each individual tag value and aggregating cells to different levels. We will first consider the construction.

1. Design

Design began by selecting a one meter grid cell scale for a base map, represented upon common grid paper with 1/10 inch grid cells. Orthophoto 'quad' maps of 1":50', 1":100', and 1":200' scales were obtained for the city of Carmel, Indiana, as typical indicators of land use. Using these maps and other aerial photographs, a large-scale base map was drawn to describe both rural and suburban land-covers and patterns in proper scale and proportions. Once the base map outline was drawn, the scene was divided into two portions: the rural areas to be represented by 18-meter cells, and the urban areas at a one-meter cell. Grid paper was taped together to form two map areas, one of 720 by 720 cells (representing the 18-meter scale rural areas) and the other 3,240 by 3,240 (representing the one-meter scale urban area). On these sheets was drawn the land-cover and patterns we wished to represent, the urban map being only a one-fourth area representation (for reasons which will become clear in the following section).

2. Encoding

After the features were sketched out, a west-to-east orientation was chosen and the process of polygon identification of the land-covers and patterns began.

Each of the features was typed as an open polygon, a closed polygon, a linear pattern, a blocked pattern, or a point. We will look at each of their types in turn. But no matter what the type, points had to be selected by line, column and tag values. This produces a three-dimensional view of the base map. Thus, typing a feature meant selecting a set of points, a type number (a switch for computer generation purposes), and a tag value.

For polygons the points chosen were perimeter points, but not all the points defining the perimeter had the same tag. That is, a polygon was defined by a polygonal line of one tag which begins it (on the 'west' side) and a polygonal line of a different tag that terminates it and begins an adjacent one. The tag was the value that would be given the interior points of the polygon (and, of course, the polygonal line) as the polygonal line was imagined to sweep west to east on the base map.

Linear features were typed by a polygonal line, a tag, and a switch that indicated whether it was to be of fixed or variable width over its length. The switch also indicated the size of the variable or fixed width associated with the feature, facilitating automated encoding of rail, river, and highway systems.

The polygon and linear techniques applied mainly to the rural areas; the urban sections consisted of mostly blocked patterns. A set of house patterns and lots were designed which included all the land uses associated with a typical urban lot. There were nine such blocked patterns laid out and stored on data cards so that each cell at the one-meter level was assigned a tag value. Moreover, each pattern was defined by a base point, usually the end of a driveway. Since the pattern matrix could be translated or rotated to any point on the base map, it was necessary to choose a house type, a base point, and an orientation to have an entire city block replicated. This was done for one-fourth of Pixeltown; the other quadrants were then obtained rotating this pattern around a pivot point which was at the center of Pixeltown.

Point features were the easiest to deal with. It was simply necessary to assign a tag to a point. Some 'point' features, such as trees, were not actually points but collections of adjacent points. Here a center point and a switch were chosen that generated the predetermined tree pattern around the center point. There were seven such extended point patterns used.

As soon as the initial design phase on the rural areas was done, simultaneous encoding of this area and the design of the urban area was begun. Encoding was done on less than 1500 punch cards for the entire rural area by specifying for each feature a type number, a polygon status number, and a tag on one card followed by as many cards as needed for the polygonal perimeter points. This was an arduous task but was accomplished with efficiency and remarkable accuracy.

3. Computer Generating

The final part of the construction was accomplished by three computer programs. The first program generated line segments from the given polygonal points. Since the design choice of points emphasized using as few points as possible to define a feature, and since the scale of features is so large compared to the one meter cell size, line segment generation involved the finding of a great many points. For the rural scene alone 366,268 points were generated from approximately 4,500 design points, and for the entire scene some 2.5 million points were generated from approximately 10,000 design points.

Points were generated by treating the base map as a Cartesian coordinate system and using the two-point expression for a straight line. To minimize the stair-casing effect in finding integral points along a line in a grid, either the vertical line coordinates or horizontal column coordinates were used as the independent variable depending on the slope of the line as being above or below forty-five degrees (at that angle, it made no difference which variable was treated as the independent one). These points and the tag associated with the perimeter were combined to form a thirteen digit number of the form LINE:COLUMN:TAG which was then written on a magnetic tape.

The second program sorted the extended polygon points first by line value, column value, then tag value. Hence the points were arranged from left to right and up to down (west to east, north to south). This meant that the third porgram could assign all cells a tag according to the initial design view. This program left the points from the linear, blocked, and point features intact but filled out the polygons by simply turning a value to an 'old' tag or a 'new' tag as it scanned the points from one polygonal line to another. Points not covered by a design feature were given a constant background value. Because higher tag values at the same points replaced lower tag values, overlaps were permitted in the design choice of points; this overlap also effected a simple check for improperly encoded points.

This computer generation resulted in eight reels of magnetic tape, each twenty-four hundred feet in length. Each reel holds 1,620 lines and 12,960 columns of a fully tagged Pixeltown scene.

The scenes were examined by using a line printer, shades-of-gray mapping program that produced for each reel a symbol map of six sheets at the 18-meter level. A full one-meter scale map measures 108 x 108 feet. Mapping can be done at any level by appropriately skipping lines or columns so that the entire map can be quickly scanned for encoding errors. Verification and correction were done from the symbol maps. The computer map matched exactly the hand-drawn base map, except for a few encoding errors, on the first generation of the rural scene. Aggregation is the next step beyond current activities of verification.

B. Aggregation to Various Pixel Sizes

After this fully tagged framework was developed, a particular data base was implemented by n-tuple association and aggregation. The n-tuple associations can be of two types: the association of reflectance or emission spectral levels. Aggregation of a sub-matrix of the base grid into a new cell size (pixel) so that adjacent matrices form adjacent new cells can be done by splitting cells, in the case that lengths not factors of 12,960 are used for the matrix dimensions, or with exact counts of cells. In the split-cell case, the last line or column on the resulting aggregated data base might be lost due to insufficient numbers of cells available to form a new larger cell. The number of one-meter cells lost in either direction would never exceed the chosen length of the submatrix, and because the line and column count of 12,960 has a sufficient list of divisors, there is little reason to do cell splitting.

Aggregating cell areas and combining n-tuples to form new n-tuple spectral levels is straightforward for reflectance spectral data; for emission spectral data there are some interesting questions.

In the reflectance case, the aggregation and replacement is done by a simple weighted linear average with the weights being the area count of one-meter cells. Since the given mean, standard deviation and ranges are the typical values for land-covers as gleaned from our studies, those of others, and the available literature, they are practical numbers derived from use. And since the object is to derive typical values to use in classification studies, where the difference of the values is important to distinguish classes, this method works quite nicely.

In the case of combining emissions spectral data, a simple weighted linear average is not appropriate. The exact relationship between area count and spectral levels in combination to form a new spectral level appears to be complicated. Although our methods allow the insertion of any function to transform area counts and spectral levels to a single spectral level for the aggregated cell, we have not as yet developed such a function for emission data.

C. Assignment of Reflectance Values

The method of assigning n-tuples is as follows: to each tag, associate a value typical of each land use/land cover category (a number within a range dictated by the typical mean and standard deviation values) signified by that tag, whether for reflectance or emission data, or indeed, for pseudo-spectral levels of the user's design. By using a random number generator, which can be set to select uniformly random numbers of the given mean, standard deviation, and range or a simple bi-modal distribution derived from the given values, each cell is given an appropriate n-tuple of spectral values, one for each band, which is written to a magnetic tape in an appropriate MSS format. In practice, the aggregation and n-tuples are set at once. One additional reflectance assignment technique, that of using only mean values, is also useful to study the split cell behavior. Split cells are so called because more than one land cover category is reflecting its own special spectral signatures back toward space within one pixel.

D. Classification of Each New Pixel Resolution

HRI has available a multi-dimensional histogram cluster analysis system[2] to classify MSS data, (similar to the system used by the Canadian

Center for Remote Sensing) that operates in an unsupervised manner, thus adding objectivity, accuracy, and uniformity to classification procedures. This system clusters the MSS data according to the natural breaks occurring in the data set and maps each spectrally distinct class.

Rather than approaching the classification questions from the perspective of how much information is gained by improving ground resolution, the following approach is taken: The question under consideration then becomes, "What is the nature and source of information loss by beginning with a 10-meter resolution, and progressing through the designated resolution levels to 80 meters?"

The land use/land cover classes built into the Pixeltown data set are in accordance with the Anderson (1976) land use and land cover classification for use with remote sensor data[3]. Of particular interest during the classification processes is to note at which spatial resolution units we pass from Anderson's level III, to level II, and then to level I. By adhering to this classification system, standards can be maintained, adding more continuity to the study.

A classification procedure that allows mapping of each spectrally distinct land use/land cover category is necessary to ensure that all possible classes have been mapped. By proceeding in this manner and classifying each aggregate level independently, (i.e., 10, 20, 30, 40, and 80 meter resolution aggregates of the one meter data base) a final classification map for each aggregate is produced. Each of these maps is amenable to interpretation.

III. MAJOR AND MINOR RESEARCH FOCUS

Several major issues will be considered as each resolution level is classified by machine processing techniques including:

a. The change of average spectral characteristics as the spatial resolution is aggregated upward;

b. Noting the break-off points in the aggregation process where required information begins to become obscured for such disciplines as land use planning, agriculture, and forestry;

c. Noting the loss, gain, or no change in the number of identifiable land use/land cover classes at each level;

d. Noting the relative cost of processing and interpreting these data at each level.

Minor topics to be investigated following the above studies will include:

a. Evaluation of all reasonable ground resolution levels;

b. Evaluation of all reasonable spectral resolution bands to guide future scanner development decisions;

c. Assessment of various classification algorithms and procedures;

d. Evaluation of geometric correction schemes, and reverse operations to measure the variances;

e. Evaluation of data compression and decompression methods;

f. Evaluation of distortion in precision registration;

g. Evaluation of projection variations for maps and geographic information system overlay;

h. The creation of functions to "distort" data at various places and in different ways;

i. Missing lines of data and evaluation of closest neighbor and cubic convolution and higher order operations to identify the most appropriate technique;

j. Modification of the data base to examine change detection techniques and their efficiency; and

k. Area calculation accuracy at each aggregate resolution level.

IV. PRELIMINARY OBSERVATIONS OF PIXELTOWN AT 20 METERS

Each homogeneous land cover class contains predictable reflectance values. Where two unique land cover classes meet, a common boundary is shared. More likely than not, pixels will overlap such boundaries. Within overlap pixels a blending of the two unique reflectance values occurs, yielding an average of the two original reflectance levels, but weighted by the area of each of the two classes present. It is typical that this artifact data is near statistically to some other valid class or classes. In such cases the overlap pixels are indistinguishable from these valid classes, thus contributing to misclassification. In some cases, however, these artifacts assume a unique spectral identity and can be mapped and accounted for independently; therefore, mapping accuracy is not degraded per se.

An interesting observation has been noted as we compare error attributed to overlap pixels between the 80m and 20m resolutions. To illustrate this point, let us envision a parcel of land one square kilometer in size. Within this parcel is a uniform or homogeneous land cover. Surrounding this parcel on all sides is a second type of uniform or homogeneous land cover. At the 80-meter ground resolution level at least 56 overlap pixels on the inside of the parcel can be expected to occur around the perimeter. This number of cells can cause a 25.2% error in mapping accuracy alone. At the 20 meter level at least 196 overlap cells will occur, accounting for only a 7.8% error in mapping accuracy because the cell size is smaller (see Table I).

When working with a randomly positioned grid such as that acquired by Landsat, overlap pixels must occur at boundaries between two or more distinct surface cover features. Even though we expect averaging to take place, no reasonable methods of observing and measuring this process have been available until now. However, as the one-meter resolution digital terrain of solar reflectance values is aggregated to the 20m resolution level in Pixeltown, these processes can be closely observed. Adjacent land parcels of statistically distinct land covers often have a clear border between them because the overlap pixels themselves form a unique class. This border effect is most common, it appears, when the differences in average spectral responses between the two parcels are relatively large; how large will soon be explored.

Very low spectral responses when averaged with very high responses yield new, or artifact data values in mid-range. Those responses with less difference behave in very strange ways. It is appropriate to examine the manner in which things become averaged. It may be useful to understand the types and probabilities of artifact data that can be expected. The combinations and permutations may prove to be too numerous to manage; however, it seems reasonable that it will be possible to properly identify these data and classify them by anticipating them.

In Pixeltown it was observed that pixels along the boundary between coniferous trees and lawns assume their merged spectral characteristics which just happened to correspond to the spectral characteristics of the deciduous trees, hemlock and sugar maple. Likewise, pixels sharing the presence of water and grass merge these reflectance levels which likewise almost perfectly match the spectral characteristics of deciduous trees. This is partially because deciduous reflectance characteristics are in the mid range of possible reflectance values and partly because of the averaging process that occurs.

Table I. For a square kilometer land parcel area, the minimum and maximum number of overlap pixels can be calculated. For the example case selected, the measured reduction in error derived from overlap can be reduced almost 31% by choosing 20m resolution over that of 80m pixels.

	CELLS INSIDE	CELLS OUTSIDE	EXPECTED ERROR INSIDE	EXPECTED ERROR OUTSIDE
80m	56	64	25.2%	28.8%
20m	196	200	7.8%	8.0%

V. SUMMARY

Most multispectral data users expect to calculate the area of all land cover classes identified from machine classification processes. Taken by itself, the reduction in error due to overlap pixels at smaller pixel sizes will have two benefits. First, area calculation can be expected to improve; and second, the overall classification accuracy will be improved.

It is likewise obvious that overlap or spectral reflectance mixing will occur when a grid-cell format is selected, regardless of the cell size (or resolution). Area calculation and improved overlap conditions appear to be possible by moving from the 80-meter resolution to the 20-meter level. Time has not yet permitted examination of a potentially disruptive characteristic cropping up with pixels of improved ground resolution characteristics. This disruptive force may be associated with the detail that is influencing matters at this scale. The more closely we observe features, the more detail. Details on the landscape have the potential of biasing the average spectral character of a cell. This bias can become stronger as resolution is improved. For example, a small lighter colored spot upon a larger dark background will have an increasing influence upon the average spectral characteristics of a pixel the smaller the field of view becomes. This phenomenon remains to be studied in the context of resolution, classification, and mapping accuracy.

It is anticipated that by late December 1981 the 10, 30, 40, and 80 meter aggregations will be completed. Immediately following, interpretation of the maps will begin, as will the cost effectiveness study. These results, as well as the results from the major and minor research topics, will be published in a special monograph series by Holcomb.

REFERENCES

1. Hyde, Richard F., Nicholas J. Vesper, et al. "An Evaluation of the Utility of Smaller Landsat Resolution elements (Resolution of Spatial Resolution." Presented at the Fifteenth International Symposium on Remote Sensing of Environment, Ann Arbor, Michigan. May 1981.

2. Hyde, Richard F., Samuel N. Goward, and Paul W. Mausel. "ISURSL Levels Classification: A Low Cost Approach to Multispectral Data Analysis." Presented at the Fourth Annual Symposium on Machine Processing of Remotely Sensed Data, Purdue University, West Lafayette, Indiana. June 21-23, 1977.

3. Anderson, James R. et al. A Land Use and Land Cover Classification System for use with Remote Sensor Data. United States Geological Survey Professional Paper 964, 1976.

CLUSTERING OF MULTISPECTRAL DATA BASED ON GENERALIZED DISTANCE CRITERION FUNCTION

S Chandrasekhar M Maruthachalam and K R Rao

National Remote Sensing Agency
#4, Sardar Patel Road
Post Box No. 1519
Secunderabad
Andhra Pradesh - INDIA - 500003

Abstract: This paper briefly discusses a clustering technique on generalized distance criterion function optimization. The criterion function used is a quadratic function based on total scatter matrix and distance of samples from the means. Initially, all the sample points to be clustered from individual cluster centres and subsequently clusters are merged depending upon change in criterion function till the desired number of clusters are obtained. The algorithm was tested on Landsat data and it was found to perform quite good when compared to least square criterion function. The results are discussed.

1. INTRODUCTION

Multispectral data is normally acquired with the help of a satellite or aircraft using scanners in a number of spectral bands or channels. Hence, each data point can be considered as a vector in multi-dimensional measurement space which is also termed as pattern space. Pattern classification aims at partitioning the pattern space into different regions wherein, each region corresponds to a specific class. This partitioning can be done depending upon whether we have apriori knowledge of pattern class or not. Based on the above concept, we have two broad techniques of classification. They are:

1) Supervised approach
2) Unsupervised approach

In supervised approach, samples that are known to belong to different classes are available and based on the information contained in them, decision function which partition the pattern space can be computed. From these computed decision functions, any unknown pattern can be classified.

Paper presented at a conference on Matching remote sensing technologies and their applications convened in London by the Remote Sensing Society, 16-18 December, 1981.

In unsupervised approach, the samples belonging to known classes are not available before hand. Hence, by mathematical formulation based on certain criterion function, one has to identify different classes that are present in the given data set. It is a common phenomenon that patterns belonging to the same class tend to form groups or clusters in pattern space. The discussion in the present paper will try to bring out clusters in pattern space. Hence, the clustering problem can be defined as given a data set the object is to partition the data set in such a way that patterns that have similar attributes are grouped together. Quantitatively, this is equivalent to stating that data set has to be partitioned in such a way that certain criterion function has to be optimized. The detailed discussion follows.

CLUSTERING CRITERION

As already explained, the basic thing in clustering is given a data set, it has to be partitioned into distinct data sets or clusters based on certain criterion function. The commonly used criterion function is the mean square error. Let us assume that the given data set is partitioned into m distant subsets $S_1, S_2, \ldots, S_m$ each containing certain number of samples. The mean of each class gives a good representation of that class. The mean square error performance index represents total error in representation of data.

The mean square error can be expressed as,

$$J_e = \sum_{i=1}^{m} \sum_{X \in S_i} \| X - \mu_i \|^2$$

where J_e = mean square error

m = number of classes

$$X = \begin{bmatrix} x_1 \\ x_2 \\ \vdots \\ x_{n-1} \\ x_n \end{bmatrix} = \text{pattern vector of dimension } n.$$

μ_i = mean vector of class i = $\frac{1}{N_i} \sum_{X \in S_i} X$

N_i = number of samples in class i.

An optimum partition is the one for which J_e is as small as possible.

Other measures of criterion functions can be formed using what are known as 'scatter matricies'. As, has already been pointed out, that, data points belonging to same class have similar attributes. In some sense scatter measures the amount of spread of data and intutively it is a known fact that amount of spread within pattern of same class have to be as small as possible and at the same time between classes have to be as large as possible. It is also shown that minimizing within group scatter is equivalent to maximizing between group scatter.

We define the following:

Mean vector of ith cluster $\mu_i = \frac{1}{N_i} \sum_{x \in S_i} x$

N_i = number of samples in cluster S_i.

If there are m clusters $S_1, S_2, \ldots, S_m$ and N total number of data points,

Total mean vector $\mu = \frac{1}{N} \sum_{x \in S} x = \frac{1}{N} \sum_{i=1}^{m} \mu_i N_i$

Scatter matrix for i^{th} cluster $S_i = \sum_{x \in S_i} (x - \mu_i)(x - \mu_i)^t$

Total within cluster scatter matrix,

$$S_W = \sum_{i=1}^{m} S_i = \sum_{i=1}^{m} \sum_{x \in S_i} (x - \mu_i)(x - \mu_i)^t$$

Total scatter matrix $S_T = \sum_{x \in S} (x - \mu)(x - \mu)^t$

If we define between cluster scatter matrix S_B as,

$$S_B = \sum_{i=1}^{m} N_i (\mu_i - \mu)(\mu_i - \mu)^t$$

then, it can be shown that,

$$S_T = S_W + S_B$$

Since, the total scatter matrix is independent of the way in which data set is partitioned, it is clear that minimizing S_W also maximizes S_B. As, the above quantities are matricies, some sort of scalar measure will be useful to measure them. Commonly used scalar measures are the trace and the determinant. The trace of a matrix which is defined as sum of diagonal terms, roughly measures the square of scattering radius. The determinant of the matrix measures the scattering volume as it is proportional to product of variances. It can be easily seen that, minimizing trace of with in group scatter S_W is equivalent minimizing mean square error J_e.

ALGORITHM BASED ON GENERALIZED DISTANCE

In this technique, the criterion function used is

$$J = \sum_{i=1}^{m} \sum_{x \in S_i} (x - \mu_i)^t S_T^{-1} (x - \mu_i)$$

where m = number of classes

μ_i = mean of N_i samples contained in partition S_i

S_T = total scatter matrix

Once, the criterion function is chosen, the clustering problem reduced to that of finding partition for which J is minimum. This is done by selecting an initial partition and move the data points from one partition to another if such a move improves the criterion function. An iterative algorithm is suggested to achieve the optimization.

The selection of initial partition is also a crucial point. As in most cases, one method is to chose all the samples in the data set as initial cluster centres, merge successively based on the change in the criterion function selected, until the desired numbers of clusters are obtained. This approach is used in our technique. Another rule that is observed is that at any stage samples belonging to that cluster from a previous stage is retained.

It can also be shown that the criterion function J changes to J* by the relation,

$$J^* = J - \left[\frac{N}{N_j+1} (\hat{x}-\mu_j)^t S_T^{-1} (\hat{x}-\mu_j) - \frac{N}{N_i-1} (\hat{x}-\mu_i)^t S_T^{-1} (\hat{x}-\mu_i) \right]$$

whenever a sample $\hat{X}$ belonging to S_i is moved to S_j. N_i, N_j are the number of samples in set S_i and S_j.

The algorithm can be written as follows:

m_{ini} = initial number of cluster centres normally equal to total number of samples.

m = number of cluster centres desired.

Begin: 1. Let $\hat{m} = m_{ini}$ and $S_i = X_i$, i = 1, 2, ..., m_{ini}

Loop: 2. If $\hat{m} \leq m$ Stop.

3. Find the pair of clusters S_i and S_j for which the criterion function J changes as little as possible.

4. Merge S_i and S_j, delete S_j and decrement $\hat{m}$ by 1.

5. Go to Loop.

DISCUSSION OF RESULTS

The above algorithm was implemented on IBM 370/155 and was programmed in FORTRAN IV. Landsat data pertaining to an approximate 512 scan lines x 512 pixels was used and all the four channels were used for clustering. The final number of clusters that was requested was 14. The results of computations are given below:

CLUSTER SIZES

4572.	3796.	3007.	2610.	1425.	1412.
1051.	211.	166.	71.	63.	55.
26.	2.				

MEAN VECTORS

CLUSTERS	CHANNELS			
	1	2	3	4
1	52.31	47.54	57.95	26.47
2	47.63	41.69	60.06	29.46
3	44.48	37.91	62.69	31.52
4	50.41	46.31	65.04	31.30
5	55.36	55.53	64.95	29.53
6	48.55	43.48	54.67	25.23
7	42.55	35.67	59.75	30.09
8	54.45	54.49	59.22	26.38
9	62.87	66.23	69.12	29.90
10	47.52	41.79	67.94	33.13
11	46.49	40.35	55.27	26.54
12	42.87	36.04	45.35	19.20
13	68.08	74.58	74.65	31.92
14	57.50	50.50	66.00	1.00

DISTANCES OF SEPARATION BETWEEN CLUSTERS

	1	2	3	4	5	6	7	8	9	10	11	12	13	14
1	0.0	8.3	14.2	8.9	11.5	6.5	15.8	7.4	24.5	14.1	9.6	20.8	35.9	27.4
2	8.3	0.0	5.9	7.6	16.6	7.1	7.9	14.9	30.3	8.7	5.9	19.4	41.5	31.9
3	14.2	5.9	0.0	10.5	20.9	12.3	4.4	20.3	34.4	7.4	9.5	21.4	45.2	35.6
4	8.9	7.6	10.5	0.0	10.6	12.5	14.3	11.9	23.9	6.4	13.0	26.4	34.7	31.4
5	11.5	16.6	20.9	10.6	0.0	17.8	24.2	6.7	13.7	16.5	20.3	32.1	25.0	29.1
6	6.5	7.1	12.3	12.5	17.8	0.0	12.1	13.3	30.9	15.6	4.0	14.5	42.3	29.1
7	15.8	7.9	4.4	14.3	24.2	12.1	0.0	22.6	37.9	11.8	8.4	18.1	48.9	36.5
8	7.4	14.9	20.3	11.9	6.7	13.3	22.6	0.0	17.9	18.2	16.7	26.8	29.3	26.7
9	24.5	30.3	34.4	23.9	13.7	30.9	37.9	17.9	0.0	29.1	33.8	44.6	11.5	33.5
10	14.1	8.7	7.4	6.4	16.5	15.6	11.8	18.2	29.1	0.0	14.4	27.6	39.3	34.8
11	9.6	5.9	9.5	13.0	20.3	4.0	8.4	16.7	33.8	14.4	0.0	13.6	45.2	31.5
12	20.8	19.4	21.4	26.4	32.1	14.5	18.1	26.8	44.6	27.6	13.6	0.0	56.0	34.4
13	35.9	41.5	45.2	34.7	25.0	42.3	48.9	29.3	11.5	39.3	45.2	56.0	0.0	41.5
14	27.4	31.9	35.6	31.4	29.1	29.1	36.5	26.7	33.5	34.8	31.5	34.4	41.5	0.0

The output was compared with imagery and it was found that clusters formed by the above algorithm matched closely with the existing classes.

ACKNOWLEDGEMENT

The authors would like to thank G. Rajaraman for his help in preparing this manuscript.

REFERENCES

1. Haralick, R.M. and I. Dinstein : "An iterative clustering procedure". IEEE Trans. on Sys., Man, and Cyb., SMC-1, 275-289 (July 1971).

2. Hillborn, C.G. Jr. and D.G. Lainiotis : "Optimal unsupervised learning multi-category dependent hypotheses pattern recognition". IEEE Trans. Info. Theory, IT-14, 468-470 (May 1968).

3. Johnson, S.C. : "Hierarchical clustering schemes". Psychometrika, 32, 241-254 (September 1967).

4. King, B.F. : "Stepwise clustering procedures". J. American Statistical Assn., 62, 86-101 (March 1967).

5. Ling, R.F. : "Cluster Analysis". Tech. Report No.18, Dept. of Statistics, Yale University, New Haven, Conn. (January 1971).

6. Meisel, W.S. : Computer oriented approaches to pattern recognition (Academic Press, New York and London, 1972).

7. Patrick, E.A. and J.C. Hancock : "Non-supervised sequential classification and recognition of patterns". IEEE Trans. Info. Theory, IT-12, 362-372 (July 1966).

8. Sammon, J.W. Jr. : "Interactive pattern analysis and classification". IEEE Trans. Comp. C-19, 594-616 (July 1970).

9. Batchelor, B.G. and B.R. Wilkins : "Method for location of clusters of pattersn to initialize a learning machine". Electronics letters, Vol.5, No.20, pp.481-483 (1969).

10. Blaydon, C.C. : "Recursive algorithms for pattern classification". Office Naval Res. Tech. Report.520. Division of Engineering and Applied Physics, Harvard University, Cambridge, MASS (1967).

11. Duda, R. and P. Hart : "Pattern classification and scene analysis". John Wiley & Sons, New York.

12. Tou, J.T. and R.C. Gonzalez : "Pattern recognition principles". Addison-Wesley Publishing Company, MASS. USA.

APPLICATION OF HIGH EFFICIENCY DIGITAL TECHNIQUES TO REMOTE SENSING IMAGE PROCESSING

Vito Cappellini

Istituto di Elettronica, Facoltà di Ingegneria
Florence University and IROE-C.N.R.
Florence, Italy

Abstract: Some high efficiency digital techniques, which can be applied to processing data given by different remote sensing systems, are presented. In particular digital filtering processing is described to perform image enhancement and noise reduction or to extract edges for different land regions separation. Data compression techniques, based on digital filtering or transformations, are also presented to reduce the tremendous amount of data connected to remote sensing images or maps. Some pattern recognition techniques, applied after filtering and edge extraction and based on prototypes or contour transformation are finally described. Some processing examples are shown.

1. INTRODUCTION

Remote sensing systems on aircraft or satellite platforms are becoming of increasing importance for resource monitoring, agriculture, cartography and pollution control. Different remote sensing systems are available: photographic cameras, multispectral scanners, infrared detectors, microwave radiometers, side-looking radars (SLR), lidars.

From the above systems different measurements and data are obtained, in general (at least at a suitable acquisition step) in form of images or maps. Digital processing techniques are required for several reasons, such as: improvement of image quality (performing corrections, reducing the noise, enhancing the gray level); reduction or compression of data to solve handling and storage problems; recognition of useful configurations and classification of different image parts.

In the following some high efficiency digital processing techniques are presented, regarding the considered aspects for a better practical utilization of remote sensing data. In particular digital filtering, data compression and pattern recognition techniques are described.

2. DIGITAL FILTERING TECHNIQUES

Two-dimensional (2-D) digital filters are to be used for the processing of images, such as obtained from remote sensing systems. A 2-D digital filter can be defined through the following relation (Cappellini et al.,

Paper presented at a conference on Matching remote sensing technologies and their applications convened in London by the Remote Sensing Society, 1981

1978)

$$g(n_1,n_2) = \sum_{k_1=0}^{N_1-1} \sum_{k_2=0}^{N_2-1} a(k_1,k_2)f(n_1-k_1,n_2-k_2) - \sum_{\substack{k_1=0 \\ k_1+k_2\neq 0}}^{M_1-1} \sum_{k_2=0}^{M_2-1} b(k_1,k_2)g(n_1-k_1,n_2-k_2) \quad (1)$$

where: $f(n_1,n_2)$ are the input data (digital samples of the analog image for a photo), $g(n_1,n_2)$ are the output data (samples of the output image), $a(k_1,k_2)$ and $b(k_1,k_2)$ are the coefficients defining the 2-D frequency response of the digital filter, N_1,N_2,M_1,M_2 are integers.

Many types of 2-D digital filters have been defined: with finite impulse response (FIR) of non-recursive type having all $b(k_1,k_2)=0$; with infinite impulse response (IIR) of recursive type. Some important design methods of 2-D FIR digital filters are: methods using "windows"; frequency sampling methods; approximation methods (Cappellini et al., 1978). Filters of this type can present good efficiency (in particular linear phase characteristics) and relatively simple software implementation: they are useful and practical for remote sensing data processing. 2-D IIR digital filters require in general a lower number of coefficients, but present stability problems and more complex design procedures.

2-D digital filters, in particular of FIR type, can be applied to remote sensing data processing to smooth the data (by means of low-pass filtering) or to perform frequency corrections or to obtain enhancement (by mean means of high-pass or band-pass filtering), extracting also boundaries or edges. In particular edge extraction can be efficiently obtained by performing enhancement as above and applying hence a 3x3 local space operator (estimating the gradient).

3. DATA COMPRESSION TECHNIQUES

Data compression is a transformation performed to reduce the amount of not useful or redundant data. Several data compression techniques have been introduced of reversible and not reversible type (reversible means that the inverse transformation or decompression can recover all the original data).

Some important data compression techniques are: prediction, interpolation, adaptive sampling, encoding, filtering, use of transformations (Fourier, Hadamard, Walsh, Haar, Slant...) (Benelli et al., 1980). All one-dimensional data compression techniques can also be applied to image processing, if the image is scanned line by line.

In particular 2-D digital filtering of low-pass or band-pass type represents by itself a sort of data compression, because a limited part of the space frequency spectrum is extracted, requiring a lower number of data to be maintained. Further 2-D low-pass digital filtering can be used as an useful pre-processing before the application of other data compression algorithms, because the smoothed data are more efficiently compressed.

For what regards the use of 2-D transformations, their interest is connected to the fact that the transformed data are in general more concentrated than the original space data and hence are more easily compressed. Some useful methods consist in dividing the transformed image data into several squares of small size (4x4, 8x8, 16x16) and in employing for each of them a minimum word length.

The interest of applying data compression techniques to remote sensing data processing is due to the fact that an extremely large amount of data is already now - and more in future - given by the sensors and systems aboard aircrafts or satellites, making difficult and sometimes impossible to handle the data and to storage them for subsequent utilization.

4. PATTERN RECOGNITION TECHNIQUES

Two main pattern recognition techniques are known. One type uses prototypes, that is reference configurations to which the input data are compared to recognize and classify them. The other type is based on statistical analysis and structural properties (decomposing complex configurations in basic elements).

The first class of techniques using prototypes is more useful in practical applications to remote sensing images. On this line, before to estimate the different configurations, an important step is represented by the enhancement of the image structure (i.e. through digital filtering) and the edge extraction. A correct edging can separate the different land parts in the examined earth image or map.

5. APPLICATIONS TO REMOTE SENSING IMAGE PROCESSING

The above briefly described techniques were extensively applied in Florence (as a cooperation among the Istituto di Elettronica, IROE-C.N.R. and I.A.T.A.-C.N.R.) to the processing of remote sensing images and maps.

In particular aircraft photos, regarding agriculture lands, were processed by means of 2-D digital filtering and local space operators to extract the edges and recognize configurations. By using gray level prototypes, three land classes were defined: land with oil and wine-grape, land with forest, other land (Cappellini et al., 1979).

Further, by applying 2-D digital filtering and suitable thresholds on gray levels, water regions were extracted from LANDSAT images estimating water resources (Cappellini et al., 1979).

A new method was also defined to compare and correlate maps obtained from different sensors (at different height or with different ground space definition) as aboard aircrafts or satellites. The method is based on the use of a suitable 2-D low-pass digital filter applied to the high definition map and subsequent decimation procedure (Cappellini et al., 1980).

Data compression techniques, in particular by using transformations, were applied to aircraft photos and satellite images (as LANDSAT ones): compression ratios (between original and compressed data) in the range 3-5 were obtained with neglegible loss of information (that is maintaining the quality of the images).

In conclusion the digital techniques above considered can be very useful to process the remote sensing data for encreasing the image quality, reducing the amount of data and recognizing important configurations.

6. REFERENCES

BENELLI, G., et al (1980): Data compression techniques and applications. The Radio and Electronic Engineer 50 (1-2) 29-53.

CAPPELLINI, V., et al. (1978): Digital Filters and Their Applications, Academic Press, London-New York.

CAPPELLINI, V., et al. (1979): Analisi delle coltivazioni di un'area del Chianti tramite telerilevamento aereo. Report Istituto di Elettronica, University of Florence.

CAPPELLINI, V., et al. (1979): Telerilevamento di aree campione mediante dati da satellite. Report GIT, R-3-79, University of Florence.

CAPPELLINI, V. and MARACCHI G.P., (1980): Digital comparison and correlation of map data obtained from different remote sensing systems: a new proposal. Intern. Conference on Image Analysis and Processing, Pavia-Italy.

Part E

STUDENT PAPERS

A TEST OF AN ATMOSPHERIC CORRECTION OF LANDSAT DATA OVER WATER SURFACES

N. MacFarlane, Department of Oceanography, University of Southampton.

Abstract

Part of the atmospheric correction algorithm being implemented in the Oceanography Department of Southampton University is the use of LANDSAT MSS Band 7 data to estimate the contribution to the radiance received in the other Bands due to aerosol scattering. The basic theory is an adaptation of similar work performed for CZCS imagery. This method is tested by attempting to remove a large atmospheric plume from two LANDSAT scenes of the Solent/Isle of Wight area. Removal of the atmospheric radiances should enhance our future sediment calibration algorithms.

Introduction

Any method for calibrating LANDSAT MSS imagery over the sea in terms of suspended sediment or any other water quality parameter should attempt to remove the effects of the atmosphere from the signal received at the satellite. This is the case whether a statistical technique or a theoretical model based on the physics of radiative processes is being employed in the calibration algorithm. In order that an MSS calibration can be performed simply and at relatively low cost by an environmental scientist who wishes to apply the data, the aim of the remote sensing specialist should be to provide as simple a calibration model as possible but based on a sound theoretical basis. The atmospheric corrections should therefore only require such atmospheric data as is readily available.

The testing of atmospheric correction algorithms normally requires measurement of many parameters synoptic with the satellite overpass, which can be a lengthy and expensive process. However, where an identifiable haze or smoke plume from an industrial complex provides a well defined limited area of atmospheric interference on an image, over water, it may be possible to test, at least approximately, the efficacy of atmospheric correction models without the need for an extensive measurement programme. In this paper, such a method is used to test an atmospheric correction algorithm which is based on simple atmospheric parameters.

Sturm (1981) has shown in his thorough account of calibration techniques that the radiance received in a wavelength range centred on average wavelength λ by a satellite-borne nadir-looking sensor can be written as

$$L(\lambda) = Lsg(\lambda) + Lhg(\lambda) + L\omega(\lambda) + Lpa(\lambda) + Lpr(\lambda) \quad (1)$$

where Lsg is the radiance due to direct sun radiance reflected from the water surface (sunglitter), Lhg is the radiance due to diffuse radiance reflected from the water (skyglitter), $L\omega$ is the radiance due to radiance leaving the water and Lpa and Lpr are respectively, the radiances scattered from particles and molecules. Since a constant view angle is to be assumed in this work, simplfied forms of Sturm's equations can be used to calculate values for Lsg, Lhg and Lpr. These will be constant across the image, the only variations being in $L\omega$ and Lpa. Rewriting equation (1) we obtain

$$L(\lambda) - Lshr(\lambda) - Lpa(\lambda) = L\omega(\lambda) \quad (2)$$

where

$$Lshr(\lambda) = Lsg(\lambda) + Lhg(\lambda) + Lpr(\lambda) \quad (3)$$

This would be used to obtain the radiance related to subsurface scattering and absorption.

To remove the effect of aerosol scattering Lpa can be subtracted from the total radiance L, or an equivalent amount from the digital number I. In the case where originally a distinct haze plume could be identified, the validity of the technique can be tested by checking to see if the resulting image shows only water and no atmospheric patterns.

The method used for calculating radiance L from digital number I is that proposed by Ahern and Murphy (1978).

$$L(\lambda) = \frac{R_2(\lambda) - R_1(\lambda)}{255} \; I(\lambda) + R_1(\lambda) \tag{4}$$

where R_1 and R_2 are the radiances corresponding to the minimum and maximum digital values (0 and 255). This test therefore consists of subtracting

$$\frac{255}{R_2(\lambda) - R_1(\lambda)} \; Lpa(\lambda)$$

from the digital numbers I.

The calculation of the radiance due to aerosol scattering, $Lpa(\lambda)$, is based on two assumptions:

(a) there is no water leaving radiance in Band 7, i.e. $L\omega^7 = 0$. The only variation in Band 7 is thus assumed to be due to variation in aerosol scattering Lpa^7 and use of equation (1) allows a direct calculation of this quantity for each pixel,

(b) the aerosol scattering radiance in Band 7 can be related to the aerosol scattering radiance in each of the other Bands by an equation proposed by Gordon (1978)

$$Lpa^i = L(i,7).\ Lpa^7 \; ; \; i = 4,5,6 \tag{5}$$

The value of $\alpha(i,7)$ used in this work was taken to be the ratio of the values for Lpa^i and Lpa^7 produced by

$$Lpa(\lambda) = \sum_{\lambda=\lambda_1}^{\lambda_2} E_o(\lambda,J).\ \tau_a(\lambda).\ \left[T_{o2}(\lambda,\theta_o).\ P(\theta_o) + T(\lambda,\theta_o).\ \rho(\theta_o).\ \cos\theta_o.\ P(\pi-\theta_o)\right].\ T_{o2}(\lambda,0) \tag{6}$$

where $E_o(\lambda, J)$ is the incident solar irradiance in the small wavelength range $\Delta\lambda$ centred on wavelength λ, $T_{o2}(\lambda,\theta)$ is the ozone layer transmittance, $\rho(\theta)$ is the Fresnel reflectivity (all at incident angle θ), θ_o

is the solar zenith angle, $P(\theta)$ is the aerosol scattering phase function - represented by one-term Henyey-Greenstein (OTHG) functions used by Gordon (1976) for scattering angle θ,

$$P(\theta) = (1-g^2)/\left[4\pi(1+g^2+2g\cos\theta)^{3/2}\right];\ g = 0.69 \tag{7}$$

and the aerosol optical thickness $\tau a(\lambda)$ is calculated from a value of horizontal visibility recorded by a coastguard station. The reliability of $\tau a(\lambda)$ in this calculation depends on the value B in the basic assumption.

$$\tau a(\lambda) = A\ \lambda^{-B} \tag{8}$$

Sturm states that B varies in the range 0.8 to 1.5 but may become less than zero under some maritime conditions, a value of 1.0 was assumed in this work. Optical thicknesses used in the transmittance calculations were obtained from the seasonal averages in LOWTRAN 2 (McClatchey and Selby, 1972).

Equation (6) has been derived by adapting Sturm's derivation of a similar equation for the Rayleigh scattering Lpr. The fact that the aerosol optical thickness is calculated from the measurement of horizontal visibility which is based on general haze levels rather than the denser plume may mean that the technique works best for the areas of low contamination.

Substitution of (5) in (2) yields:

$$Lw^{i} = L^{i} - \alpha(i, 7).\ L^{7} - (Lshr^{i} - \alpha(i, 7).\ Lshr^{7}) \tag{9}$$

and hence by considering equation (4) an image can be obtained showing $(a.Lw^{i} + b)$ by calculating

$$I_i - \alpha(i,7)\,\frac{R_2^7 - R_1^7}{R_2^i - R_1^i}\,I_7$$

This image will have data values which are still suitable for input to most present image processing algorithms, i.e. they are still in the 0-255 range and are not radiance values which are normally less than 1.0.

The LANDSAT image of 1st January 1979 from the Solent/Isle of Wight area - path 217, row 25 - (fig. 1 shows part of Band 4) shows a large atmospheric plume emanating from the area surrounding Fawley power station and petrochemical complex and traversing the sediment laden waters north of the Isle of Wight. Figure 2 shows the same part of the image after the equivalent aerosol scattering component has been removed. Whilst the original data had been processed for destriping at the Earthnet receiving station, Fucino, this data was transformed back into pre-standard processing values using the radiometric look-up tables, see Benny (1980), the aerosol correction was applied and only then was the output statistically destriped. Deprocessed data should be nearer to the actual information recorded by the sensor and should therefore produce more accurate results; this data cannot always be obtained because the receiving station at Kiruna carries out a resampling procedure which invalidates the look-up tables.

The atmospheric correction has reduced the size of the values along the axis of the plume and the data values on the sides of the axis have become similar to the values away from the plume. The technique, therefore, has been successful in the areas of lower contamination but although the plume has been reduced in size and continuity it is still visible. The contrast stretch in the photographic process is now actually larger for the sea pixels since the land areas have been masked out. This image was designed to strictly test for the existence of the plume; the land areas are no longer needed for identification of the area. Apart from exaggerating any remainder of the plume this has also increased the number of different visible radiance levels in the sea, making the image more visually noisy and the identification of the sea patterns more difficult.

Figure 3 and figure 4 show, respectively, part of the LANDSAT image of 23rd February 1979 before and after the above correction process was applied. The technique has been almost completely successful at the southern end of the plume but has left traces of the plume

IMAGE MADE BY SPACE DEPT. RAE FARNBOROUGH 15 OCT 81 001227

FIGURE 1   9/1/79 SOLENT ORIGINAL BAND4

IMAGE MADE BY SPACE DEPT. RAE FARNBOROUGH 15 OCT 81 001227

FIGURE 2 9/1/79 DESTRIPED ATMOS.CORRECTED BAND4

IMAGE MADE BY SPACE DEPT. RAE FARNBOROUGH 15 OCT 81 001227

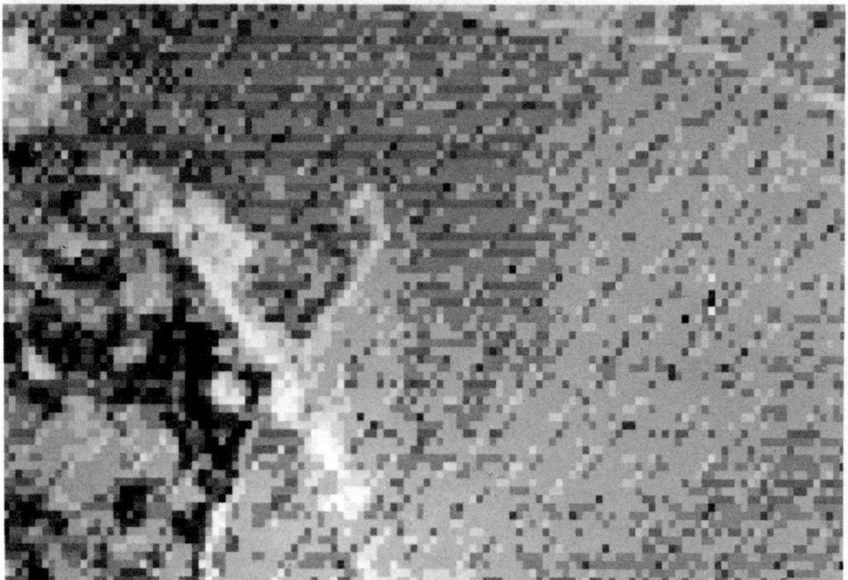

FIGURE 3 23/2/79 SOLENT ORIGINAL BAND4

IMAGE MADE BY SPACE DEPT. RAE FARNBOROUGH 15 OCT 81 001227

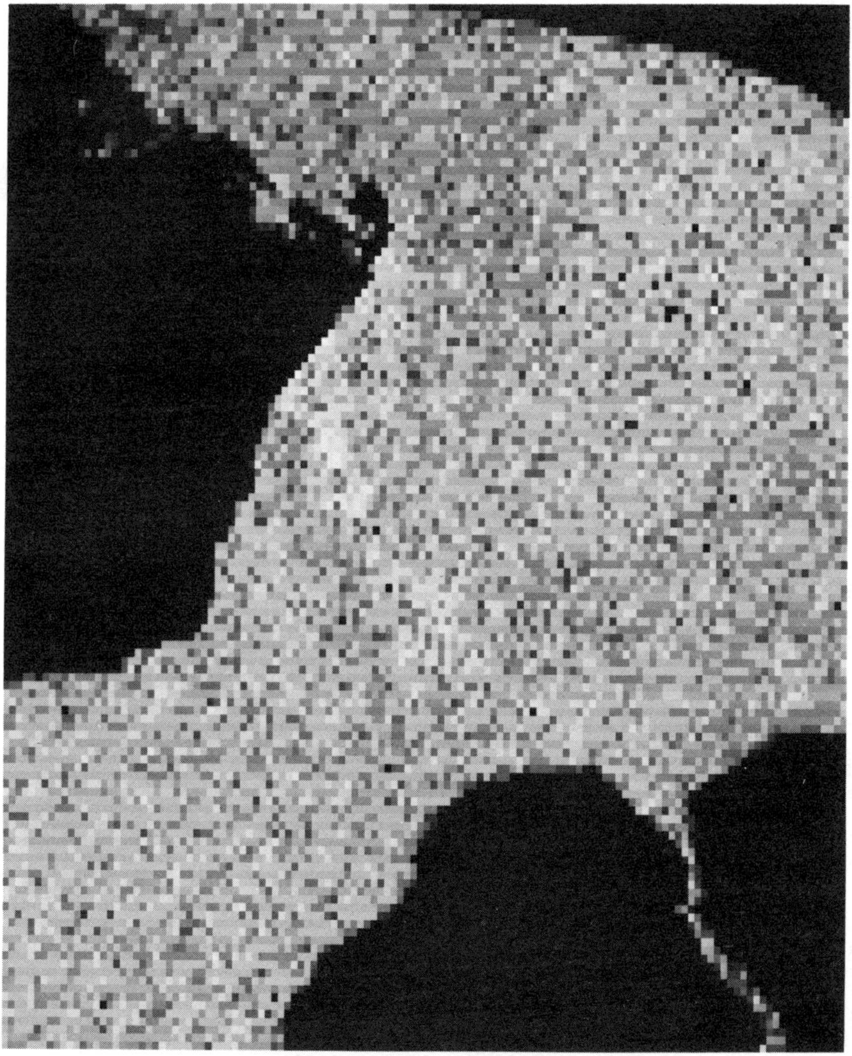

FIGURE 4 23/2/79 DESTRIPED ATMOS. CORRECTED BAND4

where the original contamination was again high. The increase in visual noisiness of the image again makes identification of water patterns more difficult.

This test strongly suggests that the equations used will enable the removal of atmospheric effects in LANDSAT data from sea areas, but it is still inconclusive as to what type of effects are being removed and what effect the equations are having on the water radiances. The technique is less successful in Bands 5 and 6 in terms of plume removal, thus suggesting that a different value of B in equation (8) will produce better results, though variation of B to produce a "best" set of the three Bands will probably be limited by subjective judgement of the images. Further work in this area should, however, enable us to get closer to the subsurface - originating radiances and increase the possibility of calculating accurate, reliable suspended sediment concentrations.

References

Ahern, F. J. and Murphy J. (1978): Radiometric Calibration and Correction of LANDSAT 1, 2 and 3 MSS data. Canada Centre for Remote Sensing, Research Report 78-4.

Benny, A. H. (1980): A study of LANDSAT MSS data tapes. Procurement Executive, MOD, Farnborough, Technical Report 80018, Feb. 1980.

Gordon, H. R. (1976): Radiative transfer - a technique for simulating the ocean in remote sensing calculations. Appl. Opt. 15, No. 8, 1974-9.

Gordon, H. R. (1978): Removal of Atmospheric Effects from Satellite Imagery of the Oceans. Appl. Opt. 17, No. 10, 1631-6.

Selby, J. E. A. and McClatchey, R. M. (1972): Atmospheric Transmittance from 0.25 to 8.5 μm; Computer Code LOWTRAN 2. AFRCL-72-0745, Dec. 1972, Environmental Research Papers No. 427.

Sturm, B. (1981): Atmospheric correction of remotely sensed data and quantitative determination of suspended matter in marine water surface layers. In Remote Sensing in Meteorology, Oceanography and Hydrology, ed. A. P. Cracknell, Ellis Horwood, 1981, pp. 147-162.

THE REMOTE SENSING OF INADEQUATELY RECLAIMED LAND

P.J. Lakin

Department of Land Surveying
North East London Polytechnic
Forest Road, London E17

Abstract: In many areas, reclamation schemes have been carried out and the results have been detrimental to the environment, degrading the land to a condition worse than if it had remained derelict. Landfill sites suffer from toxic waste overflows and underground fires, and leachates discharging from decomposing filling materials can be harmful to adjacent areas. These effects cannot be fully appreciated on the ground, so an overview, using remote sensing, can locate these phenomena, identify their nature and provide information to aid environmental protection projects. This paper outlines and evaluates the techniques used, with reference to several examples from the South Essex (Thurrock) region.

1. INTRODUCTION

Reclaimed land can be put to various after-uses as agriculture, forestry, building, industry, recreation, or just landscaped to improve the aesthetic appearance. The method used for a reclamation scheme is dependant upon several factors, the major concerns being the most suitable after-use, the available funding and expected return of income from the chosen after-use, availability of materials or locality of tipping sites and geographical aspects such as topography and drainage. In addition to these factors, the social and environmental impact of reclamation are of major importance when it comes to deciding the after-use.

Remote Sensing techniques may be employed at any stage of the project, from providing information to aid the planning stage, through the operations stage, insuring the correct specifications are adhered to, and finally to the monitoring stage, checking the rate of settlement and effects on the surrounding areas (Bullard 1981).

With limited funds and a lack of the necessary environmental information, local authorities and private concerns are very hesitant when faced with a reclamation scheme. Quick profits are constantly being sought, usually at the expense of the land. The cheapest methods of reclamation are used, with cost being the only factor taken into account. The method used may be incorrect and some disasterous effects may appear sometime after the contractor has completed. For example, differential settlement will cause humps and hollows to form, followed by waterlogging. The leaching of toxic wastes or

Paper presented at a conference on <u>Matching remote sensing technologies and their applications</u> convened in London by the Remote Sensing Society, 16-18 December, 1981.

minerals can be quite harmful to adjoining areas, and underground fires may result from combustion within the filling material, especially within domestic or industrial wastes or refuse. These problems lend themselves to be detected by remote sensing technologies, as they cause changes in the emitted and reflected energy of the ground from that expected of the undisturbed surface.

2. THURROCK CASE STUDY

Research at the Department of Land Surveying, North East London Polytechnic has been concerned with Thermal Infra-red Linescan (TIRLS) imagery and panchromatic aerial photography of the Thurrock area (South Essex) where problems of dereliction and abandonment are very acute. Large quarries have been dug during the last century, producing chalk, sand, gravel and clay. Some areas have been reclaimed satisfactorily-the clay pits at Grangewaters now provide the public with a water recreation centre, surrounded by a Country Park. Other quarries are being used with the minimum of reclamation work - oil storage tanks can be found in some quarries, these unsightly constructions screened from everyday sight by careful positoning and tree-planting.

Many of the sand and gravel quarries have been filled with domestic and industrial refuse, some successfully under rigid control, as at Bush Farm (DOE/MAFF 1977), although it is some of these 'Landfilled' areas which cause problems. The pits at South Ockendon have been filled and covered with topsoil, the work being completed more than five years ago. One quarry (site 76) was so well reclaimed that it won a Sand and Gravel Association (SAGA) award. Since then, as the refuse has broken down, differential settlement has left the surface very irregular and unusable,except that part used for horseyculture(Coleman 1979) or by squatting gypsies. In places the earth cover is so thin and has been eroded so that refuse has worked itself to the surface, many of the hollows are waterlogged and methane is issuing from the reclaimed surface;thus the land is being constantly degraded to such an extent that another major reclamation scheme will have to be considered.

The problems of the resultant reclamation can be kept to a minimum if action is taken at an early stage. Regular monitoring for several years after reclamation is desiable so as to locate these problems when they occur. If the total area to be monitored is large enough,as that around South Ockendon,Thurrock, then airborne remote sensing imagery can be justified for monitoring. The imagery provides an overview of the area, and it locates problems of overspill. Projects for different purposes, traffic control, land-use surveys or heat loss surveys, for example, may all be included in one flight to enable the cost to be spread(Collins and Ballam 1975).

Panchromatic aerial photography covering the Borough of Thurrock was commissioned in 1976. This imagery provided the Local Authority with a data base for many of these projects, especially in the land reclamation division. In February 1980 and again in February 1981, strips of TIRLS imagery were aquired of certain priority areas.

3. THE USE OF THERMAL INFRA-RED LINESCAN IMAGERY

Inadequately reclaimed land is often attributed to the misuse of the filling materials and topsoil during reclamation, resulting in an over or under-compaction of the subsoil and unevenly spread topsoil. Many soil characteristics may be determined from TIRLS imagery (Myers and Heilman 1969, Reeves et al.1975), although TIRLS only detects temperature differences over a very thin layer of the surface. By careful analysis many subsurface conditions may be interpreted through their effect on surface temperature. Ambiguities surround the detection of absolute temperatures the observations affected by the soil air humidity,rate of evaporation of the soil moisture and wind conditions(speed,direction and temperature), all of which vary considerably, even over small areas (Gurney and Blythe 1980). Therefore, unless measurements are taken at ground level at the time of the aircraft pass, then relative temperatures only can be established. Some idea of the magnitude of the temperatures can however be obtained by comparison with the blackbody values in the scanner.

Soil is a very complex medium, and the amount of radiation emitted and the soil temperature are dependant upon many physical factors within the soil itself.These include the soil's structure, its topography (slope, aspect, altitude and latitude, all of which directly affect the amount of incident sunlight), vegetation cover, moisture content and the rate of evaporation, chemical or physical erosion (resulting in leaching), microbiological activity and the rate of decomposition of organic matter. The soil topography is probably best examined using a stereoscopic overlapping pair of aerial photographs(see section 4). An areal estimate, and perhaps the type,quantity and, to some extent, the quality of the vegetation may be possible, giving clues to the underlying soil, but for more detailed analysis, especially whilst dealing with water content and quality, then TIRLS imagery is recommended. From such information, details of water for irrigation,drainage and crop planting schemes can be proposed. Soil water has a high specific heat; up to five times that compared to other soil constituents; this implies that drier soils warm at a faster rate, and that the degree of soil saturation is the foremost controlling factor in determining soil temperature and associated chemical and biological activity. Excess water content tends to reduce availiable oxygen, thus limiting microbiological growth. Nutrients can also be removed by leaching as the excess water is drained. Too little water can lead to no activity, denying the aerobic organisms of their required oxygen and warmth. An optimum amount of water when held in the soil capillaries can be of great advantage to crop or vegetation growth as organic matter is encouraged and many nutrients are held in solution and are therefore easier for the plants to absorb.

Organic matter in the soil has a direct effect on many of the soil's characteristics - rate of energy absorption and emittance, heat capacity, moisture content capacity, nutrient availiability, soil nature, rate of chemical weathering and biological breakdown. According to Reeves et al. (1975), "a rapid and easy assessment of this parameter (organic content) may be of great utility to the soil surveyor and land-use planner." To be of any use, the topsoil must contain a certain amount of organic matter, and new topsoils tend to be much less fertile than the same material which has had time to develop and aquire its nitrogen-fixing organisms (necessary for plant growth) and other nutrients. Crop yields are often improved by the addition of organic matter and nutrients, in the form of fertilizers, which ,either natural or man-made,tend to breakup the soil,providing improved aeration,stabilization of the soil structure and improved plant-water relationships. The amount of organic material is variable to the climate, vegetation cover, land-use,microbiological and organic activity and finally the temperature of the soil, which affects all the above variables. Organic activity increases with a rise in temperature; from 40°C to 80°C the rate of activity increases eightfold (Miller, Turk and

Foth 1958). The chemical processes and activities are also influenced by the temperature. Soil with a high organic content can hold more water and can thus absorb more energy giving the soil a darker colour, and a different rate of emission. By monitoring the newly reclaimed land using TIRLS imagery, detailed analysis may provide information to determine the varying rates of organicactivity and indicate the locations and amounts of of fertilizer to be added to improve the soil before the damage becomes acute.

Apart from the capability of determining moisture and the possibility of interpretation for organic content, TIRLS can also locate seepage zones, areas of toxicity and leaching, the prediction of landslips (Tanguay and Chagnon 1972), the checking of the efficiency of drainage systems and run-off conditions. Underground fires due to the internal combustion of land-fill refuse give indications at the surface of these subsurface phenomena.

4. THE USE OF AERIAL PHOTOGRAPHY

The basic cause of a poor reclamation scheme is often the lack of information availiable at the planning stage. Some of this information can be obtained by the planners from the stereoscopic interpretation of aerial photographs. Stereoplotting provides accurate and objective survey information (Keech 1979), starting with a base plot of the area, which includes topography, contours and other site details, such as buildings, roads and vegetation cover. In addition to this information concerning the site as an entity ,a synoptic overview of the neighbourhood can form the basis for an Environmental Impact Analysis to aid a landuse or other planning scheme (Bullard 1981). Landuse surveys are usually carried out efficiently using panchromatic aerial photography, despite interpretation ambiguities (Rhind and Hudson 1980) and another set of photography at a later date can be used to monitor changes without additional ground work. Specific information regarding the reclamation scheme may be added to the topographical base-map: the landform, together with type, condition and character of the existing surface cover. Volumes of heaps and holes can be measured photogrammetrically to determine the amounts of materials that require movement during the landscaping process. Often the cause of abandonment may be apparent on the photographs, a fact which may influence the method of reclamation. For example, an adjoining quarry may have a dominating effect on the site,with various pollutants overspilling into the area, sterilizing the soil. The amount of information availiable on aerial photographs is directly related to the scale; that is, there is a minimum size of feature that may be easily identified. Using modern cameras, low flying height (implying larger scale photography) and stereoscopic precision plotters, the required resolution of the imagery for most reclamation projects presents no problems, apart from the adverse weather and the high costs of flying. Thurrock Borough Council has used aerial photography at 1/10,000 scale, which is quite satisfactory for these multipurpose requirements.

It is well known that the use of colour aerial photography can extend photo-interpretation (Reeves et al. 1975). For reclamation work, the only advantage colour has over panchromatic photography is that any slight discolouration on the surface due to toxic waste overspills and leaching are more easily identified. Unfortuneately, the advantages of colour aerial photography are offset by the extra costs and the additional requirements of clearer weather.

By extending the range of sensitivity of the film emulsions to include the near-infrared part of the spectrum (projected as red, causing modifications to the emulsions: green appears blue, red appears green - thus implying 'false colour' infrared film or 'FCIR'), the usefulness of colour film increases substantially, despite the extra cost. Until recently,FCIR photographs have been taken as verticals which provided the user with stereoscopic aerial photography, which, although initially more difficult to interpret, may be used for constructing the base sheet, as well as providing extra information, unavailiable on panchromatic. Information concerning the vegetation cover, types and condition (the use of FCIR for disease detection is well known), and soil characteristics -type,chemical nature, salinity and nutrient levels, moisture, drainage, leaching and pollution can be interpreted using FCIR rather than panchromatic. This implies that FCIR should be very useful in detecting areas of inadequate reclamation, where problems of leaching, the dumping of toxic wastes and insufficient drainage would be highlighted. One initial problem with FCIR, as the name implies, is the fact that objects appear in colours different to those as normally recognised by the eye (healthy crops appear bright red), thus hindering interpretation to some degree, but this is soon overcome by familiarity. The Multispectral Scanner (MSS.) on Landsat, which provides FCIR imagery, detects the same parts of the spectrum that have been examined, but the resolution has proven to be inadequate so far for reclamation planning.

5. CONCLUSION

Current research at the Department of Land Surveying, North East London Polytechnic,is investigating, with partial finacial support from Thurrock Borough Council, the correlation between TIRLS imagery, the amounts of organic matter in the topsoil, and the depth of this topsoil cover when placed over refuse in a landfill site. TIRLS seems to be the optimum sensor to use for this kind of work, as much of the research has been concerned with water content and quality in the reclaimed sites as well as the problems of bad drainage (many of the completed landfill sites are water logged and unusable), and leaching, associated with the inadequately reclaimed land. It is hoped that from this project a monitoring technique using TIRLS will be developed, which could benefit local authorities in their environmental protection and improvement schemes.

REFERENCES

BRADSHAW A.D., and CHADWICK M.J., (1980): The Restoration of Land, Blackwell, Oxford, 317pp.

BULLARD R.K., (1981): Abandoned Land Detection, Reclamation Proposals and Monitoring aided by Remote Sensing. Unpublished PhD. thesis. University of Sheffield.

COLEMAN A., (1979): Land Use Problems, paper presented at the LAND Council's inaugural conference, Land-use in the 1980s, London.

COLLINS W.G., and BALLAM C.J., (1975): Economic advantages of Aerial Photography for Mapping Potential Tipping Sites, Solid Wastes, 65(10) p485-496.

CURTIS L.F., (1976): Remote Sensing for Soil Moisture: User Requirements and present prospects. Proceedings of the 28th Symposium of the Colston Research Society, Remote Sensing of the Terrestrial Environment, Bristol.

DOE., MAFF., SAGA., (1977): Joint Agricultural Land Restoration Experiments Department of the Environment, London.

GURNEY R.J., and BLYTHE K., (1980): Experimental Design for TIRLS Surveys. Paper presented at the Seventh annual conference of the Remote Sensing Society, Geological and Terrain Analysis Applications of Remote Sensing, Plymouth.

KEECH M.A., (1979): Aerial Photography in Land Use Planning, in Land Use and Landscape Planning (ed. LOVEJOY D.J.,), Leonard Hill, Glasgow.

LAMERS J.G., (1980): Application of TIRLS to map slaking phenomena. Paper presented at the Seventh annual conference of the Remote Sensing Society, Geological and Terrain Analysis Applications of Remote Sensing, Plymouth.

MILLER C., TURK L., and FOTH H., (1958): Fundamentals of Soil Science(3rd edition), Wiley and Sons, New York. 526pp.

MYERS V.I., and HEILMAN M.D.,(1969): Thermal Infrared for Soil Temperature Studies, Photogrammetric Engineering, 35 p1024-1032.

RHIND D.W., and HUDSON R., (1980); Land Use, Methuen, London, 272pp.

TANGUAY G., and CHAGNON J., (1972): Thermal Infrared imagery at the St.Jean -Vianne landslide, First Canadian Symposium on Remote Sensing, Ottawa, p387-402.

Part F

ABSTRACTS FOR POSTER CONTRIBUTIONS

UNIVERSITY OF DUNDEE SATELLITE IMAGE DATA ACQUISITION AND ARCHIVING FACILITY

P E Baylis
Department of Electrical Engineering
and Electronics, University of Dundee

Abstract: The University of Dundee maintain a facility for data reception and archiving of high resolution image data from various environmental satellite scanners. These include ITOS/NOAA VHRR, METEOSAT VISSR, TIROS-N/NOAA AVHRR and NIMBUS 7 CZCS. Details of archive data products available are given, followed by a technical description of the station.

1. INTRODUCTION

The first images received at Dundee were from Automatic Picture Transmission TV vidicons on ESSA 2 and NIMBUS 2 in 1966. During 1975 construction of an ITOS/NOAA Very High Resolution Radiometer receiving system was completed. With the aid of a grant from the Natural Environment Research Council, daily archiving of VHRR data commenced on August 23rd, 1976. A total of 2500 VHRR recordings were taken from NOAA 5 during the period up to February 28th, 1979 when the mission was terminated.

A METEOSAT Primary Data User Station for the reception of Visible and Infra-red Spin Scan Radiometer digital data was constructed prior to launch of METEOSAT 1 in November 1977. A receiving system for Advanced Very High Resolution Radiometer digital data reception from TIROS-N was commissioned in time for the launch of TIROS-N on October 13th, 1978. Archiving commenced soon after launch and continues now with NOAA 6 and NOAA 7.

The Dundee station was upgraded to receive NIMBUS 7 Coastal Zone Colour Scanner data during the first half of 1979. Daily archiving of CZCS data commenced on August 3rd, 1979.

In December 1979 a Ramtek 512 x 512 pixel 8 bit TV frame store hosted by a Data General Nova 2 minicomputer was interfaced with the station video data systems. In November 1981 a DEC PDP 1134 computer was installed in order to improve the station data handling capability.

Prime funding is now in the form of a contract between the University

Paper presented at a conference on Matching remote sensing technologies and their applications convened in London by the Remote Sensing Society, 16-18 December, 1981

of Dundee and NERC Scientific Services. The station is operated 14 hours a day, seven days a week. Automation of night-time data reception is presently being implemented.

2. SPACECRAFT AND SENSORS

This section contains a summary of the spacecraft and imaging sensors of concern to the Dundee station.

2.1 VERY HIGH RESOLUTION RADIOMETER

Spacecraft:	ITOS/NOAA (second generation series, terminated 28th February 1979)
Orbit:	1500 Km, sun-synchronous
Equator crossing time:	0830 LST southbound
Scan rate:	400 lines per minute
IFOV at nadir:	0.9 Km
Spectral channels:	(1) 0.6 - 0.7 μm
	(2) 10.5 - 12.5 μm
Coverage (overhead pass):	Along track 8000 Km, across track 4400 Km
Grey scale resolution:	Approx. 100 steps covering VHRR dynamic range
Video data link:	Analogue

The last of the second generation of NOAA polar orbiters was NOAA 5, launched July 29th, 1976.

2.2 ADVANCED VERY HIGH RESOLUTION RADIOMETER

Spacecraft:	TIROS-N/NOAA (third generation, series commenced October 13th, 1978)
Orbit:	840 Km, sun-synchronous
Equator crossing time:	Two spacecraft in orbit at once; 0730 LST southbound and 1500 LST northbound
Scan rate:	360 lines per minute
IFOV at nadir:	1.1 Km
Spectral channels:	(1) .55 - .68 μm (.55 - .90 μm TIROS-N)
	(2) .725 - 1.10 μm
	(3) 3.55 - 3.93 μm
	(4) 10.5 - 11.5 μm
	(5) 11.5 - 12.5 μm (NOAA 7)
Coverage (overhead pass):	Along track 6000 Km, across track 3000 Km
Video data link:	10 bit binary words, 665.4 Kb/s.

First of the third generation of NOAA polar orbiters was TIROS-N, launched October 13th, 1978. AVHRR failed in November 1980. NOAA 6 was launched June 27th, 1979 and is still operational. Replacement for TIROS-N is NOAA 7 launched June 23rd, 1981. It is fitted with the first AVHRR to have channel 5, intended to improve atmospheric correction of sea and land surface temperature measurements.

2.3 COASTAL ZONE COLOUR SCANNER

Spacecraft:	NIMBUS 7 (Last of the NIMBUS series, launched October 24th, 1978)
Orbit:	970 Km, sun-synchronous
Equator crossing time:	1200 LST northbound
Scan rate:	484.84 lines per minute
IFOV at nadir:	.8 Km
Spectral channels:	(1) .433 - .453 μm (4) .660 - .680 μm (2) .510 - .530 μm (5) .700 - .800 μm (3) .540 - .560 μm (6) 10.5 - 12.5 μm
Coverage (overhead pass):	Along track, typically 6 - 11 minutes Across track 1500 Km
Video data link:	8 bit binary words, 800 Kb/s

NIMBUS 7 is a NASA research satellite. CZCS is switched on only as required by approved experiments.

2.4 VISIBLE AND INFRA-RED SPIN SCAN RADIOMETER

Spacecraft:	METEOSAT
Orbit:	35,700 Km, geostationary
IFOV at nadir:	2.5 Km chan. 1 5.0 Km chans. 2 and 3
Spectral channels:	(1) .4 - 1.1 m (2) 5.7 - 7.1 m (3) 10.5 - 12.5 m
Video data link:	8 bit binary words, 166.6 Kb/s.

The METEOSAT programme is managed by the European Space Agency. METEOSAT 1 was launched on November 23rd, 1977 and failed on November 24th, 1979. METEOSAT 2 was launched on June 19th, 1981.

3. STATION ROUTINE

The station is operational seven days a week throughout the year with two overlapping shifts per day, covering the period approximately 0800 to 2200 GMT. Work is divided between tracking selected orbits and archive retrieval.

All archived data are stored on high density instrumentation tape unprocessed, as received from the spacecraft and will be kept indefinitely.

3.1 VHRR AND AVHRR

Basically, at least two satellite passes per day closest to overhead the U.K., have been recorded since archiving commenced in August 1976. With NOAA 5, a N-S pass was recorded each morning and a S-N each evening. With TIROS-N, recordings were taken from two successive S-N passes each afternoon up until AVHRR failure on November 1st, 1980. Since the launch of NOAA 6 (June '79) recordings have been taken each morning from the closest to overhead U.K. N-S pass. However, after the failure of TIROS-N AVHRR recordings were taken each evening from two successive S-N NOAA 6 passes instead of the TIROS-N passes, in addition to the NOAA 6 morning passes. However, since the launch of NOAA 7 (June '81) afternoon recordings have been resumed and the evening NOAA 6 recordings discontinued.

In addition to the above, extra passes are recorded on occasion, by special request to support oceanographic research activities. Typical examples are JASIN, MARSEN, EURASEP, KONTUR etc. In particular, early a.m. recordings were taken from NOAA 7 during the N. Sea KONTUR project (Sept. 14 - Oct. 22 1981) in order to enable groups interested in sea surface temperature measurement to validate the new AVHRR five channel instrument.

3.2 CZCS

The CZCS is usually, but not always, switched on over European waters during the S-N daytime passes of NIMBUS 7. Normally it is switched off at about latitude 60° N, but sometimes is left on up to the northern horizon in polar regions. Since August 3rd, 1979 the Dundee station has tracked NIMBUS 7 every day for 2-3 successive passes, recording CZCS data whenever it is switched on.

3.3 VISSR

Recordings from METEOSAT are taken only by prior request. Routine archiving is not performed at Dundee since an archive is maintained by the spacecraft managers at Darmstadt in the Federal Republic of Germany.

4. PRODUCTS AVAILABLE FROM THE DUNDEE ARCHIVE

4.1 HARDCOPY

Primary output from the VHRR/AVHRR/CZCS archive is hard copy photofacsimile imagery, from any chosen channel, linearised to correct earth curvature distortion, but not earth rotation de-skewed. Linearisation means that the scale of the images is very nearly (but not exactly) constant in all directions and at all points in the image. However, in the case of CZCS further distortion in the image is caused by tilt of the scan mirror to avoid sun glint.

Images are basically in one of two categories:

(a) Full pass (x1) with or without latitude/longitude and land outline grids. These images are contact printed from full size negatives. They can also be supplied as Xerox copies, of reduced quality, which are quite readable in terms of major meteorological features and are much less costly.

(b) Sectorised enlargements. These are first generation products derived directly from the archive tape and are selected sectors of the full pass image. Sectorisation is by electronic stretching and the resulting image represent the best quality available from the archive. Gridding of sectors is not at present available, but they are linearised.

A summary of VHRR/AVHRR/CZCS sizes and areas covered is given in Table 1.

VHRR/AVHRR and CZCS images can be grey scale stretched to any desired extent to bring out such features as sea surface temperature gradients in IR images.

4.2 BROWSE FILES

A complete set of VHRR/AVHRR/CZCS x1 contact prints is kept on the station to indicate the contents of the archive. However, not all scanner channels are included. For CZCS channel 5 is used and for VHRR/AVHRR VIS. and IR from the closest to overhead U.K. passes are included. For other passes either VIS. or IR is selected. Periodically the AVHRR browse file prints are photographed on to 35 mm film and then contact printed 36 to a page, either as positive prints or positive transparencies as required. Both full size prints and miniatures are distributed to users on the mailing list. Back numbers dating back to August 1976 are available. Full size x1 contact prints are also distributed on a weekly batch basis.

The station browse file may be consulted by visitors, alternatively browse file searches can be undertaken on behalf of data users.

4.3 COMPUTER COMPATIBLE TAPES

CZCS or AVHRR scenes can be transcribed from the archive tapes on to CCT against specific requests.

Data on CCT are unprocessed, as received from the spacecraft and therefore contain the full scan width and all channels. The format used is 1600 BPI/PE and approximately 30 cm of CCT is required for each scan line of CZCS or AVHRR.

VHRR CCTs are not available. CCTs of any video data, provided they are correctly formatted, may be sent to Dundee for hard copying in black and white photo-facsimile with a maximum line density of 118 lines/cm. (300 LPI) up to size 25 x 25 cm.

TABLE I

Magnification	Lines/cm	Size cm	Total Lines	Grid	Area Coverage H x W Km
VHRR x 1	246	25 x 21	6250	Optional	5625 x 4400
VHRR x 2	118	25 x 25	3000	No	2800 x 2800
VHRR x 4½	59	25 x 25	1500	No	1350 x 1350
VHRR x 6¼	39	25 x 25	1000	No	900 x 900
AVHRR x 1	177	24 x 16	4275	Optional	4700 x 3000
AVHRR x 1½	118	25 x 25	3000	No	3300 x 3300
AVHRR x 3	59	25 x 25	1500	No	1650 x 1650
AVHRR x 4½	39	25 x 25	1000	No	1100 x 1100
AVHRR x 9	20	25 x 25	500	No	550 x 550
AVHRR x 18	10	25 x 25	250	No	275 x 275
CZCS x 1	118	25 x 17½	3000	No	2400 x 1682
CZCS x 2	59	25 x 25	1500	No	1200 x 1200
CZCS x 3	39	25 x 25	1000	No	800 x 800
CZCS x 6	20	25 x 25	500	No	400 x 400
CZCS x 12	10	25 x 25	250	No	200 x 200

5. TECHNICAL DESCRIPTION OF STATION EQUIPMENT

Most of the equipment was designed and constructed in-house. Bought-out equipment include the magnetic tape transports and computers. Some items, such as microwave front ends and telemetry receivers, can be constructed at Dundee under contract for other users wishing to set up a ground receiving station.

The AVHRR receiving system is described with brief references to METEOSAT VISSR and NIMBUS 7 CZCS as appropriate.

5.1 ANTENNAS AND TRACKING

The main tracking antenna is a spun aluminium reflector, 3.6 m diameter fixed to an elevation on azimuth mounting. Azimuth range is about 560° and elevation, 0-180°. Position is determined by a pair of incremental shaft encoders with a resolution of .1°. These encoders have a built in reference position which produces a pulse whenever the shaft rotates through it. They were aligned with the known azimuth and elevation of a local building.

The drive motors are variable speed D.C. shunt wound, thyristor controlled. Maximum azimuth slew rate is 11°/sec. and elevation 2°/sec.

Tracking near zenith is accomplished by commanding the antenna to go up to and over 90° if the maximum elevation required is above 88°. The very high azimuth slew rate demanded by antennas with an elevation range of only 0-90° is avoided. Nevertheless, the antenna boresight tracking trajectory is deliberately modified from true satellite position when the satellite is near zenith in order to minimise tracking errors due to limited slew rate. If the antenna was commanded to follow the true satellite trajectory, azimuth slew rates above 11°/sec. would be demanded and tracking errors would be greater. By a combination of 0-180 elevation capability, a -3 dB beamwidth of 3.5° and optimised tracking trajectories near zenith, the Dundee antenna can maintain signals well above margin for near overhead passes.

A NOVA 1210 with 20 K of non-volatile memory is programmed in machine language to compute look angles at a rate of approximately four complete computations per second during the pass. A special interface is connected between the computer I/O data bus and the motor thyristor control and shaft encoder BCD outputs. The computer compares calculated azimuth and elevation look angles with actual antenna position. The difference is used to command motor speed and direction. Just before a pass is due, the operator enters via a teletype keyboard, which satellite is being tracked together with the longitude and time of equator crossing. The computer calculates horizon time and commands the antenna to the start position. When actual time equals horizon time, tracking commences. Minor timing corrections can be made during the pass via the keyboard, the effect being noted on an S-meter.

The METEOSAT antenna is identical to the tracking antenna except that the mounting is such that tracking is limited to a few degrees by hand.

5.2 FRONT ENDS

There are two front ends fixed to the tracking reflector. One is at the focus and is designed for 1.7 GHz AVHRR data reception and the other is strapped on as close to the focus as possible for 2.2 GHz CZCS data. Elevation offset for CZCS is 7.5°. Each front end consists of a circular waveguide feed, bandpass transistor amplifier and down converter. Estimated G/T is 10.1 dB/°K for AVHRR and 13.0 dB/°K for CZCS. The METEOSAT front end is similar to the AVHRR.

5.3 TELEMETRY RECEIVERS

A common receiver is used for AVHRR and CZCS but a second one is used for VISSR. First I.F. from all front ends is centred on 112.5 MHz which is bandpass amplified and down converted to 10.7 MHz in the receivers. After passing through a bandpass hard limiter the signal is demodulated by a carrier recovery phase lock loop phase demodulator. An FM discriminator is included in the AVHRR/CZCS receiver for use when the FM down-link from NIMBUS 7 is in operation. Output of the demodulator is bipolar Bi-ϕ-L which is passed to the appropriate bit conditioner, there being a separate one for AVHRR, CZCS and VISSR.

5.4 BIT CONDITIONER

A phase lock loop recovers 2 x clock rate from the input Bi-ϕ-L signal after it has been lowpass filtered and then squared. The bipolar

Bi-ϕ-L is then converted to bipolar NRZ-L in a four quadrant multiplier before being integrated bit by bit. Integration of each bit gives maximum possible discrimination against noise in the process of deciding if a 0 or 1 is being received. 2 x clock is divided by two to produce clock which is used both to convert Bi-ϕ-L to NRZ-L and to control the integrator. Phase ambiguity introduced by the divide by two circuit is resolved with the sync. pulse detector in the frame synchroniser. If a sync. pulse is not detected within a frame period, the clock phase is reversed.

5.5 FRAME SYNCHRONISER

NRZ-L from the bit conditioner is clocked into a shift register long enough to accommodate the sync. words. Parallel outputs of the register are monitored by gates arranged to detect the presence of sync. words in the register. Detection of a sync. word sets the phase of a $\div$ n counter (where n is the number of bits per data word), the output of which operates an n bit data latch connected to the last (first in) n bits of the register. In this way the serial bit stream is sliced up into n bit words. A word counter that recycles at the frame rate is used to control the channel decommutator.

A separate frame synchroniser and decommutator is used for AVHRR, CZCS and VISSR since the frame structure, word lengths and sync. words are different.

5.6 DECOMMUTATOR

Each decommutator is a dedicated hard wired circuit that strips off the required channel pixel words as selected by a panel switch. The words from one channel scan are temporarily stored in a RAM from which, in the case of AVHRR and CZCS they are read out at a variable rate to correct earth curvature distortion. Selection of the required words is aided by the word counter in the frame synchroniser and the linearisation algorithm is stored in a PROM. Image sectorisation is accomplished by reading out a selected portion of the RAM at a reduced clock rate. Output from the RAM is passed to the video processor and/or the Ramtek image processor as required.

5.7 VIDEO PROCESSOR

The video processor is used to select which part of the satellite scanner dynamic range is to be matched to the photo-facsimile black/white range. This process is achieved by a RAM look up table which is addressed by each input pixel word. Contents of the RAM determine the processor transfer function which is loaded by a special circuit controlled by two front panel analogue potentiometers. In effect the potentiometers generate two separate numbers A and B both in the range $0-(2^n - 1)$. Any input pixel value $<$ A produces a saturated black level output and any value $>$ B, saturated white. In the range A to B, the output rises linearly. A third front panel control applies gamma correction to the linear part of the transfer function as required. Output of the RAM is passed to a photo-facsimile hard copy machine via a D/A converter.

Sector enlargement factors are determined by the facsimile machine line densities, which are adjustable to 118, 59 or 39 lines /cm. Alternatively, higher enlargement factors can be achieved by use of the Ramtek 512 x 512 pixel TV frame store. Data are fed into the store from the archive tape transport, magnified under software control and then fed from the store to the facsimile machine. The latter method has the advantage that the operator has interactive control over image location and contrast before committing to hard copy. Provided no more than 512 scanner lines are processed this way, no spatial resolution is lost.

5.8 TAPE TRANSPORTS

All data are archived unprocessed on instrumentation recorders.

The NOAA 5 VHRR data link was analogue, in the form of a 99 KHz subcarrier frequency modulated by video. The subcarrier was recorded on an S.E. Labs 7000A 14 track machine running at 76.2 cm/sec. One track of a 1400 m tape was adequate for one pass so that each tape accommodated 14 passes.

In the case of AVHRR and CZCS data, NRZ-L is converted to DM (Miller Code) before recording on an S.E. Labs 7000 M 14 track machine. With AVHRR the tape speed and packing density are 7.5 cm/sec and 8.85 Kb/cm respectively and with CZCS 152.4 cm/sec and 5.32 Kb/cm respectively. One reel of 1400 m tape can accommodate 28 AVHRR passes or 14 CZCS. All tapes are 2.54 cm width.

The station has two S.E. Labs 7000 M machines and associated electronics, so that data reception and archive retrieval work can be carried on independently.

An S.E. Labs 8000 computer compatible tape transport can accept raw data from the high bit density 7000 M archive machines, so that CCTs of selected AVHRR or CZCS scenes can be transcribed into standard computer format as required. In the case of AVHRR data, groups of 3 10 bit words are assembled into groups of 4 8 bit words and recorded in blocks of 3700 bytes. With CZCS the 8 bit words are recorded in blocks of 2476 bytes.

The same CCT transport can accept CCTs of video data written in blocks of 4096 bytes, one block per scan line. Output is hard copied in a black and white photo-facsimile machine which has optional line densities of 118, 59 or 39 lines per cm on 25 x 25 cm paper.

In due course the CCT transport will become a peripheral of the PDP 1134 allowing much greater flexibility in reading and writing CCTs.

6. CONCLUSIONS

The continuing requests for data from the Dundee archive indicate a requirement for the facility. The greatest use of the archive data is in oceanography where the requirement is for sea surface temperature and water colour. Although the station is not primarily geared up to provide real-time data, images can be produced and dispatched to users very quickly in support of particular projects.

A university electrical engineering department provides a good environment for the station because of the excellent facilities for in-house design and construction of specialised electronic equipment. Much of the equipment and software are the results of student effort at both undergraduate and postgraduate level.

BIBLIOGRAPHY

The following references may be useful to anyone requiring more detailed information about the satellite systems and station equipment discussed in this paper.

1. SPILKER, J. J., "Digital Communications by Satellite", Prentice-Hall, 1977.
2. NOAA Technical Memorandum NESS 95 "The TIROS-N/NOAA A-G Satellite Series", NOAA/NESS Washington D.C.
3. NOAA Technical Report NESS 75 "Guide for Designing RF Ground Receiving Stations for TIROS-N", NOAA/NESS Washington D.C.
4. "METEOSAT Ground Segment Handbook", ESA/ESOC Darmstadt, Germany, 1977.
5. MATHER, J. R. "Study and Specification of a Low Cost/High Performance METEOSAT Primary Data User Station" Dept. of Electrical Engineering and Electronics, University of Dundee, 1981.
6. "NIMBUS G Data Systems Description" NASA Goddard Space Flight Centre.
7. BAYLIS, P. E., MATHER, J. R. and BRUSH, R. J. H., "A METEOSAT Primary Data User Station" ESA/ESOC, Darmstadt, Germany, 1980.
8. BRUSH, R. J. H., "An Algorithmic State Machine Decommutator for METEOSAT Primary Data User Stations", The Radio and Electronic Engineer.
9. BAYLIS, P. E. and BRUSH, R. J. H., "NOAA SR/VHRR Receiving and Archiving Facility", U.K. Remote Sensing Society, 1977 Conference Proceedings.
10. BENTLEY, W. E., "Squeeze More Data onto Mag. Tape", Electronic Design, 21 October, 1975.
11. EMI Engineering Handbook, "Modern Instrumentation Tape Recording", EMI Technology Inc. 1978.

MICROWAVE REMOTE SENSING OF CORN

F Benincasa: G Maracchi: S Paloscia: P Pampaloni: G Zipoli
CNR-IATA, Firenze, Italy

The physical parameters of corn have been investigated with remote sensing techniques during the growth stage from June to July 1981. For this Research concerned with ESA-JRC SAR 580 Campaign, experimental equipment for ground based remote sensing of the crops has been arranged by IATA-CNR of Florence.

The equipment consists of two microwave radiometers at 9.8 and 36.6 GHz mounted on a 10m tower together with an IR (8-14 m) radiometer, many micrometeorological sensor placed over vegetation and an automatic acquisition system with a digital cassette recorder. 45 readings were taken every 20 minutes during the day and one hour at night. The radiometric data were taken at the two linear polarizations V and H with an angle of incidence variable between 10 and 70 degrees from nadir, while IR radiation was measured at a fixed angle of 45 degrees. The other 'in situ' measured parameters were the following:

Air temperature and humidity under and over vegetation (5+5 points)
Soil temperature, moisture and heat flux
Solar radiation global and net
Wind velocity (direction and intensity at two heights)
Evapotranspiration computed from the energy budget (Bowen Ratio)
Stomatal resistance of the leaves
Plant heights and water content

In order to evaluate the potential of microwave radiometry in agriculture many correlations have been made between microwave data and the measured physical parameters. One objective was to establish from which layer, between soil and the top of vegetation the microwave radiation originates, whether this layer is the same for the two frequencies, whether it changes during the day and finally whether and in which way it depends on the physical and the growth stage of the culture.

The results are reported in diagrams and give interesting information on the general problem of the interactive mechanism of the electromagnetic waves and vegetation. The most important application should be the prediction of evapotranspiration. During this study the flight of the SAR 580 aircraft took place on June 19 preceded by a survey made with an airborne IR multispectral scanner. When these data are available they will be compared with radiometric and ground data.

CALIBRATION OF DIGITAL SAR IMAGES IN SUPPORT OF THE EUROPEAN SAR-580 CAMPAIGN

D G Corr   Systems Designers Ltd, Fleet, Hants, UK
S Bruzzi   Earthnet, ESRIN, Frascati, Italy

ABSTRACT

An important part of the recent European SAR-580 Campaign was the collection of radar data from calibrated corner reflectors. The purpose of these data is to provide a means by which the values given in digital SAR image products can be related to absolute values of radar cross-section. The objectives of the calibration task are to investigate the extent to which this is achievable to quantify the experimental errors and to provide calibration information for use by experimenters. The calibration activity is being undertaken by Systems Designers Ltd working at Esrin, Frascati.

Digital SAR images can be produced by processors at RAE Farnborough and at DFVLR Oberpaffenhofen. Associated with each calibration site image, there is also an image of simulated point targets. These are derived from the use of a calibration signal (CALSIG).

The calibration software locates and extracts the image data for real and simulated targets. The peak pixel intensity is determined by interpolation. A model of the radar system and image processor is then used to relate these peak values to the known target scattering cross-sections. The simulated data are used to help to relate data from different flights or from different processors. The calibration software is under development and project completion is expected by July 1982.

TOPOGRAPHIC MAPPING FROM SPACE IMAGERY

Progress and Potential

A Report on a Remote Sensing Society Workshop

I. J. Dowman
Department of Photogrammetry and Surveying
University College London

Abstract: This paper consists of a report on the discussion at the workshop and incorporates short papers by G. Davison (RAE), M. Faulkner and S. Fraser (OS) and P. Carter (Harwell). The main aspects presented are a review of applications to date; the needs for the future and an assessment of the potential of currently planned sensors to meet the need. The amount of work done so far has been limited because of insufficient resolution (Landsat) or limited imagery (Skylab). The Metric Camera Experiment of Spacelab, the Large Format Camera on the Shuttle and SPOT will considerably extend the range of possibilities. A particular problem which exists is the determination of heights and some developments on this topic are discussed.

A major discussion point is just what are the requirements for topographic mapping, how great is the need for geometrically correct images and how far space imagery can meet the requirements The main needs are:

- base maps for thematic mapping;
- small scale maps to give a synoptic view;
- national map series at 1:50 000 scale and smaller where none exist;
- revision of national map series.

Some points relating to this discussion will be presented.

INTRODUCTION

A remote Sensing Society Workshop was held at University College London on 10th April 1981 to review the present situation regarding topographic mapping from satellite imagery and to discuss future plans and requirements. This paper gives a summary of the short papers given by several speakers and a report of the discussion. A list of participants is given in the appendix. The objectives of the meeting can be briefly stated as:

1. To present information about the work done to date using satellite imagery and imagery relevant to satellites, such as Infra red linescan and radar, for topographic mapping.
2. To present information about the plans for future satellites.
3. To discuss requirements of UK users which could be satisfied by satellite imagery.

The morning session was devoted to the first two objectives and the afternoon session to the third.

Paper presented at a conference on Matching remote sensing technologies and their applications convened in London by the Remote Sensing Society, 16-18 December, 1980.

SATELLITES AND RESULTS

A review of satellites and results of analysis of the imagery was given by Professor G. Petrie of Glasgow University. Three principal types of imagery have been used: Skylab 190B, Landsat MSS and Landsat RBV. Skylab imagery is particularly interesting because it is photographic and of a similar scale and resolution to the photographs which will be obtained from cameras in the first U.S. shuttle missions. Landsat MSS has been widely used for thematic mapping because of its multispectral nature but Landsat RBV has been less used. Tests made on the imagery have been of two types - analysis of geometrical properties and evaluation of the information content. Some results of geometric analysis are given in Table 1; it can be seen that after transformation quite high accuracy can be obtained in plan. Generally no useful height information can be obtained, although some work (Dowman and Mohammad 1981) has indicated that there is some potential here. The accuracy of position is not the only consideration in using imagery for topographic mapping because the required detail has to be visible and the whole image has to be rectified. Detectability is particularly important and can be assessed in mapping tests; in order to make direct comparisons the image must be geometrically transformed.

Several comparative tests on mapping have been carried out. Work on Landsat MSS in the Sudan has shown that detail seen on existing maps such as roads and settlements cannot be seen on the imagery but that information seen on the imagery such as land use, water courses and an impression if relief is not shown on the maps. More information is visible on Landsat RBV and Skylab photographs.

Some examples are available of using Lansat imagery for National Mapping. The Directorate of Overseas Surveys (DOS) has produced maps of Antarctica at a scale of 1: 250 000 and 1:500 000 and have recently produced a 1:1 000 000 map of Tanzania in collaboration with the Regional Centre for Services in Surveying and Mapping in Nairobi. The Earthsat Corporation have produced maps of Libya at 1:500 000 scale. DOS have found that the information content of Landsat MSS is limited and have supplemented it in Tanzania with RBV imagery. Hunting Surveys have used Landsat MSS for maps in the Arabian Gulf area. The conclusion from Petrie's paper and the following discussion was that the limit of use of Landsat MSS is 1:500 000 mapping although there may be specific applications at larger scales.

Three speakers then presented reports of work carried out in the U.K. First G. Davison from the U.K. Remote Sensing Centre, spoke on The Geometric Transformation of Satellite imagery.

GEOMETRIC TRANSFORMATION AT RAE

pace Department of the Royal Aircraft Establishment (RAE) first became involved in the processing of data from earth resources satellites in mid-1977. The principlal type of data received is Landsat multi-spectral imagery, which, although corrected for earth curvature and rotation, satellite attitude errors and sensor non-linearities, still contains significant geometric distortions, with the centre of the scene only located to an accuracy of 1km.

When the UK Remote Sensing Centre was formed at RAE in April 1980 it was decided to provide software facilities for transforming Landsat imagery into the British National Grid. To determine the transformation required the user selects a number of ground control points (i.e. features which are readily identifiable on both the image and the map). The first task of the UK Remote Sensing Centre was therefore to compile a ground control point library for Great Britain. This has been done using the Ordnance Survey 1: 50 000 scale maps which give coordinates in the British National Grid. The choice of

TABLE I: RESULTS FROM SENSORS ALREADY LAUNCHED

Mission	Launch Date	Alt. (km)	Sensors	Spectral Bands µm	IFOV/ Resolution	Accuracy			Application
						Plan	Height	Comment	
LANDSAT 1 and 2	1972 1975	920	MSS	.5 - .6 .6 - .7 .7 - .8 .8 -1.05	80 m	50 - 80 m	50 - 200 m	Accuracy after geometric transformation	Thematic mapping Resource mapping monitoring etc. Mapping at 1:500 000
			RBV		80 m				
SKYLAB	1973	435	S 190 A Camera	.4 - .7	$29 \ell pmm^{-1}$ 100 m	40 - 60 m	150 - 180 mm	Between lens shutter IMC	Multi spectral camera
			S 190 B Camera	.4 - .88	$25 \ell pmm^{-1}$	140 m	500 m	Focal plane shutter	Earth terrain camera
SOYUZ 22	1976	250	Zeiss MKF6	6 bands in range .48 - .84	$80\ pmm^{-1}$				Earth observation
HCMM	1977	600	Radiometer		500 m				
LANDSAT 3	1978	920	MSS	as LS 1+ 2 + 10.5 - 12.4	80 m	As LANDSAT 1 and 2			
			RBV	0.5 - .75	40 m				
SEASAT	1978	800	SAR	L band	25 m	> 50 m	-		Oceanographic applications

coordinate systems is not a constraint, however, as facilities exist for transforming between the National Grid and other map projections (e.g. Universal Transverse Mercator, Lambert conical and latitude, longitude and height relative to any geoid). There are 204 1:50 000 map sheets covering Great Britain and at least five control points have been selected on each map giving an overall total of 1200.

To obtain the map coordinates of a GCP a 25*25 patch of image centred on the control point is output as a photographic negative enlarged to a scale of 1:50 000. The centre of each patch is marked with a cross and a line is drawn at the angle of the satellite orbit to grid east-west. Each patch is laid over the appropriate portion of the map and translated until the best fit is obtained. A pin is then pressed through the centre of the cross to mark the exact position of the control point on the map, and the coordinates are read off.

For each GCP, the map coordinates (Northing, Easting and height) and the image coordinates (column and row) are stored, and when a scene has been completed (i.e. around 100 control points found), a least squares fit is performed to determine the transformation between image and map coordinates. The residuals in this fitting process give an indication of the transformation accuracy and yield a typical r.m.s. error of 50m.

The selection of ground control points is a repetitive, labour intensive task and so an automated procedure has been developed so that any subsequent scene of the same area can be dealt with using software alone. In this procedure a library of n*n sub-images centred on each control point is formed, and correlation techniques are then used to identify the corresponding points in the new scene. Experiments have shown that the optimum balance between accuracy and computing time is obtained by setting n = 19.

At present the ground control point library only contains data relating to Landsat MSS imagery, but the filing systems have been designed so that they can easily be extended to cover other types of data.

Software has also been written to enable users to obtain transformed imagery by specifying the orientation and pixel sizes for the scene required. In the transformation process the image is resampled with the software providing a choice of three possible techniques namely:- nearest neighbour, linear interpolation and cubic convolution.

For many applications, users require some form of classified imagery. Conventional raster scans are an expensive way (in both cost and computing time) of presenting this data, and software has been generated to convert classified imagery into map files (i.e. line diagrams). The transformation can then be applied to the map file, and the geometrically transformed data output to a graph plotter, together with an overlay grid.

TESTS ON GEOMETRICALLY TRANSFORMED LANDSAT IMAGERY

The Ordnance Survey has carried out tests on geometrically transformed data and these were described by Col. S. Fraser.

Landsat satellite imagery has potential for the depiction and update of coastal information on OS small scale topographic maps. Because the pixel size is approximately 79m x 56m the largest OS scale at which such information could be used is likely to be 1:250 000, 1:50 000 maps may therefore be used as a standard to assess accuracy. Co-ordinates from these maps can be measured to about 20 metres.

Following liaison with the Remote Sensing Centre, prints and negatives of Landsat imagery geometrically transformed to National Grid were supplied to Ordnance Survey. The imagery, was part of Landsat 2 scene 219/024 Band 7 and covered the Severn Estuary. The negatives were produced at a scale of approximately 1:500 000 on the RAE Linoscan equipment. The prints were photographic enlargements by RAE to 1:250 000. A print of untransformed imagery of the same scene was also supplied (OSM).

The procedure was designed to test the repeatability of measurement on the imagery and the accuracy of the imagery compared with 1:50 000 maps. Forty points of detail were selected on the imagery. These points were identified on the ten 1:50 000 sheets covering the area. A Wild A10 plotting table fitted with EK22/teletype equipment was used as a coordinatograph providing a direclty printed record of co-ordinates which were read to 0.01 mm from the imagery. The Haag-streit Rectangular Coordinatorgraph was used to give NG coordinates to 10m from the 1:50 000 maps. A least squares linear transformation was used to fit in turn each set of co-ordinates to the National Grid. Scale, rotation and shift parameters were obtained and the residuals listed.

Analysis of 5 sets of measurements to two documents provide an indication of repeatability of measurements. The standard error of single measurement on a print at scale 1:250 000 was ±17m and on a negative at scale 1: 500 000 was ±28m. Following transformation to National Grid co-ordinates and elimination of points with Eastings and Northings residuals 3 or more times the standard error, the results shown in Table II were obtained.

TABLE II

Document	Nominal scale	Computed	Error	σ	No. of iterations
01A Print	1:250 000	1:251 154	0.46%	53m	1
02B Print	1:250 000	1:250 630	0.25%	44m	2
031 Neg	1:500 000	1:500 627		47m	3
04L Neg	1:500 000	1:500 146		61m	1
OSM Print	1:250 000	1:251 260	0.5%	402m	0

These limited tests indicate that the accuracy of the transformed imagery is satisfactory for 1:250 000 revision.

A study of the direction of the residuals on the unrectified print indicates local systematic error. The overall accuracy is inadequate for 1:250 000 revision but the untransformed image could be useful if detail is fitted to map detail over small areas. The majority of topographic information availabl from Landsat imagery is readily available from other sources, and for the time being at least Ordnance Survey has not identified a viable application of the imagery.

MONITORING URBAN GROWTH

The third speaker was Dr. P. Carter from the Harwell Image Analysis Group which has been monitoring urban growth using Landsat imagery for the Department of the Environment. The Instantaneous Field of View (IFOV) of Landsat is 79 x 79m and thus each ground element covers a mixture of features such as roofs, garden paths and roads and consequently the scattered radiation intensities recorded are average values from all the categories present.

In order to be able to compare the results of automated classification of Landsat data with ground-truth, and also to combine the data for several scenes over an area, the Landsat digital data is geometrically corrected by the use of selected ground-control points to the Ordnance Survey National Grid coordinate system with a cell size of 50m. Typical errors are ±50m after rectification. After the determination of the spectral-signatures for the ground categories of interest by using areas of known ground-truth the Landsat data is automatically classified into these categories. Typical categories are water, inner urban, suburban, woodland, heathland, grassland and agricultural.

In order to check the accuracies of the classified data the Department of the Environment have visually interpreted aerial photography over a number of test sites and from the resultant 'developed-area' maps the results have been transferred into a digital base based on the O.S. coordinate system with a 50m cell size.

The results for the test sites (Reading, Northampton, Preston, South Manchester and Sheffield) were that urban areas were overall delineated with about 70% accuracy and rural areas about 90%. The main errors of omission in the urban areas were due to the older more select residential areas having a predominantly vegetation spectral signature, while the main errors of commission in the rural areas were due to the fact that bare soil, etc. has for the present Landsat a spectral signature similar to urban areas.

Therefore the classification of a single Landsat scnene is not accurate enough for resurveying completely developed areas and growth cannot be reliably delineated by simply subtracting the results from two Landsat scenes. New growth is relatively a very small amount equivalent to about 0.1% of the total area per year.

However we can derive information on urban growth if we adapt the requirements to the data available. The errors of omission can be eliminated by using the D.O.E. 1969 Developed area data base (obtained from aerial photography) as a starting point and only looking for growth behind the existing boundary. The errors of commission can also be reduced by limiting the area of search to the region around the existing boundary (this is where most of the growth takes place), only accepting areas greater than 2.5 hectares and checking for consistency on more than one scene.

After the application of these elimination rules urban growth can be defined to within an accuracy of 80% while the commission errors are considerably reduced. Thus the Landsat data is a very cost effective way of collecting intelligence information on urban growth at national and regional level. The application of textural parameters and use of RBV imagery would further improve accuracies. The potential of future satellites, sparticularly the Landsat D Thematic Mapper for land use applications of this type is considerable.

RADAR IMAGERY

Dr. P.H.A. Martin-Kaye of Hunting Geology and Geophysics Limited gave a review of the characteristics of radar imagery, the type of purposes to which mapping radars have been applied and problems that are encountered in interpretation.

Side-looking airborne radar (SLAR) has been used in numbers of resource mapping projects, the largest being Projecto RADAM of Brazil (9 million sq. km) and Nigeria (1 million sq. km). Other programmes have been flown for Peru, Colombia, Venezuelan Nicaragua, Togo, in parts of Indonesia and elsewhere, aside from substantial cover in North America. Both real and synthetic aperture systems have been used for airborne work. Synthetic Aperture Radar (SAR) imaging of the Earth has also been undertaken from Space by the SEASAT satellite.

An extremely useful property of mapping radars is cloud penetration enabling large areas to be rapidly and consistently imaged. This feature, amongst other factors, renders it a particularly useful tool for regional inventory type surveys over areas that pose problems for the photographic aircraft.

The semi-independence of resolution from range of radar (SAR) together with the cloud-penetration capability renders SAR very apt for satellite systems although there are problems with the very large quantities of data which are produced.

The imagery is capable of interpretation in much the same manner as air photography, and upon the basis of texture, tone, shape and context of features.

However, layover distortions due to the distance ranging basis of acquisition of radar images creates increasing interpretative difficulty with increasing severity of relief. In mountainous terrain, apart from the layover problem, data can also be lost in highlight and shadow.

Col. S.E.G. Fraser of the Ordnance Survey spoke about the possible application of radar to map revision. The basic large scale mapping of Great Britain is now virtually complete. Likewise nationwide map coverage is available at small scales (1:50 000, 1:250 000 and 1:625 000). Thus remote sensing has come too late to be of assistance in the primary mapping of the country. It may, however, offer potential for map revision, which is now the primary survey task of the Ordnance Survey.

The resolution of currently available satellite imagery renders it unsuitable for scales of mapping larger than 1:250 000. Improved resolution (in excess of 5 metres) is required before radar imagery is likely to be of value to O.S. This may be achieved by improvingthe resolution of satellite imagery or by using an airborne platform.

The forthcoming SAR 580 campaign offers an opportunity to acquire airborne synthetic aperture radar imagery with a resolution of about 3m x 3m. It also provides an opportunity to test the all weather capability of SAR, which in a cloud covered country like Britain is of great potential value.

In conjunction University College London, O.S. will be participating in the SAR 580 campaign. Imagery from two of the proposed sites (Northampton GB5 and the east of Scotland GB1) will be used. This imagery will cover areas mapped at all O.S. basic scales (1:1250 urban, 1:2500 rural and 1:10 000 mountain and moorland).

The overall objective of the experiment will be to assess the information content and ease of interpretation of the imagery and the geometry and optimum flying and processing conditions for such imagery.

INFRA RED LINESCAN

Dr. D. Williams of Clyde Surveys Ltd. gave a brief report on this company's use of Infra Red Linescan and showed that this imagery could be matched to maps to indicate the location of areas of different heat emisison from the ground or from buildings. No geometric rectification had been applied to this imagery.

DISCUSSION

In the discussion on radar imagery several speakers stressed the advantages of radar and H. Lewis (Huntings) estimated that whilst complete air photograph coverage of the United Kingdom could take up to five years and cost £1.75m, radar imagery could be obtained in a shorter time at much less cost. Speakers also emphasised the problems still remaining in the use of radar imagery and the considerable work necessary to make full use of the imagery.

THE NEED FOR NEW MAPS

Dr. I. Dowman of University College London gave a review of the requirements for mapping which could be satisfied with space imagery. It is generally agreed that a realistic and desirable aim is that space imagery should be used for the compilation of 1:50 000 maps with 20m contours. This is the requirement laid down by the United States Geological Survey (USGS) for a mapping satellite and the French have a requirement of 1:100 000 mapping from SPOT with 40m contours. A study carried out for the European Space Agency (ESA) to identify organisations in Africa interested in the use of the Spacelab Metric Camera indicated that these saw their main need as the production of 1:50 000 topographic map and

and the revision of existing mapping. In addition the replies to the Call for for Experiments with the Metric Camera on Spacelab have come from 30 different countries with many experiment proposals involving 1:50 000 mapping. Other reports from S.E. Asia and S. America have similar conclusions.

It is uncertain whether these requirements can be met by the missions currently planned and proposed; these are shown in Table III. The photographic systems both have an image scale of 1:820 000, very similar to the Skylab S190 B photographs and there are likely to be problems of interpretation as discussed by Petrie earlier, also the height accuracy is limited and even in the best configuration of the Large Format Camera (LFC) photographs with a base to height ratio of 1:1.2 only 10m accuracy is possible allowing at best 30m contours. SPOT is designed for mapping and the production of topographic information but even after the most intensive level of processing only 1:100 000 maps and ortho-photographs are definitely planned and 40m contours. Landsat D, whilst having an IFOV of 30m is not intended as a mapping system. There are no significant geometric distortions in the photographic systems and the distortions in SPOT can be more easily corrected than those in Landsat but processing is still needed.

The Atlas programme sponsored by West Germany and, in the case of Atlas 1 by ESA (the Metric camera on the first Spacelab mission) aims at developing a photographic system in three stages. Atlas 1 uses an existing camera, the Zeiss RMK 30/23 in the Spacelab module. Atlas 2 will use a modified camera, probably a Zeiss camera with 600m principal distance and image motion compensation so that high resolution photography is obtained and Atlas 3 will use a new camera developed to produce very high resolution images with a suitable base to height ratio. An elliptical orbit would be used to allow a perigee over the area of interest at about 200km and an apogee at 500km which would permit a parking orbit to be entered when the area was cloud covered.

The United States have two push broom stereo systems under consideration. Mapsat is promoted by the USGS and is designed for mapping at the specifications quoted above, if the applications are met the satellite should ably provide accurate, distortion free maps very quickly. Stereosat is aimed at geological investigation using similar methods.

It appears then that the French, Germans and Americans are aware of a need for a mapping satellite and have plans to satisfy the needs, in the case of the Germans and Americans the money has yet to be found and experience form LFC, Spacelab and SPOT needed to further investigate some of the problems. The United Kingdom's only direct link with any of these projects is representation on the ESA metric camera working group; does this indicate a lack of interest or do we feel that impossible aims are being set?

UNITED KINGDOM REQUIREMENTS

In the discussion on U.K. needs it emerged that there are two - one is to extend and update conventional topographic mapping at 1:50 000 scale in areas where U.K. organisations are working and the other is to provide a small scale data base as a basis for planning and thematic mapping.

The requirement for 1:50 000 mapping is well established but mapping at this scale will not be possible for a number of years. Discussion concentrated on the necessity for maps at smaller scales. Lewis argued a worldwide need for up to date mapping at 1:250 000 scale which he claimed the mapping community had neglected bacause of its concentration on basic mapping and control surveys. He felt that because of the pressing need of administrators and others for such maps, their production and maintenance should be treated quite separately from that of larger scales. Because resolution and geometric accuracy are of lesser importance than up-to-dateness satellite data can provide a major input. Several

TABLE III PLANNED AND PROPOSED SENSORS.

Mission	Launch Date	Alt. (km)	Sensors	Spectral Bands µm	IFOV/ Resolution	ACCURACY			Application
						Plan	Height	Comment	
Photographic									
Large Format (LFC) on Shuttle	1983	250	ITEK Camera f = 305 mm Format 460 x 230 mm with IMC	Visible	80ℓpmm^{-1}	10 - 20 m	10 - 20 m	No geometric correction needed	Mapping; map revision orthophotos; thematic mapping
Metric Camera on Spacelab (ALTASA)	1983	250	Zeiss RMK Camera f=305mm format 230 x 230 mm	Visible	40ℓpmm^{-1}	10 - 20 m	20 - 30 m	No geometric correction needed	"
ATLAS B	1984	250	Modified Camera	Visible				Objective is to develop high resolution camera with IMC to give 10 m plan and height accuracy	"
ATLAS C	1986	200/500	New camera	Visible					"
Digital									
LANDSAT D	1982	705	Thematic Mapper	0.45 - .52µm 0.52 - .60µm 0.63 - .69µm 0.76 - .90µm 1.55 - 1.75µm 2.1 - 2.35µm	30 m	20 m	?	Based on Landsat 1,2,3 0.5 - 1.0 pixel after geometric transformation	Thematic maps reserve studies monitoring etc.
				10.4 -12.5 µm	120 m				
			MSS	As LANDSAT 2					
SPOT	1984		HRV	0.5 - 0.9	10 m			Different levels of product to various specifications	Mapping, map revision, orthophotos. Thematic mapping.
				0.5 - .59 0.61 - .69 0.79 - .90	20 m				
SIR - A	1981	200	SAR	L band	40 m			Same sensor as SEASAT	
MRSE on Spacelab	1983	250	SAR Scatterometer Passive Microwave Radiometer	X band	25 m				Land and oceanographic applications.
ERS - 1	1985/86	675	Active Microwave Radar Altimeter	C band	30 m				Ocean applications
MAPSAT	?	913	3 camera CCD arrays	.47 - .57 .57 - .70 .76 -1.05	10 m	10 m	< 10 m	Intensive processing using epipolar geometry	1:50 000 mapping with 20 m contours (USGS)
STEREOSAT	?	705	3 camera ccD arrays	?	15 m				geological mapping

organisations, including the Royal Aircraft Establishment and the Experimental Cartography Unit at the Natural Environment Research Council Scientific Services have expressed a need for 50m contours of the United Kingdom in digital form and as the Ordnance Survey cannot provide this at the moment it could be more economical to obtain it directly from satellite imagery than by digitising the data held by the Ordnance Survey. This argument applies even more forcibly in overseas country where no large scale data exists anyway. Mr. O'Brien of D.O.S. supported the view that satellites could economically provide generalised data and useful mapping in areas such as the Sudan. It may be that land surveyors and photogrammetrists need to take a fresh look at requirements in this area and listen to the way that users are thinking.

Assuming that there is a need for such a data base, whether in digital or graphical form, there are many problems still foreseen. If a photographic system is used - will it be possible to obtain photography of the desired areas within a reasonable period of time? Will the photographs be too expensive? Is it indeed possible to obtain sufficient resolution and sufficiently accurate height information? At present optimistic answers are possible to all these questions but only time will tell whether optimism will be transformed into reality.

If digital systems are used there is still the question of resolution and height accuracy but there is also the problem of how to cope with the vast amount of data. Davison (RAE) and Carter (Harwell) believe that we can handle it but that more pre-processing on the satellite was needed together with more sophisticated processing on the ground. It was pointed out that enormous advances had been made and that progress was not slowing down. The question was also raised of whether all imagery would be available; in particular data from SPOT. It was also felt that if true costs were charged to users the data would only be used if a commercially successful product was possible.

CONCLUSIONS

1. Landsat imagery is limited to 1:500 000 mapping with no height information.
2. Radar imagery has considerable potential because of high resolution and all weather capability but much more work is required.
3. There is no certainty that currently planned satellites and sensors will be able to produce data sufficient for economical mapping at scales greater than 1:100 000 with 50m contours.
4. There is a significant feeling that there is a requirement for satellite imagery to provide maps at 1:50 000 with 20m contours for unmapped areas.
5. There is a need for small scale maps where up to dateness is more important than accuracy.
6. Technology is able to cope with the data processing requirements but high costs may deter commercial users.

APPENDIX

List of Participants

Dr. A. Benny, R.A.E. Farnborough.

Mr. E.M. Byrne, Map Research and Library Group, M.C.E.

Mr. A. Bryant, Directorate of Overseas Surveys.

Dr. P. Carter, Image Analysis Group, A.E.R.E. Harwell.

Mr. O.W. Cheffins, Clyde Surveys Ltd.

Mr. L. Collin, Dept. of Geography, University College of Wales, Aberystwyth.

Mr. M. Coulson, Dept. of Geography, University College of Swansea.

Mr. G. Davison, R.A.E. Farnborough.

Dr. I.J. Dowman, University College London.

Captain C. Esslemont, School of Military Survey, Hermitage.

Col. S. Fraser, Ordnance Survey.

Dr. E. Kapokakis, University College London.

Mr. H. Lewis, Hunting Surveys Ltd.

Dr. P. Martin-Kaye, Hunting Geology and Geophysics Ltd.

Dr. M. Mohammad, University College London.

Mr. C.I.M. O'Brien, Directorate of Overseas Surveys.

Professor Gordon Petrie, University of Glasgow.

Mrs. B. Rose, Department of the Environment.

Mr. T. Welsh, Easams Ltd., Camberley.

Mr. C. Whitehead,

Dr. D. Williams, Clyde Surveys Ltd.

REMOTE SENSING OF TOXIC MINERAL VEINS

BENEATH PERMANENT PASTURE

D C Ferns and Prof J Barber
Remote Sensing Group
Department of Pure and Applied Biology
Imperial College of Science and Technology
Prince Consort Road
London SW7 2BB

Several workers have examined the effect of heavy metals on the spectra of natural and semi-natural vegetation communities. This study however is in the agricultural area of Callington, Cornwall, where much of the land is permanent or semi-permanent pasture, traversed by mineralised dykes, many of which exhibit soil geochemical anomalies. The mineral rights for the study area are owned and partially exploited by South-West Consolidated Minerals Limited.

Several remote sensing techniques will be utilised to best discriminate these mineralised veins by their expression on the reflectance of the surface vegetation, which may suffer physiological change due to mineral toxicity. Laboratory experiments have shown that p.p.m. values of copper, lead and zinc similar to those in the field area, can be detected by high resolution spectrometry.

Ground spectroradiometer transects over soil geochemical anomalies will be operated with both a high resolution instrument - the 903 band Barringer Refspec, and the lower resolution Milton Radiometer representing the four Landsat bands. Biomass measurements will be taken from spectra at several growth stages, and full details of nitrogen application and other fertilizers will be recorded. Imagery has been selected not only with the purpose of defining mineral veins, but also to assess the land use changes on a regional and 'per field' basis, as the area contains both permanent pasture and pasture in long rotation. Landsat data has been processed by both electro-optical and digital methods, and recent aerial photography includes black and white, colour and infrared imagery of the area.

A 'Local Toxicity Indices' has been derived accounting for both foliar and root metal content, which it is hoped will improve the regression analysis between vegetation reflectance and soil geochemical data.

The study is continuing and results will be presented at a later date.

Acknowledgements
SOUTH WEST CONSOLIDATED MINERALS LIMITED
ROBERTSON RESEARCH INTERNATIONAL LIMITED

Poster presented at a conference on Matching remote sensing technologies and their applications convened in London by the Remote Sensing Society, 16-18 December, 1981

USE OF A GIS SYSTEM FOR REMOTE SENSING IMAGERY SELECTION AND COVERAGE PLOTTING

C A Kitcho
Woodward-Clyde Consultants
San Francisco

Large remote sensing projects often require several phases of imagery selection and coverage plotting. Imagery selection includes obtaining listings of available imagery or aerial photography, selecting scenes, and plotting areal coverage of the scences for final selection. This process has been done manually in the past, and can be time consuming. For this reason, Woodward-Clyde Consultants has developed a remote sensing imagery selection sytem (RISS) that allows real time searches, selection, and automated coverage plotting, using an inhouse geographic information system.

This system keys the geographic coordinates from the listings,converts the geographic boundaries and plots them on a map base. The system was developed in two steps: (1) implementation of an imagery search system from two separate imagery and aerial photography data bases, which accesses the information via terminal; and (2) development of a linking program to direct imagery coverage coordinates to the plotting hardware, and then to plot coverage on base maps produced from digital terrain tapes.

Creation of RISS is an example of a future trend in maximising the capabilities of an operational GIS system for remote sensing applications.

APPLICATION OF LANDSAT IN ESTIMATING CARRYING CAPACITY OF WOODED RANGELANDS IN TANZANIA

I R Lane, Edinburgh School of Agriculture

On semi-arid grassland plant biomass can be estimated with Landsat to within ± 250kg/Ha. Many rangelands in the tropics however have a variable woody cover. As forage production is variable between years a technique for its estimation on an annual basis would be valuable in these ecozones.

The research schedule was: (1) estimate woody cover in late dry season; (2) estimate total cover at end of rainy season: (3) derive cover of herb layer by subtraction: (4) relate satellite estimate to ground truth of forage production; (5) repeat (2-4) each year.

In 1979 vegetation was surveyed in a semi-arid ecozone in Tanzania, Ecotypes were grassland, semi-deciduous woodland, deciduous bushland and thorn shrubland. Woody types had grass understorey. The type of woody species had a differential effect on grass cover and productivity; this has to be taken into account in estimating carrying capacity by satellite imagery.

A revised schedule is outlined in the poster, and relationships between Landsat imagery and vegetation characteristics for the site are illustrated.

A QUANTITATIVE STUDY OF THE SPECTRAL HABIT OF ARID LANDFORMS IN SOUTHERN TUNISIA

TJ Munday, University of Reading

The Tunisian Pre-Sahara is dominated by the presence of three large closed drainage basins known as the Chotts. They are similar in form to the sebkhas and garaets that occur throughout N Africa, and to the playas of Iran and the West United States.

An in situ spectral sampling and ground measurement programme was conducted over the Chott Djerid and its immediate surroundings. The primary aim of the study was to determine the quantitative relationships between spectral reflectance and ground parameters pertaining to these areas. The secondary aim of this research was to highlight those problems involved in the collation of such information in arid and semi-arid regions.

Ground spectra were sampled using a Milton Multiband Radiometer, recording spectral data in Landsat MSS bandpass equivalents.

Statistical Analyses were performed on these data. Results showed that several ground variables were important in determining reflectance over the different surface crusts of Chott Djerid. The ratio of % soluble salts to the % detritals in the crust was the single most important determinant of reflectance. The most prominent saline mineral is halite while the detritals are predominantly quartz. Percent non-structural water, and the presence of a surficial cover of aeolian derived material on the crusts were also significant in determining spectral reflectance. The latter assumes greater importance on the periphery of Chott as does the effect of a vegetation cover.

The results of this work give substantial encouragement to the utilisation of Landsat MSS data in the study of Chott landforms. Multi-temporal analysis in particular could provide a significant source of information on these forms, to hydrogeologists and geomorphologists alike.

REMOTE SENSING FOR LAND MANAGEMENT DECISIONS

C Purcell, Ertec Western Inc, Long Beach, California
L W Anderson, US Bureau of Reclamation, Denver, Colorado

A study of approximately 8300 squarekilometres of northwestern Arizona, USA, was conducted for the United States Bureau of Land Management to evaluate the use of remote sensing in the assessment of potential geology, energy, and mineral (G-E-M) resources. The G-E-M resources investiaged were divided into four resource categories: 1) locatable mineral (gold, silver, etc), 2) salable industrial (sand, gravel, etc), 3) leasable energy (oil, geothermal, etc), and 4) water (surface and ground). A three-phased approach was used to test the adequacy of the remote sensing data: 1) interpretation of drainage patterns, land forms, cover types, and linear features entirely from computer enhanced Landsat imagery 2) interpretation of geomorphology, stratigraphy, lineaments, structure, and geodynamics from a combination of Landsat and large and small scale stereo aerial photographs, and 3) literature, supplemented with limited field data.

Data on which to base land management decisions were ranked in to the following four categories: 1) unsatisfactory, 2) marginally satisfactory, 3) satisfactory, and 4) highly satisfactory. At no time was Landsat data alone satisfactory for making management decisions. Overall, highly satisfactory data was provided on the potential for salable, leasable, and water resources from a combination of literature, remote sensing, and field data. However, only marginally satisfactory data was compiled for identifying the potential locatable mineral resources. A much expanded field program would be necessary to provide satisfactory data on locatable mineral resources.

THE MANIPULATION OF RASTER-BASED TOPOGRAPHIC DATA ON A PARALLEL PROCESSOR

T.A. Adams, Department of Geography, University of Durham

K.A. Smith, DAP Support Unit, Queen Mary College, University of London

The raster image of an 1:1250 Ordnance Survey topographical map consists mostly of 'empty space'; that is, the feature code boundaries are only a small percentage of the image. An efficient method of storing the raster image is described which allows the image to be processed on the ICL Distributed Array Processor (DAP). The DAP is essentially a 64x64 array of processing elements which can all execute the same instruction simultaneously.

The poster describes algorithms to do the following manipulations:
(i) the compression of the map by factors of 2, 4 and 8;
(ii) the retrieval of all polygons from the map;
(iii) the vectorization of the polygons;
(iv) the determining of a spatial feature code for each polygon.

INTEGRATING THE USE OF LANDSAT TO DEVELOPMENT STUDIES

J W Trevett: G Deane, Hunting Surveys and Consultants, Borehamwood,UK

The poster display will illustrate three actual projects where remote sensing techniques have been used to provide additional data for use in development studies.

1. Costa Rica The Coto Sur area comprises the Pacific Lowlands region within the Costa Rica-Panama Frontier Project. Its total area is 61,200 ha of which 6,500 ha belong to Banana Companies, leaving a net area of 54,700 ha for which semi-detailed survey is required.

Interpretations from 1:20,000 scale aerial photographs together with field work (minimum of 5 observations per km^2) are being used to map the Project Area. In addition, enhanced Landsat multispectral scanner images have been generated using the Hunting Image Processor and Analysis System(HIPAS). On these 1979 Landsat images, areas where the vegetation cover has changed since the aerial photography was flown can be located. This up to date information is essential for both navigation in the field and preparation of the final thematic maps.

Soil maps will be produced at a scale of 1:50,000, showing soil series, associations and phases. Land capability is being determined and crop aptitude charts prepared. Appropriate measures for soil conservation and management are being defined.

2. Indonesia A study to evaluate the agricultural potential for transmigrant settlers in an area of 30,000 km^2 of southeast Sulawesi Province. Within the framework of the Provincial Government's development plans proposals were drawn up for new settlement areas and for the improvement of existing schemes. The plans involved the selection of appropriate cropping patterns, and farm technologies. Agricultural supporting services required for the successful implementation of the plans were also drawn up. The relationship between cropping intensity, farm technology and holding size were examined, which played an important role in the formulation of the final agricultural proposals. Settlements including supporting infrastructural facilities were planned.

The Landsat imagery was used to examine one of the potential development areas where initial field work indicated that there was a potential difference in soils which was reflected in the vegetation cover mainly grassland. The extent of these differences were impossible to determine on the ground because of the terrain. Examination of Landsat on HIPAS showed that a vegetation difference did in fact occur and these data are being used by the field teams to assist them in mapping development units.

3. Iraq An area of intense irrigation development has many new irrigation ditches built which have not been recorded on any survey documents. To re-map would be expensive and security constraints preclude obtaining air photographs. Landsat was examined on HIPAS to determine whether canals could be identified using analysis. The enhancement of the standard tape imagery brought some success but not sufficient to recognise irrigation channels outside the main government farms. Subsequent band ratioing and enhancement intensified the expression of irrigation channels which could be used, in conjunction with the existing maps to identify some new structures.

INTEGRATION OF FUTURE AEROSPACE REMOTE SENSING FOR CARTOGRAPHIC AND THEMATIC APPLICATIONS: AN ITALIAN SIMULATION PROGRAM

S Vetrella, Istituto Di Aerodinamica, Napoli, Italy

This paper is intended to provide an overview of the research and applications developed at University of Naples in the field of aerospace remote sensing.

Several papers have already emphasised the need to integrate different remote sensing systems in order to optimize the analysis procedures and to get better results.

The future satellites (Landsat D, SPOT, ERS.1) and Space-Shuttle/Spacelab experiments (particularly active microwave systems) need a new approach of data analysis due to high geometric resolution, high repetition rate and radiometric accuracy.

The stereoscopic view of the Earth from space is a new area of research in which it is necessary to integrate morphological and textural land informations with remote sensing data through a digital analysis system.

The Synthetic Aperture Radar can be useful for several applications, but data analysis procedures and spectral signatures are to be deeply investigated.

The development of new technologies for operational systems requires simulation programs to identify, the requirements and characteristics of future systems, using the existing aerospace platforms and sensors.

This paper describes also the point of view of an Italian research group and the simulation program that cover four years of activity.

VEGETATION MAPS OF THE SWEDISH MOUNTAINS BASED ON HIGH ALTITUDE INFRARED COLOUR IMAGES

L Wastenson, Department of Physical Geography, University of Stockholm

The first mapping of vegetation over large areas in Sweden (66 00 km^2) will result in 22 Maps in the scale of 1:100,000 which will cover the Swedish mountains. The series will be complete in 1983. Up to now 10 sheets have been published. The work is done at the Department of Physical Geography of the University of Stockholm and is financially supported by the National Swedish Environment Protection Board. The mapping is based on air photo interpretation of infrared film and a limited field work. Ten alpine vegetation types (heaths and meadows), three subalpine types of birch forest, three types of coniferous forest and six types of mire are distinguished. On each map there is a circle diagram showing the vegetation types in percent of the mapped area. There is also a general description of the vegetation within the map, and small scale maps of topography, bedrock and interesting botanical areas, eg one map showing where to find rich vegetation.

COMPUTER ANALYSIS OF MULTI-DATE LANDSAT DATA FOR MAPPING LAND USE, FOREST CLEAR CUTTING AND MIRES

L Wastenson: Wolter Arnberg; L Boresjo och Margareta Ihse
Remote Sensing Laboratory, Department of Physical Geography, University of Stockholm

A brief summary is given of experiments of land-use mapping from LANDSAT-2 data and from airborne MSS in the Linköping region, Sweden. The results from a single-data analysis have been compared with a multi-date analysis.

Methods for monitoring forest clearings during the period 1975-1979 using digital multi-temporal Landsat imageries are also presented. The last study deals with methods to use digitized map information in the multi-date Landsat classification process and methods to select the most informative combination of wave length bands and registration dates. In the first application of these methods we have tried to classify mires.

THREE-DIMENSIONAL HISTROGRAMS AND QUANTITATIVE IMAGE PROCESSING FROM LOW RESOLUTION SENSORS

M Wilson A Henderson-Sellers* and N A Hughes
Department of Geography
University of Liverpool

* GISS, NASA, 2880 Broadway, New York, USA

Monitoring of system albedo by meteorological satellites provides a new and unexpected method of environmental assessment which has yet to be fully developed. System (or planetary) albedo is the ratio of reflected radiation from the Earth-atmosphere system to the total incident solar radiation. It is a function of surface and atmospheric state. Variability of the observed fluxes is clearly almost as important as the radiance values themselves. The value of albedo for the geographical area around the UK for instance, is 35% with a standard deviation of 16%. System albedo variations are both a cause and a consequence of environmental variability.

Monitoring system albedo is clearly of considerable interest, but both its evaluation and any analysis of its variability must be undertaken with care. Variations in system albedo as measured by low resolution satellite-borne scanning radiometers are likely to be associated primarily with changes in cloud cover and type and changes in the surface reflectivity. The restrictions imposed by the geometry of the satellite-sun-surface configuration and limited waveband response of the scanning radiometers are more easily overcome if analysis can be made user-interactive; two methods of system albedo analysis described are: i) the IDP-3000 at RAE, Farnborough has been used in this study to produce colour sliced images of albedo; ii) a rapid-response graph plotting video display unit which permits cloud and surface classification by interactive analysis of 3-D displays of channel irradiances. Greater use of meteorological satellites to monitor both system and surface albedo changes on a regional scale is urged and quantified analysis methods recommended.

REMOTE SENSING SOCIETY

PRELIMINARY NOTICE AND CALL FOR PAPERS

for the

ANNUAL TECHNICAL CONFERENCE

on the theme

REMOTE SENSING AND THE ATMOSPHERE

to be held in the University of Liverpool

15 — 17 December 1982

REMOTE SENSING AND THE ATMOSPHERE

University of Liverpool 15-17 December 1982

The purpose of the meeting will be to promote the dialogue between those who sense the atmosphere and those who are obliged to remove atmosphere distortion before interpreting surface data. There will be some general sessions to permit discussion of other research. There will also be an opportunity to offer material in poster sessions.

Those proposing to present papers or poster sessions should address their comments specifically to:-

Atmospheric observations and/or inversion techniques with reference to:

- Atmospheric dynamics, state and constituents
- Climatology and climatic variables
- Minimisation/removal of atmospheric effect
- Air pollution monitoring from remote platforms
- Planetary atmospheres
- Papers by students

The conference will extend over 3 days including 2 nights residence in Liverpool.

Dr A Henderson-Sellers
Geography Department
University of Liverpool
PO Box 147
Liverpool L69 3BX
England

Tel No (051)709 6022 X2707
Telex 627095